Lecture Notes in Physics

Volume 998

The series Lecture Notes in Physics (LNP), founded in 1969, reports new developments in physics research and teaching - quickly and informally, but with a high quality and the explicit aim to summarize and communicate current knowledge in an accessible way. Books published in this series are conceived as bridging material between advanced graduate textbooks and the forefront of research and to serve three purposes:

- to be a compact and modern up-to-date source of reference on a well-defined topic;
- to serve as an accessible introduction to the field to postgraduate students and non-specialist researchers from related areas;
- to be a source of advanced teaching material for specialized seminars, courses and schools.

Both monographs and multi-author volumes will be considered for publication. Edited volumes should however consist of a very limited number of contributions only. Proceedings will not be considered for LNP.

Volumes published in LNP are disseminated both in print and in electronic formats, the electronic archive being available at springerlink.com. The series content is indexed, abstracted and referenced by many abstracting and information services, bibliographic networks, subscription agencies, library networks, and consortia.

Proposals should be sent to a member of the Editorial Board, or directly to the responsible editor at Springer:

Dr Lisa Scalone
Springer Nature
Physics
Tiergartenstrasse 17
69121 Heidelberg, Germany
lisa.scalone@springernature.com

More information about this series at https://link.springer.com/bookseries/5304

Fulvio Ricci · Massimo Bassan

Experimental Gravitation

 Springer

Fulvio Ricci
Dipartimento di Fisica
Sapienza Università di Roma
Rome, Italy

Massimo Bassan
Dipartimento di Fisica
Università di Roma, Tor Vergata
Rome, Italy

ISSN 0075-8450   ISSN 1616-6361   (electronic)
Lecture Notes in Physics
ISBN 978-3-030-95595-3   ISBN 978-3-030-95596-0   (eBook)
https://doi.org/10.1007/978-3-030-95596-0

This Springer imprint is published by the registered company Springer Nature Switzerland AG
The registered company address is: Gewerbestrasse 11, 6330 Cham, Switzerland

To our patient wives Simonetta and Stefania

Preface

Our current knowledge on the gravitational interaction is based on the intellectual breakthroughs of Galilei, Newton and Einstein. The theory of General Relativity summarizes the development, over centuries of our knowledge, in this field of Physics and is the stable reference for all experimental efforts. These aim at highlighting the dynamic aspects of the theory of gravitation and try to test its limits.

Einstein's "General Theory of Relativity" was presented to the scientific community in 1916 and for a long time was confirmed only by a limited number of experimental evidences. Its mathematical complexity made its spread difficult to the scientific community and it was only in the second half of the twentieth century, following impressive technological developments, that the experimental evidences multiplied and reached a level of precision that highlights the great robustness and completeness of Einstein's theory. Important advances in instrumentation opened up the field to new astrophysical observations that can only be interpreted in the light of General Relativity and allow us to establish the absolute prevalence of this theory in the interpretation of gravitational phenomena.

Precision radio-astronomical observations on binary systems of compact stars have opened the chapter of gravito-dynamics, indirectly showing the existence of gravitational waves. The direct detection of this new type of radiation allows now to investigate the dynamics of emission processes, opening the way to a new field of astronomy. The experimental effort, started by Joseph Weber in the 60s of the twentieth century, has now reached a level that made it possible to attain in 2015 the first direct detection: other sensational discoveries have followed and more will take place in the coming years. However, gravitation still remains conceptually separate from other fundamental interactions and the problem of framing gravity in a wider context of unification of fundamental interactions remains open. In recent years, experimental research in this direction is also gaining strength and proceeds, free of theoretical prejudices, in a constant effort to expand our level of knowledge of Nature.

We have tried to summarize this effort in the chapters of this book, which aims to contribute to the field development by highlighting the fundamental role of the Galilean experimental method in the process of understanding gravitational phenomena.

The book is composed of two main sections: the first six chapters present a survey of key experiments that were (or will be) cornerstones in our understanding of gravitational phenomena in the stationary or slowly changing regime.

From classical gravity to general relativity and post-Newtonian parametrization, we sketch the theory and we go in deeper detail for the experimental verifications.

The following chapters, instead, deal with gravitational waves, i.e. transient, rapidly changing gravity. Here too, we summarize the theoretical background and discuss in detail the detectors, both operating and planned, and the technologies that make them possible.

The pulsars, which we deal with in Chap. 12, actually are laboratories for both gravito-statics (measurement of relativistic effects in the orbits) and gravito-dynamics, as key element of the searches for ultra-low-frequency gravitational radiation.

Finally, in the last two chapters, we describe two practical applications of this science: the Sagnac gyroscope and the GPS navigation system, with attention to the required relativistic corrections.

The emphasis is always on the experimental methods that made these progresses possible: some of these techniques were developed for the specific purpose, like the torsion pendulum (in its numerous varieties) or the superconducting gyros of GP-B; some were adopted and expanded: it is the case of Mössbauer's spectroscopy (for the Pound-Rebka measurement of red-shift), Michelson interferometry (for gravitational wave detectors) or Bayes' statistical inference. The reader will find the tools for a first approach to these experiments and the related techniques, and references to pursue the topics in further detail.

What the reader will not find in this book. We had to make choices, and reluctantly left out some fascinating new topics and techniques: the astronomical studies of galactic black holes that are providing important insight into our understanding of black hole physics and of space-time in strong field regime; all wonderful gravity experiments that rely on atom interferometry and Bose-Einstein condensates; the squeezing and all quantum optics techniques that are finding application in the interferometers.

This textbook is intended for graduate or advanced undergraduate students. The reader is expected to have confidence with basic undergraduate physics. General relativity is summarized, but in too quick a fashion, and previous knowledge of the theory is recommended. For some technical topics, like modulation, noise and power spectra, we have provided a quick recap in the first paragraphs of the relevant chapter, or in the Appendices.

A few words about notation: SI units are used throughout the book, so our general relativity equations are rich in G and c. In rare exceptions, we gave in to tradition and reluctantly cited cm or other non-SI units. Greek indexes run from 0 to 3, Latins from 1 to 3 and the signature $(+, -, -, -)$ is adopted. Finally, about frequencies: it was our intention to reserve f for the frequency of the signals, typically in the audio band, and ν for carrier frequencies, most often optical. In all cases, it always intended $\omega = 2\pi \cdot$ frequency.

This book derives from the lecture notes of the course of "Experimental Gravitation" that we both taught in our respective universities. We are grateful to all our students, who provided motivation for these notes and stimulated the study in further depth of some topics. We take the opportunity here to express our gratitude to friends and colleagues who offered help, advice and proof-reading: at the risk of forgetting someone, we mention Pia Astone, Massimo Bianchi, Marta Burgay, Valerio Ferroni, David Massimo Lucchesi, Mehr Un Nisa, Paola Puppo, Giuseppe Pucacco, Alberto Sesana, Fabio Spizzichino, Angelo Tartaglia, Jean Yves Vinet and Massimo Visco. A special mention for Stefano Braccini, who taught this course at the University of Pisa: sadly, he could not join us in this venture.

Gravitation is the oldest and, at the same time, the most lively and vibrant field of experimental physics: the awarding of three Nobel prizes in the last decade (2011, 2017, 2020) is the proof, if ever needed. Our wish is to provide, with this textbook, young and perspective experimental physicists with some of the necessary tools to continue in this search.

Rome, Italy   Fulvio Ricci
November 2021   Massimo Bassan

Table 1 A very limited selection of constants, conversion factors and astronomical quantities. Several constants, including c, k_B, $\hbar$, e, etc., have been defined *exact* since 2018: this has implied redefining metre, second and kilogram in terms of these new "yardsticks". Note that G, the most important (for this book) constant, is the least well determined, with a relative uncertitude. $u_r = 1.5 \cdot 10^{-5}$

Name	Symbol	Value	Units	Notes
Fundamental constants				
Boltzmann constant	k_B	$1.3806489 \cdot 10^{-23}$	J/K	Exact (SI 2018)
Planck's constant	$\hbar$	$1.054571817 \cdot 10^{-34}$	J s	Exact
Speed of light (...and of GW)	c	$2.99792458 \cdot 10^8$	m/s	Exact
Gravitational constant	G	$6.67430(15) \cdot 10^{-11}$	m^3/s^2kg	$\sigma_G/G = 1.5 \cdot 10^{-5}$
Elementary e.m. charge	e	$\pm 1.602176634 \cdot 10^{-19}$	C	Exact
Derived constants				
Supercond. flux quantum	ϕ_0	$2.067833... \cdot 10^{-15}$	Wb	Exact
Fine structure constant	α	$1/137.036...$		$\sigma_\alpha/\alpha = 1.5 \cdot 10^{-10}$
Non-SI units and conversion factors				
Modified Julian Date	MJD	Days since midnight of November 17, 1858		
(Julian) year	yr	$3.15557600 \cdot 10^7$	s	365.25 days
Sidereal day	$sid.day$	86164.09	s	0.99727 solar day
Astronomical unit	AU	$1.495978707 \cdot 10^{11}$	m	Exact (IAU 2012)
Parsec	pc	$\sim 3.08567758 \cdot 10^{16}$	m	
		$648000/\pi$	AU	Exact (IAU 2015)
Light-year	ly	$\sim 9.460730 \cdot 10^{15}$	m	
		~ 0.3066	pc	
		~ 63198	AU	
Mega electron volt	MeV	$1.60217 \cdot 10^{-13}$	J	$10^6/e$
Kilogram	kg	$5.61 \cdot 10^{29}$	MeV	$10^6 c^2/e$
A few data about the solar system				
Distance from galactic centre	d_{gal}	$2.35 \cdot 10^{20}$	m	
Sun speed around the galaxy	v_{gal}	$2.2 \cdot 10^5$	m/s	
Mass of the Sun	$M_\odot$	$1.99 \cdot 10^{30}$	kg	
Mean radius of the Sun	$R_\odot$	$1.3909 \cdot 10^9$	m	
Sun quadrupole moment	$J_{2,\odot}$	$2.21 \cdot 10^{-7}$		
Mass of the Earth	$M_\oplus$	$5.9736 \cdot 10^{24}$	kg	
Mean radius of the Earth	$R_\oplus$	$6.378137 \cdot 10^6$	m	
Earth quadrupole moment	$J_{2,\oplus}$	$1.0826 \cdot 10^{-3}$		
Mass of the Moon	$M_\mathbb{C}$	$7.349 \cdot 10^{22}$	kg	
Mean distance Earth-Moon	$d_\mathbb{C}$	$6.378137 \cdot 10^6$	m	
Eccentricity of Moon orbit	$e_\mathbb{C}$	$5.54 \cdot 10^{-2}$		

Contents

Acronyms

ADC	Analog-Digital Converter
AGN	Active Galactic Nuclei
AGWB	Astrophysical Gravitational Wave Background
AM	Amplitude Modulation
ASI	Agenzia Spaziale Italiana
BAE	Back Action Evading
BH	Black Hole
BS	Beam Splitter
CARM	Common mode of the Fabry-Perot arms
CCSNe	Core Collapse Supernovae
CERN	European Centre of Nuclear Research
CMB	Cosmic Microwave Background
CNES	Centre National d'Études Spatiales
CODATA	Committee on Data of the International Science Council
CPT	Charge Parity Time
CWB	Coherent Wave Burst
DAC	Digital-Analog Converter
DARM	Differential mode of the Fabry-Perot arms
DFACS	Drag-Free Actuation Control System
DL	Delay Line
DM	Dispersion Measure
DSP	Digital Signal Processor
e.m.	electromagnetic, electromagnetism
ECEF	Earth Centred—Earth Fixed
ECI	Earth-Centred Inertial
EEP	Einstein Equivalence Principle
EGM	Earth Gravitational models
EMRI	Extreme Mass Ratio Inspiral
EOM	Electro-optical Modulator
EP	Equivalence Principle (EP)
EPTA	European Pulsar Timing Array
ESA	European Space Agency

ETRS89	European Terrestrial Reference System, updated 1989
FFT	Fast Fourier Transform
FIR	Finite Impulse Response
FM	Frequency Modulation
FOG	Fibre Optic Gyroscopes
FP	Fabry-Perot
FSR	Free Spectral Range
FT	Fourier Transform
FWHM	Full Width Half Maximum
GEM	Gravito-electric Magnetic
GLONASS	GLObal NAvigation Satellite System
GNSS	Global Navigation Satellite System
GONG	Global Oscillation Network Group
GP-B	Gravity Probe B
GPS	Global Positioning System
GR	Einstein's General Relativity
GRB	Gamma Ray Burst
GRS	Gravitational Reference Sensor
GW	Gravitational Wave
IAU	International Astronomical Union
IERS	International Earth Rotation and Reference System Service
IIR	Infinite Impulse Response
IPTA	International Pulsar Timing Array
ISM	Interstellar Medium
JAXA	Japan Aerospace Exploration Agency
LASER	Light Amplification by Stimulated Emission Radiation
LBI	Long Baseline Interferometry
LEO	Low Earth Orbit
LI	Lorentz Invariance
LLI	Local Lorentz Invariance
LLR	Lunar Laser Ranging
LNH	Large Number Hypothesis
LPF	LISA Pathfinder
LPI	Local Position Invariance
LT	Lense-Thirring
LWA	Long Wavelength Array
MASER	Microwave Amplification Stimulated Emission Radiation
MICH	Michelson differential mode
MOT	Magneto-optic Trap
MSE	Mean Square Error
MSP	Millisecond Pulsars
NASA	National Aeronautics and Space Administration
NAVSTAR	Navigation System for Timing and Ranging
NG	Non-gravitational
NS	Neutron Star

OB	Optical Bench
PDH	Pound-Drever-Hall
PK	Post-Keplerian
PNS	Proto-Neutron Star
PPN	Post-Newtonian Parametrization
PR	Pound and Rebka
PR	Power Recycling
PRCL	Power Recycling Cavity Length
PSD	Power Spectral Density
PTA	Pulsar Timing Array
Pulsar	PULSAting Radio source
QND	Quantum Non-demolition
QSO	Quasi-stellar Object
RAAN	Right Ascension of the Ascending Node
RLG	Ring Laser Gyroscopes
RM	Recycling Mirror
SASI	Standing Accretion and Shock Instability
SEP	Strong Equivalence Principle
SGWB	Stochastic Gravitational Wave Background
SI	Système International d'unités
SLR	Satellite Laser Ranging
SMBH	Super Massive Black Hole
SMBHB	Super Massive Black Hole Binary
SN	Supernova
SNR	Signal-to-Noise Ratio
SQUID	Superconducting Quantum Interference Device
SR	Signal Recycling
SR	Special Relativity
SSB	Solar System Barycentre
TDI	Time Delay Interferometry
TM	Test Mass
TOA	Time Of Arrival
TT	Transverse and Traceless
UFF	Universality of Free Falling
VLBI	Very Long Baseline Interferometry
WD	White Dwarf
WEP	Weak Equivalence Principle
WFSM	Weak Field and Slow Motion
WK	Wiener-Kolmogorov

Acronyms of Experiments and Satellites

AURIGA	Antenna Ultracriogenica Risonante per l'Indagine Gravitazionale Astronomica
CHAMP	Challenging Minisatellite Payload
FAST	Five hundred meter Aperture Spherical Telescope
GAIA	Global Astrometric Interferometer for Astrophysics
GG	Galileo Galilei
GOCE	Gravity Field and Ocean Circulation Explorer
GRACE	Gravity Recovery and Climate Experiment
HIPPARCOS	HIgh Precision PARallax COllecting Satellite
JUICE	Jupiter Icy Moons Explorer
KAGRA	Kamioka Gravitational Wave Detector
LAGEOS	LAser GEOdynamic Satellite
LARES	LAser RElativity Satellite
LIGO	Laser Interferometer Gravitational Observatory
LISA	Laser Interferometric Space Antenna
LOFAR	LOw Frequency ARray
LSC	LIGO Scientific Community
LVK	LIGO/LSC Virgo KAGRA
MeerKAT	Karoo Array Telescope
MESSENGER	MErcury Surface, Space ENvironment, GEochemistry, and Ranging
MRO	Mars Reconnaissance Orbiter
MWA	Murchison Widefield Array
NANOGrav	North American Nanohertz Observatory for Gravitational Waves
NAUTILUS	Nuova Antenna Ultra bassa Temperatura per Investigare: nel Lontano Universo le Supernovae
PPTA	Parkes Pulsar Timing Array
RHESSI	Reuven Ramaty High Energy Solar Spectroscopic Imager
SKA	Square Kilometre Array
SOHO	Solar and Heliospheric Observatory
SOI-MDI	Solar Oscillation Investigation and Michelson Doppler Imager

STEP   Satellite Test of the Equivalence Principle
VERITAS   Venus Emissivity, Radio Science, InSAR, Topography, and Spec-
                troscopy

Classical Gravity

The cosmological vision of the ancient world was dominated by the philosophy of Ptolomeus (100–178 AD), describing the Earth as the centre of the universe, with several celestial spheres spinning around it with different angular speeds. Although a heliocentric vision had been proposed as early as the third century BC by Aristarchus of Samos, the Ptolemaic universal order, so well depicted by Dante in the Divine Comedy, was only reversed by Nicolaus Copernicus in 1532, when he wrote in his famous text *De Revolutionibus Corporum Coelestium: The Sun is set at the centre of all the things. Which other position could this source of light have in the cosmos, a wonderful temple, if not the centre from which it can illuminate every single thing at the same time? Therefore, the Sun is not improperly named by some "the lamp of the universe", "its corresponding mind" by others, and even "the metronome" by some others.*

It is interesting to notice the time line of the scientific discoveries and how they have been progressively integrated to constitute the framework of classical gravitational theory. The key scientists of such a revolution are depicted in Fig. 1.1.

1.1   Kepler and Newton

Johannes Kepler assumed the Copernican point of view in his analysis of the planetary motion. This analysis was based on data, incredibly accurate for the times, collected by Tycho Brahe over more than 20 years of observation. In *Astronomia Nova*, published in 1609, Kepler announced his first two laws, describing the motion of a generic planet

1. *All the planets move along elliptical orbits with the Sun occupying one of the two foci.*

© The Author(s), under exclusive license to Springer Nature Switzerland AG 2022
F. Ricci and M. Bassan, *Experimental Gravitation*, Lecture Notes in Physics 998,
https://doi.org/10.1007/978-3-030-95596-0_1

Claudius Ptolemy
(100 - 178 A.D.)

Nicolas Copernicus
(1473 - 1543)

Tycho Brahe
(1546 - 1601)

Johannes Kepler
(1571 - 1630)

Galileo Galilei
(1564 - 1642)

Isaac Newton
(1642 - 1727)

Fig. 1.1 Scientists who contribute to define the classical theory of Gravitation

2. *During equal time intervals the line segment connecting the Sun to a planet sweeps equal areas.*

The third law was published 10 years later (1619) in *Harmonices Mundi* and, unlike the previous two, states a scaling law for all planets orbiting the same central mass.

3. *The ratio of the square of the orbital period to the cube of major semi-axis is a constant for all the planets.*

These laws describe the kinetic properties of the planetary motion: they are exact, as long as we deal with a simple two-body (Sun and planet, or Earth and Moon) problem. Deviations from the ideal behaviour occur when other perturbations are considered, as we shall see in the following chapters. It is customary to define an orbit via the *Keplerian Elements*: six parameters that describe the shape and size of the ellipsis, its orientation in space and the planet position, as shown in Fig. 1.2:

- The orbit inclination i with respect to a reference plane (the solar ecliptic for planets, the equatorial plane for Earth satellites).
- The longitude of ascending node Ω defines the position of the "node", i.e. the intersection of the orbital plane and the reference plane. It is an angle measured from a reference direction (the Vernal point or the Greenwich meridian). When using celestial coordinates, the right ascension takes the place of the longitude.
- The argument of pericentre (or periastron) ω specifies the orientation in such plane of the major semi-axis, measured from the line of nodes.
- The major semi-axis a is the one information about the size of the orbit.
- The eccentricity e tells us, instead, about its shape.
- Finally, the position of the planet on the orbit at a given time or *epoch t_0* is given as the angle between the satellite, the central body and the periastron. This angle is

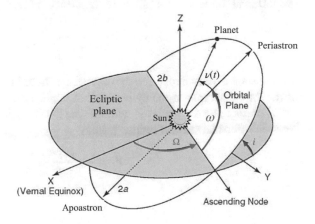

Fig. 1.2 The six Keplerian Elements

historical called *anomaly*.[1] $\nu(t_0)$ In celestial mechanics, it is chosen as the sixth Keplerian parameter.

For a Keplerian two-body system, they are all constant. We shall see in Chap. 5 how general relativity introduces perturbations resulting in secular shifts for some of these elements. Binary systems in strong gravitational field (Sect. 12.3) will also require five additional *post-Keplerian* parameters.

Kepler's laws were essential for Newton's formulation of the Law of Universal Gravitation. Nevertheless, in order to derive them, it is necessary to find a relation between the description of the motion and the cause generating it. To fill this gap, the formulation of Galileo's inertia principle was crucial: *a body upon which no force is acting either remains in rest or moves with a constant speed*, along with the formulation of the second law of dynamics by Newton.

$$\vec{F} = m\vec{a} \tag{1.1}$$

The curved trajectories of the planets imply the existence of a force, whose expression must be compatible with the kinetic properties stated by Kepler. These logic steps are today straightforwardly evident, if we just apply the elementary concepts of classical mechanics of our modern era. With the simplifying assumption of neglecting the small eccentricity of the planet orbits (i.e. assuming circular orbits), Kepler's third law suggests that, for all the planets, the square of the orbital period T is proportional to the third power of the orbital radius, r:

$$T^2 = Kr^3$$

Therefore, recalling that the centripetal acceleration is $a_c = \frac{4\pi^2}{T^2}r$ and applying (1.1), the corresponding attraction force that bends the planets' trajectories turns out to be proportional to the inverse square of the orbital radius:

$$F_c = m_P \frac{4\pi^2}{Kr^2}$$

The constant K must depend on the mass of the attracting body m_S: it is a consequence of the third principle of the dynamics that implies the equality of the force that the star S exerts upon the planet P with the one that P exerts upon S. Hence, we derive Newton's third law in its familiar form:

$$\vec{F} = -G \frac{m_S m_P}{r^3} \vec{r} \tag{1.2}$$

---

[1] There is nothing anomalous about the anomaly. It comes in three varieties: true, eccentric or mean. We shall define them in Sect. 1.3.

1.2   Newton's Gravitational Constant G

The value of the gravitational constant G was first measured by Newton from astronomical observations of planetary motion; these measurements have been repeated in recent days, benefitting from improved accuracy in the value of the orbital radii. However, we invariably end up measuring $GM_\odot$, as there is no way to disentangle this product from orbit observations, and the mass of the Sun is not well known. Therefore, these measurements provide results of limited accuracy, with relative uncertainty no better than 10^{-2}.

From the data available at the time, Newton managed to approximately calculate the Universal Gravitation constant, finding

$$G \simeq 6 \cdot 10^{-11} \ \frac{\text{Nm}^2}{\text{kg}^2}.$$

We need to wait till the first experiment of the modern experimental gravitation, based upon the torsion balance, to obtain a measurement of G of acceptably accuracy. In 1798, as the conclusion of an elegant series of laboratory measurements, (Cavendish 1798) "weighted the Earth". Only in 1811, S. Poisson introduced the gravitational constant (that he called f), that, using Cavendish data, was evaluated at

$$G = 6.74 \cdot 10^{-11} \ \frac{\text{Nm}^2}{\text{kg}^2},$$

which differs by only 1% from today's accepted value.

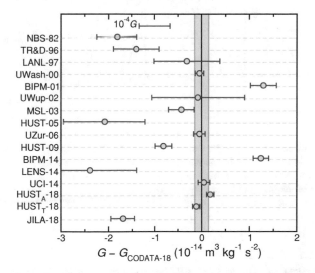

Fig. 1.3 16 different a measurements of the gravitational constant G, with their 1 σ error bars, performed between 1982 and 2018. These results were considered by the CODATA committee, in a weighted average, for the 2018 recommended value. Credits: reprinted figure with permission from Tiesinga (2021). Copyright 2021 by the American Physical Society

Since then, hundreds of experiments have been carried on to determine G to better precision (Rothleitner and Schlamminger 2017), but still today there is a spread of results limiting the accuracy. The CODATA 2018[2] value is

$$G_{CODATA-18} = (6.67430 \pm 0.00015) \cdot 10^{-11} \, \text{m}^3 \, \text{kg}^{-1} \text{s}^{-2} \qquad (1.3)$$

with a relative incertitude of $1.5 \cdot 10^{-5}$, to be compared with that of other fundamental constants, in the 10^{-9} range (Fig. 1.3).

1.3 Motion on a Keplerian Orbit

The motion of celestial bodies along a Keplerian, elliptic orbit is a classical exercise of analytical mechanics and can be found in most first-year physics textbooks. We summarize here the results that will be needed in other chapters. The relevant points and distances are shown in Fig. 1.4.

Although the description is quite general, we shall refer, to simplify definitions, to a planet orbiting the Sun. The only assumptions are to deal with a two-body problem,

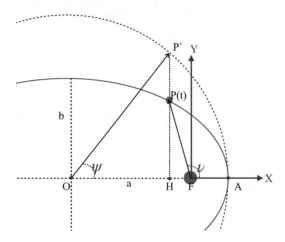

Fig. 1.4 Relevant point to define the anomalies along an elliptical orbit: O is the ellipse centre, a, b are the semi-axes. F is the focus, where the attracting body (the Sun) is, A is the periastron (perihelion), P the position of the orbiting planet. The red circumference is tangent to the ellipse at the extremes of the major axis. The X- and Y-axes of the Solar Barycentre Coordinate System are shown in red. The *true anomaly* is the angle $\nu = AFP$, the *eccentric anomaly* is the angle $\psi = AOP'$

---

[2] The Committee on Data of the International Science Council (CODATA) releases every 4 years an update on the value of several hundreds physical constants. The 2018 release (Tiesinga 2021) has revolutionized the SI unit system, defining several fundamental constants (c, $\hbar$, k_B, ϕ_0, e, eV...) as *exact*. This means redefining metre, second and kilogram in terms of these new "yardsticks". The values of constants are also available at https://physics.nist.gov/cuu/Constants/index.html.

Table 1.1 The six Keplerian parameters that define an orbit and the position of the planet (or satellite) along it

Symbol	Definition for planets and stars *(for Earth satellites)*
i	Inclination over the ecliptic *(over the equatorial plane)*
Ω	Right ascension *(Longitude)* of ascending node
ω	Argument of periastron *(of perigee)*
a	Major semi-axis
e	Eccentricity
$\nu(t_0)$	True anomaly at epoch t_0

and $M_\odot \gg M_{planet}$, so that we can take the system centre of mass to coincide with the Sun centre. The orbit is described by the simple equation in polar (r, ϕ) form

$$r(\phi) = \frac{a(1 - e^2)}{1 + e \cos \phi} \tag{1.4}$$

where the attracting body (i.e. the Sun) is at the *main focus* F ($r = 0$) and ϕ is measured from the semi-major axis directed towards the perihelion A.

The above equation is of little help when we need to describe the position of an orbiting body versus time, i.e. to follow its motion along the orbit. We thus need a different way of expressing the trajectory. First, we must revisit the definition of *anomaly*: as mentioned above, it comes in three varieties:

- **Mean Anomaly:** it is simply Ωt, the angle described by a body that hypothetically travels on a circular orbit with the same period $T = 2\pi / \Omega$ of the orbit considered. It corresponds to a constant, mean angular velocity.
- **True Anomaly:** it is the angle, seen from the focus F, between the perihelion and the planet position, the angle $\nu = AFP$ in Fig. 1.4. The true anomaly is the sixth Keplerian parameter listed in Table 1.1.
- **Eccentric Anomaly:** With reference to Fig. 1.4, consider the point P', intersection of the line HP, perpendicular to the semi-major axis and passing at the body position P, with the circumference tangent to the orbit (i.e. with radius equal to the semi-major axis length a): the eccentric anomaly is the angle $\psi = AOP'$, between the semi-major axis and the line connecting P' with the ellipse centre C.

The relevance of the eccentric anomaly depends on *Kepler's equation* that offers an explicit time dependence:

$$\psi - e \sin \psi = \Omega t \tag{1.5}$$

There is no analytic solution for this equation, but nowadays it is not a worry to numerically obtain $\psi(t)$.

Then, we introduce Cartesian coordinates, as shown in Fig. 1.2, with X-Y lying on the ecliptic plane and the origin at the Sun position F. If we were to place the origin at the centre of mass of the gravitating system (a negligible difference under our assumptions) we would have the Solar Barycentric Coordinate System.

All this considered, and recalling the definition of orbit inclination i, we can write the time-dependent equations of motion for the orbiting body:

$$X = a(\cos \psi(t) - e) \cos i,$$
$$Y = a\sqrt{1 - e^2} \sin \psi(t) \tag{1.6}$$
$$Z = a(\cos \psi(t) - e) \sin i$$

where the time dependence is in $\psi(t)$, through Eq. 1.5. It looks awkward, but it is the best we can do to describe the motion of a planet or satellite.

1.4   The Gravitational Field and Potential

Classical gravitational theory is linear (i.e. the force is proportional to the source "charge"), thus the superposition principle applies to it. We therefore deduce that the acceleration field on a point mass identified by the position vector $\vec{r}$, generated by n point masses m_k, with $k = 1, \ldots, n$, placed at positions $\vec{r}_k$ can be expressed as

$$\vec{a}(\vec{r}) = -G \sum_{k=1}^{n} m_k \frac{\vec{r} - \vec{r}_k}{|\vec{r} - \vec{r}_k|^3} \tag{1.7}$$

Just as for electrostatic, we define a gravitational *field* $\vec{g}$ as the force acting on the unit test mass.

The reader will notice that this field coincides with the gravitational acceleration felt by the test mass: this identity is the cornerstone of the principle of equivalence (see Chap. 3), the founding concept of general relativity.

This acceleration field is conservative and a potential Φ can be associated to it, such that $\vec{a} = -\vec{\nabla}\Phi$, where

$$\Phi(\vec{r}) = -G \sum_{k=1}^{n} \frac{m_k}{|\vec{r} - \vec{r}_k|} \tag{1.8}$$

The Newtonian force field of point like mass has the same central behaviour and the same $1/r^2$ dependence of the electrostatic field due to a point charge. It follows that we can prove the analogous of Gauss' theorem so that, when we deal with a mass distribution described by the volumetric density function $\rho(\vec{r}\,')$, we have

$$\vec{\nabla} \cdot \vec{a} = 4\pi G \rho \tag{1.9}$$

Hence, the general solution for the gravitational field is found by solving the Poisson equation:

$$\nabla^2 \Phi = 4\pi G \rho \tag{1.10}$$

However, the attractive character of the gravitational force (there is no negative mass) entails significant differences in the observed effects, for instance, the lack of gravitational screening.

In the case of a spatially continuous mass, distributed over a volume swept by the variable $\vec{r}'$, the general solution of Eq. (1.10) takes the form

$$\Phi(\vec{r}) = -G \int \frac{\rho(\vec{r}')d^3r'}{|\vec{r} - \vec{r}'|} \tag{1.11}$$

that also coincides with the continuum limit of Eq. (1.8).

1.5 Measurement Units in Experimental Gravitation

The symbol g denotes a mean value of the acceleration produced by gravity on the Earth surface, at sea level:

$$g = \frac{GM_{\oplus}}{R_{\oplus}^2}$$

Actually, the value of the gravitational acceleration changes from place to place, depending on the latitude, altitude and local geological structure: it is indeed the modulus of the vector field introduced in Eq. (1.7), i.e. $g = |\vec{a}(\vec{r})|$. In aero-spatial engineering, the symbol g is used also to indicate the measurement unit of the acceleration, implicitly assuming a reference value of

$$1\,g = 9.80665 \text{ ms}^{-2}.$$

This practice, widely spread among engineers, causes confusion since the same symbol g denotes the gram, a measurement unit of mass. In locations with a given latitude p, the conventional value of the gravitational acceleration at sea level can be derived from the internationally accepted relation (*International Gravity Formula*):,

$$g = 978.0495[1 + 5.3024\, sin^2(p) - 0.0059\, sin^2(2p)] \text{ cm s}^{-2}$$

We can add a final correction: $-3.086 \cdot 10^{-3}\, h$, where h is the elevation in metres, that can easily be derived by series expansion of Newton's law in the vicinity of $r = R_{\oplus}$.

The unit *Gal*, written with capital *G* to avoid confusion with the unit of volume *gal* so appreciated in the Anglo-Saxon world (the *gallon*), was introduced in honour of Galileo Galilei. It has been adopted for the measurements of local variations in the Earth's gravitational acceleration (gravity anomalies). A Gal is equal to $1 \text{ cm s}^{-2} = 1.0197 \cdot 10^{-3}$ g, that is, 1 mGal $\sim 10^{-5} \text{ ms}^{-2} \sim 10^{-6}$ g (Table 1.2).

Table 1.2 Acceleration and gradient units, expressed both in SI units and in terms of $g = 9.80665$ m s^{-2}. In some cases, the approximation $g \simeq 10$ms^{-2} is made

1 Gal	$1.02 \cdot 10^{-3}$ g	10^{-2} m s^{-2}
1 mGal	10^{-6} g	10^{-5} m s^{-2}
1 E	10^{-10} g m^{-1}	10^{-9} s^{-2}
	10^{-7} Gal m^{-1}	
1 mE		10^{-12} s^{-2}

To get acquainted with the new unit of measurement let us note that the mean acceleration produced by gravity is equal to $\bar{g} = 9.81 \cdot 10^5$ mGal and that it varies from $9.832 \cdot 10^5$ mGal at the Equator down to $9.781 \cdot 10^5$mGal at the poles. Variations due to the land's orography are of the order of $10-100$ mGal.

The unit of measurement Eötvös is used in Geophysics to measure the gradient of gravitational acceleration along an axis. An Eötvös is defined as $1\ E = 10^{-7}$Gal/m: in International Systems (SI) of Units it is 10^{-9} s^{-2}. The gravity gradient along the vertical is the largest component, being $\sim 3000\ E$ at ground level (this means that g decreases by $\sim 3 \cdot 10^{-6}$ ms^{-2} per metre of elevation). The horizontal components change roughly half as much and the off-diagonal gradient terms $\partial^2 \Phi / \partial x_j \partial x_k$ are of the order of $100\ E$. Anomalies in the gravity gradient can exhibit values of up to $10^3\ E$ in mountain regions.

1.6   Multipole Expansion of $\Phi(\vec{r})$

When we need to compute the potential in points very far from the source, we choose the origin of coordinates within or close to the source (see Fig. 1.5) and we have $|\vec{r}| \gg |\vec{r}\,'|$.

We can then expand the denominator of Eq. (1.11) in Taylor series about the origin $(x_1', x_2', x_3') = 0$

$$\frac{1}{\sqrt{\sum_i (x_i - x_i')^2}} = \frac{1}{r} - \sum_{i=1,3} \frac{x_i x_i'}{r^3} + \frac{1}{2} \sum_{i,j} (3x_i' x_j' - r'^2 \delta_i^j) \frac{x_i x_j}{r^5} + \cdots \quad (1.12)$$

where $r = |\vec{r}| = \sqrt{x_1{}^2 + x_2{}^2 + x_3{}^2}$ and same for r'.

Using this expansion, the potential Φ of Eq. (1.8) can be expressed as

$$\Phi(\vec{r}) = -\frac{GM}{r} + \frac{G}{r^3} \sum_k x_k D_k + \frac{G}{2} \sum_{k,l} Q_{kl} \frac{x_k x_l}{r^5} + \cdots \quad (1.13)$$

This expansion is useful, as it allows to factorize contributions due to the source (M, D_k, $Q_{kl} \ldots$) and terms due to the position of the test particle (x_k, $x_l \ldots$) while in Eq. 1.11 they are mixed in the $1/|\vec{r} - \vec{r}\,'|$ term.

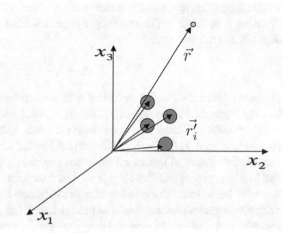

Fig. 1.5 The generic vector $\vec{r}_i'$ indicates the positions of one of the point masses m_i $(i = 1 \ldots N)$ that generate the field, while the vector $\vec{r}$ points to the position P where the gravitational potential must be computed. If the source masses are localized in a finite region far from P, we choose the origin of the reference frame within that region

Let us identify the first few terms of this multipole expansion:

$$M = \int \rho(\vec{r}\,')d^3x'$$ (1.14)

is the total mass of the system, the monopole term;

$$D_k = \int \rho(\vec{r}\,')x_k'\,d^3x'$$ (1.15)

is the mass dipole: it can always be set to vanish by choosing the centre of mass as the origin of the coordinates;

$$Q_{ij} = \int \rho(\vec{r}\,')(3x_i'x_j' - r'^2\delta_i{}^j)d^3x'$$ (1.16)

is the quadrupole term, which vanishes in the case of a spherically symmetric mass distribution. For a rotation ellipsoid, the shape of most astronomical bodies, the quadrupole moment can be expressed in terms of the equatorial and polar radius, $R_e < R_p$, respectively:

$$Q_{i \neq j} = 0; \qquad Q_{xx} = Q_{yy} = -\frac{1}{2}Q_{zz};$$

$$Q_{zz} = \frac{2M}{5}(R_e^2 - R_p^2) = I_p - I_e$$ (1.17)

where I_p and I_e are the moments of inertia with respect to the rotation axis $\hat{z}$ and to an equatorial diameter, respectively. Moreover, if the deviation from sphericity is small, $\Delta R = R_e - R_p \ll R$, with R the mean radius, we can further simplify the expression:

$$Q_{zz} \simeq \frac{4MR}{5}\Delta R$$

In the following, we adopt the convention of normalizing the quadrupole moment, dividing it by $2MR^2$. The resulting dimensionless parameter is called the J_2 factor of a quasi-spherical mass:

$$J_2 \equiv \frac{Q_{zz}}{2MR^2} \tag{1.18}$$

Newton proved that, on the outside a spherical symmetric mass system, the potential has the form $1/r^2$, analogous to that of a point source: we now know that it is a manifestation of Gauss' law. A deviation from the spherical symmetry would modify this dependence on distance. It follows that Kepler's prediction of the planets orbiting around the Sun in elliptical orbits is not rigorously true if the quadrupolar term of the Sun's gravitational field is accounted for, with the corresponding alteration of the $1/r^2$ behaviour. The quadrupolar perturbation leads to a variation of the orbital period of a planet and the resulting precession of the orbital perihelion. This effect is therefore expected within the classical gravitational theory and its magnitude depends on the value of the quadrupolar contribution of the field source.

In the case of the Sun (symbol $\odot$), the experimental results have been under debate for decades: in 1967, Dicke and Goldenberg (1967, 1970) derived the Sun eccentricity from measurements of the solar radiation intensity: they projected its image on a rotating disc with a slit and a photocell beyond the slit. An occulting disc was used to obscure most of the Sun surface, allowing to focus attention on the peripheral regions, similarly to what is done in coronagraphy. These measurements, significantly influenced by the local emission activity on the solar surface, yielded a value

$$J_{2,\odot} = \frac{Q_{zz}}{2M_\odot R_\odot^2} \simeq \frac{2}{5} \frac{\Delta R}{R_\odot} = 2.4 \pm 0.2 \cdot 10^{-5} \qquad \text{- Dicke, 1961}$$

This first result was much larger than what measured a few years later, in 1975, using the same technique, by Hill and Stebbings: $J_{2,\odot} = 1 \pm 4 \cdot 10^{-6}$.

The difficulty in assessing the quadrupole mode of the Sun is the lack of knowledge about the motion of its interior: does the Sun rotate as a solid body, or are there internal layers rotating at different speeds?

An alternate way to derive the value of the quadrupole moment was developed during the 1980s: the method is based on the analysis of the frequencies of the Sun vibrational modes. The vibrations are usually observed on the solar surface by measuring the Doppler shift of solar absorption lines formed in the lower part of the solar atmosphere. There are numerous ground-based and space observatories dedicated to solar oscillations. The most prominent projects are the GONG (Global Oscillation Network Group) with six observing stations around the globe and the SOI-MDI (Solar Oscillation Investigation—Michelson Doppler Imager) aboard the space solar observatory SOHO. These observations have provided detailed information about the internal structure and rotation of the Sun. The Sun is modelled as a quasi-spherical elastic body, and its vibrational modes are described by expanding the displacement field in spherical harmonics, Y_{lm}, where l is the harmonic index and m is the azimuthal index (see Eq. 1.19, next page).

The normal modes with the same l resonate in principle at the same frequency, but the degeneracy of modes with different azimuthal m index is resolved if the body is rotating. Therefore, one can deduce, from the frequency splitting of these modes, the angular speed of the inner layers of the Sun, address the issue of the deformation of the star and finally compute the quadrupole moment (Duval 1984).

The currently accepted value, based on more recent helioseismometric observations (Mecheri 2004) from satellites, is

$$J_{2,\odot} = (2.21 \pm 0.02) \times 10^{-7} \qquad \text{SOHO, 2015}$$

Knowledge of $J_{2,\odot}$ is crucial for all the experiments measuring precession to verify a gravitational theory. Indeed, all these experiments rely on the subtraction of this classic contributions. We shall return to discuss the Sun quadrupole moment, and its implications for gravitational theories, in Sect. 6.6.2.

The Earth too has a non-zero quadrupole moment:

$$J_{2,\oplus} = 1.082 \ 10^{-3}$$

caused by the difference between its polar and equatorial radii, namely, $\Delta R = 21.4$ km. This in turn corresponds to an eccentricity of the Earth's rotational ellipsoid of $\epsilon = \Delta R/R = 3.35 \cdot 10^{-3}$ (IERS 2020).

For a complex and structure-rich solid as the Earth, the quadrupole is just the most basic correction to the spherical approximation. As indicated by the dots in Eq. 1.13, the expansion can be carried out to an arbitrary degree. However, carrying out the Taylor series expansion beyond the quadrupole term becomes increasingly cumbersome: a more manageable, and widely used, approach consists in expanding these corrections in terms of spherical harmonics:

$$\Phi(\vec{r}) = \frac{GM}{r}\left(1 + \sum_{n=2}^{N_{max}}\left(\frac{R}{r}\right)^n \cdot \right.$$
$$\left. \cdot \sum_{m=0}^{n} \bar{P}_{nm}(sin\phi)[\bar{C}_{nm}\cos m\lambda + \bar{S}_{nm}\sin m\lambda] \right) \tag{1.19}$$

where we made explicit the spherical harmonics Y_{lm} as the product of Legendre polynomials $\bar{P}_{nm}$ in the latitude ϕ and sinusoidal functions of the longitude λ. $\bar{C}_{nm}$ and $\bar{S}_{nm}$ are the normalized, unit-less, spherical harmonic coefficients or Stokes' coefficients. For the Earth, $R \equiv R_{\oplus} = 6378136.3$ m is the mean equatorial radius of the *reference ellipsoid*, defined in Sect. 14.4. This expansion can be used for any celestial body: recent and future space missions measure the first multipoles for other planets: BepiColombo for Mercury, Cassini for Saturn, Juno for Jupiter...

1.7   Mapping the Earth Gravitational Potential

In the last few decades, a large effort has been devoted, mainly by the geophysics and geodesy communities, to measuring the multipoles of the Earth gravitational field to higher and higher order, to generate a detailed model of the Newtonian field of our planet. This is achieved by measuring the magnitude of the perturbations induced on the trajectories of artificial satellites orbiting the Earth: these geodetic surveys started as early as 1958 (just 1 year after the very first satellite, Sputnik) with Sputnik 2 that measured the Earth oblateness. They continue, with a variety of technologies and with increasing accuracy, till our days. The method is, in principle, simple: track the satellite position, interpolate positions to calculate its orbit and evaluate deviations from a pure Keplerian orbit. Then, solving an inverse problem, derive perturbations to the spherical symmetry of the source mass, i.e. the multipoles, that cause such deviations. The main techniques of satellite tracking are as follows:

– **Satellite Laser Ranging (SLR)**: The satellite is tracked by a network of laser ranging stations on the Earth: each station shines light pulses on it, and measures the travel time of the reflected beam, deriving the distance between each station and the satellite. Orbits are determined with an accuracy better than a cm.

This is a powerful technique that, by looking at the position of satellites in the sky, derives a wealth of information about the Earth: non-sphericity (multipoles), the European Terrestrial Reference System (ETRS89), drift of continental plates, motion of the tracking stations, motion of the North pole, changes of rotation period and of the Length-Of-Day, definition of Universal Time and more. Besides, the gravitational perturbations of satellite orbits are measured: effects of Moon, Sun and other planets, oceanic tides and solid tides. Last but not least, general relativity perturbations. Indeed, the most studied geodetic satellites are the two LAGEOS[3] : they are completely passive, almost spherical (60 cm in diameter) objects, with the surface covered with retro-reflectors. They are still in operation and are expected to orbit the Earth for the next 8 million years. Analysis of their orbits has led to the first measurement of the Lense-Thirring relativistic precession, as we shall see in Chap. 5.

– **Global Navigation Satellite System (GNSS)** is based on a network of synchronized radio emitters satellites: it can locate, within a few metres, any object equipped with a receiver. The first and most famous of these systems is the US network GPS (Global Positioning System). Today we also have the Russian GLONASS, the European Galileo and the Chinese BEI DOU networks. Operation of GNSS networks, and their implications with special and general relativity, are the subject of Chap. 14.

---

[3] LAser GEOdynamics Satellite (NASA 1976) and (NASA-ASI 1992).

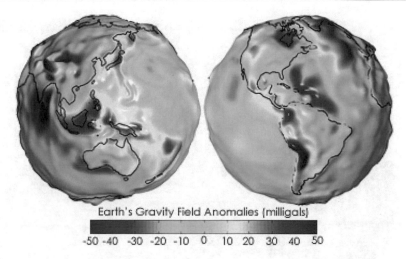

Earth's Gravity Field Anomalies (milligals)

-50 -40 -30 -20 -10   0   10   20   30   40   50

Fig. 1.6 Map of the Earth's gravity anomalies, from GRACE data. Credits: NASA

– **Gravity Gradiometry**: A gradiometer consists of a pair of accelerometers at short distance d along an axis and measures the change, over d, of one component of the gravitational field $\vec{g}$. As $d \ll R_\oplus$, this difference is representative of one component of the gravity gradient. By combining the data of three orthogonal pairs, one can verify Laplace's equation

$$\sum_k \frac{\partial^2 \Phi}{\partial x_k^2} = 0$$

More complex instruments can measure the *full* gravity gradient tensor $\frac{\partial^2 \Phi}{\partial x_j \partial x_k}$: this is actually the tidal tensor (see Sect. 1.8), symmetric and, being solution of the Laplace equation, traceless. Thus, it has five independent components and a minimum of five accelerometer pairs is needed to measure it. Gravity gradients can be integrated and, together with altitude data gathered by other instruments, return a detailed gravitational map of the land and sea that the satellite flies over. The foremost example of these surveyors was the satellite GOCE.[4]

– **Satellite to Satellite Tracking**: Two satellites in Low Earth Orbit (LEO) track their relative position by exchanging a radio link. Variations in their distance are due to irregularities of the gravity field. The twin satellites mission GRACE[5] was so successful that a new mission, GRACE Follow-On, was launched in 2018, with identical instrumentation plus a laser link between the two satellites, accomplishing the first spacecraft to spacecraft interferometer. We shall return to this in Chap. 11 dedicated to the LISA space mission. This wealth of information is used, by a large geodesy and geophysics community, to evaluate and keep track of geodynamic phenomena

---

[4] Gravity Field and Ocean Circulation Explorer (ESA, 2009–2013).
[5] Gravity Recovery and Climate Experiment (NASA-DLR 2002–2017).

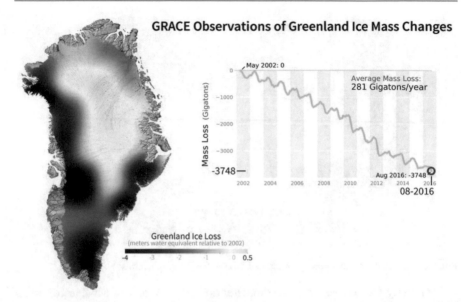

Fig. 1.7 Ice loss in Greenland, as measured from Grace data: in the period 2002–2016, 280 gigatons of ice per year were lost, causing global sea level to rise by 0.8 mm per year. Credits: NASA

(Fig. 1.6), such as crustal motion, Earth rotation and polar motion, to measure gravity field parameters, tidal Earth's deformations, coordinates and velocities of SLR stations, and other substantial geodetic data. Impressive data were recorder, just to give an example, on the constant reduction of Arctic ice Fig. 1.7. Earth Gravitational models (EGM) are periodically updated, with ever-increasing accuracy and detail: the EGM2008, largely based on data from GRACE, GOCE and the German CHAMP,[6] provides Stokes' parameters of Eq. 1.19 up to $N_{max} = 2160$, with a spatial resolution of 5' x 5' minutes of arc, requiring $\sum_{\ell=0}^{N_{max}} (2\ell + 1) =$ over 4.6 million coefficients.

1.8   Gravitation and Tides

The tidal phenomenon is another powerful manifestation of the gravitational field described by Newtonian theory, of its obvious consequence that the force of gravity exerted on an extended object is larger on the object regions nearest to the gravity source and weaker where it is farther. On the Earth surface, for example, gravity is slightly stronger at one's feet than at the head. Thus, the tide is a differential force, a secondary effect of the gravitational field.

Tides are responsible for various well-known phenomena, including sea tides, tidal locking, breaking apart of celestial bodies and formation of ring systems within

---

[6] Challenging Minisatellite Payload (DLR 2000–2010).

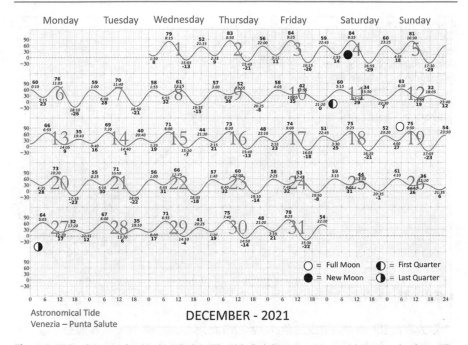

Fig. 1.8 Tides forecast for Venice (Italy). The tide heights are expressed in cm and refer to the Zero Mareografico (tidal-wave) of Punta Salute (Z.M.P.S. 1897). The average sea level has been constantly raising since 1897 and is currently about 31 cm higher than this reference. The predictions apply for normal weather conditions. Credits: Comune di Venezia

the Roche limit.[7] The most relevant manifestation of this phenomenon is the rise and fall of sea level caused by the combined effects of the gravitational forces exerted by the Moon and the Sun, and the rotation of the Earth. Tides vary on timescales ranging from hours to years due to a number of factors, which determine the lunitidal interval. To make accurate records, tide gauges at fixed stations measure water level over time. These gauges are designed so to be insensitive to variations caused by waves, with periods shorter than minutes (Fig. 1.8). The classic theory of gravitation opens the way for a detailed understanding of tidal prediction. It explains, just to give the most basic example, why tides appear twice a day. However, the dynamics of tides is more complex than what gravitation itself can predict: the actual tide at a given location is determined by an accumulation effect of those forces acting on the body of water over many days, by the depth and shape of the ocean basins (their *bathymetry*), and by the coastline shape.

---

[7] The Roche limit, named after Édouard Roche (1820–83), is the minimum distance to which a large satellite can approach its primary body without tidal forces overcoming the internal gravity holding the satellite together. If the satellite and the primary body are of similar composition, the theoretical limit is about 2.5 times the radius of the larger body. Inside the Roche limit, orbiting material disperses and forms rings, as we can see around Saturn. Its rings may be the debris of a demolished moon. Artificial satellites are too small to develop substantial tidal stresses.

Fig. 1.9 A sketch of
Newton's mechanism that
explains the ocean tides.
Dotted ellipses represent the
differential acceleration on
circles of particles. The
figure is not to scale, and the
distortion of the tidal field is
greatly exaggerated

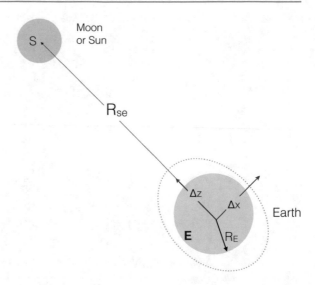

1.8.1   Modelling Tides

To introduce the mathematical study of the tides, we consider a spherical test mass E
(the Earth), of mass M_E and radius R_E moving under the effect of the gravitational
force due to a source S (the Moon or the Sun) with mass M_S.[8] We call this motion
free fall, as it is solely determined by gravity and by the initial conditions. We assume
the Sun to have spherical symmetry; Sect. 1.4 tells us how to correct this hypothesis,
if needed. In the case of an infinitesimally small elastic sphere, the effect of a tidal
force is to distort the shape of the body without any change in volume. The sphere
becomes an ellipsoid with two bulges, pointing towards and away from the other
body. Larger objects distort into an ovoid, and are slightly compressed, which is
what happens to the Earth's oceans under the action of the Moon.

To evaluate this phenomenon, we position the origin of a reference frame at the
centre of the mass E, so that the source S is at distance R_{se} along the $\hat{z}$-axis (Fig. 1.9).

We focus our attention on two points positioned at distance $\pm\Delta z$ ($\Delta z \ll R_E \ll$
R_{se}) from the centre of E (and distant $R_{se} \pm \Delta z$ from S). The acceleration in these
points can be computed:

---

[8] In general, and elsewhere in this book, the Earth is labelled with the traditional astronomical
symbol ⊕, the Sun with ⊙ and the Moon with ☾. Here, for sake of clarity and generality (our source
S can be either the Sun or the Moon), we use E and S, respectively.

$$\vec{a}_{ext}(0,0,\pm\Delta z) = -\frac{GM_S}{\left(R_{se}\pm\Delta z\right)^2}\hat{z} \simeq -G\frac{M_S}{R_{se}^2}\left(1\mp2\frac{\Delta z}{R_{se}}\right)\hat{z} =$$

$$\equiv \vec{a}_0 \pm \delta a_z\hat{z} \qquad (1.20)$$

The residual term $\delta a_z = \pm 2(GM_S\Delta z/R_{se}^3)$ is the tidal acceleration along the axis joining the centres of the two masses: while the whole body "falls" in the field of the source mass M_S with a common acceleration $a_0 = -GM_S/R_{se}^2$, the particles on its surface feel an additional acceleration δa_z due to the non-uniformity of the field. A similar calculation carried on for two particles positioned on the equatorial plane normal to $\hat{z}$, say at $\pm\Delta x$ from the centre, shows that they feel a relative acceleration that is half as large, and with opposite sign, with respect to that along $\hat{z}$:

$$\delta a_x(\pm\Delta x, 0, 0) = -\frac{GM_S}{\left(R_{se}^2 + \Delta x^2\right)}\frac{\pm\Delta x}{R_{se}} \simeq \mp G\frac{M_S\Delta x}{R_{se}^3}$$

So, the tidal force pushes away the particles along the axis pointing to the external mass M, and pulls closer together those on the plane normal to it. We shall encounter again this behaviour in Chap. 7, when dealing with gravitational waves.

We can now generalize the analysis, computing the gravitational acceleration in a generic point near the surface of E. We shall neglect the common acceleration a_0, but include, in the potential, the local gravity of body E and the tidal term due to S. The equation of motion of a particle in a gravitational potential $\Phi(\vec{x})$ is

$$\frac{d^2x^i}{dt^2} = -\hat{x}_i \cdot \vec{\nabla}\Phi(\vec{x}) = -\frac{\partial\Phi(\vec{x})}{\partial x^i} = -\sum_{k=1}^{3}\delta^{ik}\frac{\partial\Phi(\vec{x})}{\partial x^k} \qquad (1.21)$$

where x^i is the component of the position vector $\vec{x}$ of the test particle in the direction $\hat{x}_i$. To describe the tidal effect of differential gravity acceleration, we focus on two test particles falling in the field of the source mass: we define the vector $\vec{\xi} = \vec{x}_2 - \vec{x}_1$ as the separation between the two particles. While the motion of the first particle is described by Eq. 1.21, for the second we have

$$\frac{d^2(x^i + \xi^i)}{dt^2} = -\sum_k \delta^{ik}\frac{\partial\Phi(\vec{x}+\vec{\xi})}{\partial x^k} \qquad (1.22)$$

that we can expand, for sufficiently small separation, as

$$\frac{d^2(x^i + \xi^i)}{dt^2} = -\sum_k \delta^{ik}\left(\frac{\partial\Phi(\vec{x})}{\partial x^k} + \frac{\partial}{\partial x^j}\left(\frac{\partial\Phi(\vec{x})}{\partial x^k}\right)\xi^j + \cdots\right) \qquad (1.23)$$

Thus, the Newtonian geodesic deviation equation for the separation vector ξ^i is

$$\frac{d^2\xi^i}{dt^2} = -\sum_{k,j}\delta^{ik}\left(\frac{\partial^2\Phi(\vec{x})}{\partial x^k\partial x^j}\right)\xi^j \qquad (1.24)$$

This last equation shows how the distance between test particles changes due to a non-uniform field. We introduce the Newtonian tidal tensor[9] :

$$T^i{}_j = \sum_k \delta^{ik}\left(\frac{\partial^2 \Phi(\vec{x})}{\partial x^k \partial x^j}\right)$$ (1.25)

In the particular case of two particles falling in the gravitational field of an isolated, spherically symmetric mass distribution, i.e. $\Phi = \Phi_E = -GM_E/r$, we have

$$T^i_j = (\delta^i_j - 3n^i n_j)\frac{GM_E}{r^3}$$

where $n_i = x_i/r$ are the components of the unit vector in the radial direction.

Note that

$$\sum_i T^i{}_i = \nabla^2\Phi = 4\pi G\rho$$ (1.26)

is the way to write the Poisson equation via the tidal tensor, and similarly the equation for ξ, the geodesic deviation of the two particles in the gravitational field is

$$\frac{d^2\xi^i}{dt^2} = -\sum_j T^i{}_j\xi^j$$ (1.27)

We can conclude by stating that the complete description of gravity is done by $T^i{}_j$, which is a fundamental tensor which can describe the gravitational interaction beyond the individual accelerations $g_i = \partial\Phi/\partial x^i$.

Note that the tidal field, i.e. a non-uniform gravity, is the reason why gravity can be replaced by a free-falling reference frame (a founding idea of general relativity) only *locally*, in a restricted region of space and, as we shall see, time.

If we are dealing with two astronomical bodies at large orbital separations and the tidal interaction between them is negligible, the bodies behave as point masses. When the tidal interaction becomes important, the bodies are deformed, and their shape deviates from the spherical symmetry. Focusing on the *passive* role of one of the bodies (body E), we see that, in the gravitational field of S, it acquires a quadrupole moment

$$Q_{ij} = \int \rho(x'_i x'_j - \delta_{ij}\frac{r'^2}{3})d^3x'$$

---

[9] We shall see in Chap. 4 how general relativity extends the concept of geodesic deviation equation. Indeed, the tensor defined in the right-hand side of Eq. 1.25 is related to the Riemann tensor of relativity by $T^i_j = c^2 R^i_{0j0}$.

as described in Sect. 1.4. This quadrupole moment is proportional to the external tidal field T_{ij}, and is traditionally expressed as

$$Q_{ij} = -\frac{2}{3}k_2\frac{R^5}{G}\cdot T_{ij}$$

where R is the body's radius, the factor of $2/3$ is conventional and the dimensionless constant k_2 is the tidal Love number for a quadrupolar deformation. The Love numbers, introduced by A.E.H. Love in 1909, measure the rigidity of a planetary body, i.e. the susceptibility of its shape to change in response to a tidal potential. In other words, they measure the ratio between the response of the real Earth and the theoretical response of a perfect fluid sphere.[10] For the elastic Earth, $k_2 \sim 3.1$. It follows that the gravitational potential near the surface of body E, at a distance r from its centre, is composed of three terms: the monopole, depending on the source mass M_E, the external tidal field due to S and the body's own response to the tidal interaction

$$\Phi_E(\vec{r}) = -\frac{GM_E}{r} + \frac{G}{2}\left[1 + 2\frac{k_2}{3G}\left(\frac{R_E}{r}\right)^5\right]\sum_{i,j}T^{ij}x_ix_j \qquad (1.28)$$

The quadrupole term is the next-to leading order term of the Taylor expansion of the potential Φ. If we include additional terms we have to consider tidal moments of higher multipole orders, and higher powers of the coordinates x_i. Thus, the generalization to higher orders of the gravitational potential outside the body leads to a more complex formula in which the coefficients weighing the high order tidal moments are

$$G\frac{1}{l(l-1)}\left[1 + 2k_l\left(\frac{R}{r}\right)^{2l+1}\right] \qquad (1.29)$$

Here k_l is the Love number for the $l-th$ configuration, a constant of proportionality between the tidal field applied to the body and the resulting multipole moment of its mass distribution.

Tidal Love numbers are introduced also in general relativity: in reference to the gravito-electromagnetic formalism, described in Chap. 5, they are catalogued as electric-type Love numbers k_{el}, having a direct analogy with the Newtonian Love number here introduced, and magnetic ones k_{mag} that are associated to a purely relativistic effect.

1.8.2 Tides Classification

Current procedure for analysing tides follows the method of harmonic analysis introduced in the 1860s by W. Thomson. It is based on the principle that the combined

---

[10] To account for dissipation, Love numbers can be complex.

motion of Sun and Moon contains a large number of component frequencies, and at each frequency there is a component of force acting to produce tidal motion. At each place of interest on the Earth, the tides respond to each frequency with an amplitude and phase peculiar to that locality. These components are, for the most part, the six basic frequencies, and their harmonics, of the Moon-Earth-Sun dynamics.

The different harmonic content of the tide-generating potential is specified using the *Doodson numbers*, a system proposed by Arthur Thomas Doodson in 1921.

It is based on six coefficients, indicating the harmonic of each of these six basic periodic angles:

1 τ: the Greenwich Hour Angle of the mean[11] Moon plus 12 h, with a period of 1.04 days. It can be expressed as $\tau = \theta_M + \pi - s$ where θ_M is *the Sidereal Time on the Earth*, whose period is one sidereal day, or 0.9973 days, and s is as follows:
2 s: the mean longitude of the Moon, with a period of one lunar month, or 27.32 days.
3 h: the mean longitude of the Sun. Its period is 1 year, 365.25 days.
4 p_M: the mean longitude of the Moon's perigee. It has a period of 3231.5 days.
5 $-\Omega$: the negative of the longitude of the Moon's mean ascending node on the ecliptic, with a period of 6798.4 days.
6 p_S: the longitude of the Sun's mean perigee: its periodicity is extremely long, 112000 years, and is not used in describing practical tides.

A given tidal frequency f is given by a combination of small integer multiples, positive or negative, of these basic frequencies:

$$f = m_1(\dot{\theta}_M - \dot{s}) + m_2\dot{s} + m_3\dot{h} + m_4\dot{p}_M + m_5(-\dot{\Omega}) + m_6\dot{p}_S$$

where the dot indicates the derivative with respect to time.

While m_1 is used to classify the kind of tide and takes the values 0 (long period), 1 (diurnal) or 2 (semidiurnal), all other parameters take small integer values, positive or negative. To avoid dealing with negative numbers, an offset of 5 is applied to these digits, so that the first negative harmonic takes the value 4 and the third positive harmonic is 8. That periodicity is therefore catalogued with the Doodson number

$$m_1(m_2 + 5)(m_3 + 5).(m_4 + 5)(m_5 + 5)(m_6 + 5)$$

As an example, the well-known principal solar semidiurnal tide S_2, which takes place exactly twice a day, has the Doodson number 273.555, meaning that it has frequency components at twice the mean lunar time, twice the lunar month and -2 times the year, and no component of the remaining three periodicities. An exhaustive table of many tides constituents, with period, Doodson coefficients and other information, can be found in https://en.wikipedia.org/wiki/Theory_of_tides.

Tidal phenomena are not limited to the oceans, but can occur in other systems whenever a gravitational field that varies in time and space is present. For example,

---

[11] *mean*: same meaning as in *mean anomaly*, see Sect. 1.1.

the shape of the solid part of the Earth is slightly affected by Earth tide, though this is not as evident as the water tidal movements.

Tidal effects become particularly pronounced near small bodies of high mass, such as neutron stars or black holes, where they are responsible for the *spaghettification* of infalling matter. When we deal with a binary system of two neutron stars, their orbital motion produces emission of gravitational waves, which remove energy and angular momentum from the system. This causes the orbits to decrease in radius and increase in frequency, and leads to the inspiralling motion of the compact bodies. When the orbital separation decreases sufficiently and the tidal effects become important, the neutron stars acquire a tidal deformation, and this affects their gravitational field and orbital motion. The effect is revealed in the shape and phase of the gravitational waves releasing information regarding the compactness of each body, as well as its equation of state.

1.9 Active and Passive Masses

When introducing Newton's law (1.2) we have overlooked some subtle conceptual aspects. First, we directly assumed that the physical property determining the gravitational attraction of a body, the gravitational mass, cannot be different from the physical property of the same body affected by this attraction. Let us examine better this concept. Newton's law implies that the mass is the source of the gravitational force. Let us denote this source by *active gravitational mass* m_a. On the other hand, we denote the physical body affected by such a force by *passive gravitational mass* m_p. These two quantities might not coincide. Consider the force exerted by the Earth upon the Moon: we could assume that the gravitational interaction is proportional to the Active Gravitational Mass of the Earth and to the Passive Gravitational Mass of the Moon. Conversely, the force exerted by the Moon upon the Earth would depend on $m_p^{(E)}$ of the Earth and $m_a^{(M)}$ of the Moon.

Moreover, in the last section, we have derived the dependence of $\vec{F}$ on the moving planet's mass using the second law of dynamics, whence the passive mass is identified with the inertial one, m_i; this latter being the physical quantity defined by Eq. (1.1), and characterizing the dynamical response of the body to the forces acting on it. The identification of the gravitational mass with the inertial one is the foundation of the equivalence principle which we will discuss in detail in Chap. 3. For the time being, we shall continue our analysis by introducing in an operational fashion the three quantities m_i, m_a and m_p and examining the implications that result from such distinction.

Consider a two-body system, the first of which has a role of a reference unitary mass, say, for example, 1 kg, $m_1^{(1\,\text{kg})}$. In the absence of external forces, the action-reaction principle states that

$$m_1^{(1\text{kg})}\vec{a}_{1\text{kg}} + m_i\vec{a}_i = 0 \qquad (1.30)$$

From this, we deduce the operational definition of inertial mass, m_i:

$$m_i \overset{def}{=} m_1{}^{(1kg)} \frac{|\vec{a}_i|}{|\vec{a}_{1kg}|} ,$$

(1.31)

that is, the ratio between the inertial masses of the two bodies can be directly obtained from the ratio of their respective accelerations.

Having thus defined the inertial mass, we can proceed to define the passive gravitational mass, assuming there is a unitary test mass of active type which attracts the mass m_p due to gravitational interaction. Hence, by applying the second law of dynamics

$$G \frac{m_a{}^{(1kg)} m_p}{r^2} \frac{\vec{r}}{r} = m_i \vec{a}$$

(1.32)

we end up with an expression for m_p that depends on the product $r^2 a(r)$, which turns out to be a conserved quantity, according to Kepler's third law:

$$m_p \overset{def}{=} \lim_{r \to \infty} m_i \left(\frac{|\vec{a}| r^2}{G m_a{}^{(1kg)}} \right)$$

(1.33)

We proceed in the same way for defining the active gravitational mass, assuming there is a unitary test mass of passive type subject to the gravitational influence of the active mass m_a. We thus derive that the operational definition of m_a is

$$m_a \overset{def}{=} \lim_{r \to \infty} m_i \left(\frac{|\vec{a}| r^2}{G m_p{}^{(1kg)}} \right)$$

(1.34)

We can now apply these definitions to the case of two interacting masses, m_1 and m_2. Applying the third law of dynamics

$$|\vec{F}_1| = |\vec{F}_2|$$

where

$$\vec{F}_1 = m_{1i} \vec{a}_1 = -G m_{1p} m_{2a} \frac{\vec{r}_1 - \vec{r}_2}{|\vec{r}_1 - \vec{r}_2|^3}$$

(1.35)

$$\vec{F}_2 = m_{2i} \vec{a}_2 = -G m_{2p} m_{1a} \frac{\vec{r}_2 - \vec{r}_1}{|\vec{r}_2 - \vec{r}_1|^3}$$

(1.36)

we conclude that the ratio between active and passive masses of each body is a constant, that is, active and passive gravitational masses are proportional quantities:

$$\frac{m_p}{m_a} = \beta$$

(1.37)

1.9.1 The Experiment of Kreuzer

The identification of active mass with its passive counterpart was verified in the elegant experiment (Kreuzer 1968). His idea was to generate a gravitational force by a Teflon cylinder (76% of fluoride) submerged in a mixture of trichloroethene and dibromomethane (76% of bromine). The two massive elements, one liquid and the other solid, are chosen because they are chemically inert and their nuclear composition is very different (Fig. 1.10).

The Teflon cylinder, which is immersed in the liquid, is connected to a motor by a thin nylon wire and a pulley system. The motor forces the cylinder to slowly oscillate back and forth (with a period of 400 s). If there is a small density difference between the fluid and the Teflon block, the gravitational force of the liquid+solid system exerted upon the mass of a torsion pendulum placed in front of the container can be measured experimentally, see Fig. 1.11 (details of the torsion pendulum as gravitational detector are discussed in the next chapter). The density of the liquid is then changed, varying the temperature, up to the point where the densities of the two materials are equal: if passive and active masses coincide, the gravitational force must vanish. Note that the density measurements of the mixture and the Teflon block depend on the passive mass. They are indeed obtained by measuring the effect of the Earth's gravitational field in samples of these materials.

As the difference in density of materials changes with temperature, comparing the measurements of density difference versus T to those of the gravitational force exerted on the torsion pendulum, we can obtain an upper limit on the fractional difference between active and passive masses by analysing the way the two functions cross zero. The two plots of Fig. 1.11 show the original data versus temperature:

Fig. 1.10 The Kreuzer experiment: diagram of the gravitational field generator formed by a Teflon cylinder immersed in a mixture of a liquid with a density very similar to the density of the solid. The cylinder is moved back and forth in the liquid by a motor connected to the pulley depicted in the figure. Credits: reprinted figure with permission from (Kreuzer 1968). Copyright 1968 by the American Physical Society

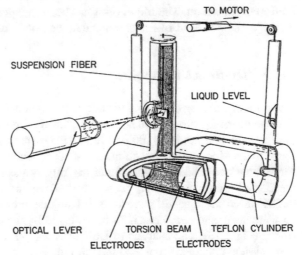

FIG. 1. Cut-away view of tank, Teflon cylinder, and torsion balance.

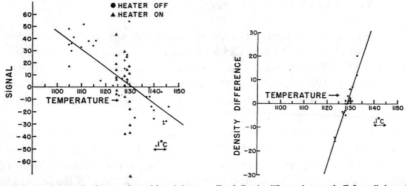

FIG. 2. Torsion balance signal averaged over 3-h periods as a function of liquid temperature.

FIG. 3. Density difference between the Teflon cylinder and the liquid as a function of liquid temperature.

Fig. 1.11 Data of the Kreuzer experiment: the right panel shows a plot of the density difference between liquid and solid versus temperature, while the left panel depicts the corresponding dependence of the signal from the torsion pendulum. Credits: reprinted figures with permission from (Kreuzer 1968). Copyright 1968 by the American Physical Society

(a) The density difference between liquid and solid.
(b) The signal from the torsion pendulum, which depends on the difference of active mass between fluid and solid.

The zero crossing of the two functions occurs at a temperature marked by a thermistor reading of 1130 Ω: analysing these data, Kreuzer claimed the ratio between active and passive masses to be 1, with an accuracy level of $5 \cdot 10^{-5}$.

A different limit, accurate but model dependent, was derived by observing the motion of the Moon. The Moon orbit has been monitored, for the last 50 years, by means of the powerful LLR technique that we briefly describe here.

1.9.2   Lunar Laser Ranging

Lunar Laser Ranging (LLR) is a technique that is conceptually very simple: evaluate the distance of the Moon by measuring the time of flight of a light beam. Short pulses of laser light, of order $10 - 100 \, ps$, are sent from a network of tracking stations on the Earth to the Moon, reflecting off arrays of corner cube retro-reflectors placed on the Moon's surface (Fig. 1.12). The round-trip time is accurately measured, from which the Earth-Moon distance may be deduced. There are five retro-reflector arrays on the Moon surface that still serve as useful targets: three left by the Apollo 11, 14 and 15 astronauts, and two, French built, deployed by the Soviet Lunokhod unmanned rovers (Fig. 1.13). The analysis of LLR data requires a sophisticated model of the solar system ephemeris that also includes all the significant effects that contribute to the range between the Earth stations and the lunar retro-reflectors. These models compute a range prediction and the partial derivatives of range with respect to each model parameter at the epoch of each measurement. The model predictions take into

Fig. 1.12 Left: the Matera Laser Ranging Observatory of ASI; photo courtesy of F. Ambrico Right: one of the retro-reflectors placed on the Moon's surface by the Apollo space program. Credits: NASA

account orbital parameters, attraction to the Sun and planets, relativistic corrections, as well as tidal distortions, plate motion and other phenomena that affect the position of both the retro-reflector and of the ground station relative to the centres of mass of the Earth and Moon. Some of these parameters are measured by other means, but most are estimated from a huge fit to the LLR data. The range measurements are corrected for atmospheric delay and a weighted least square analysis is performed to estimate the ∼170 parameters in the model, most of which are initial conditions and masses of solar system bodies. LLR data are often combined with other spacecraft and planetary tracking data to further constrain the estimates or remove degeneracy.

Many observatories around the world routinely make centimetre-level range measurements. In the last 50 years, the ground station technology has improved to the point that the range measurements have a precision of ∼2 cm, limited by the lunar retro-reflectors. Once the data are fitted with a computed orbit, the post-fit residuals (difference between observed and computed) are at the millimetre level. The continuous record of laser range measurements between Earth and Moon, dating back to 1969, has provided an unprecedented set of data by which to understand dynamics within the solar system and to test the fundamental nature of gravity.

This technique is analogous to the Satellite Laser Ranging (SLR), mentioned in Sect. 1.7, that is used to track the motion of artificial satellites covered with retro-reflectors.

1.9.3 The Bartlett-Van Buren Moon Experiment

The Moon test of the equivalence of active and passive mass is directly related to the violation of action-reaction law of the Newtonian dynamics. Bartlett and Van Buren modelled the Moon as a system made of two components: a spherical mantle of density $\rho_{Fe} = 3350$ kg/m^3, whose centre is displaced $t = 10$ km apart from the geometrical centre of the spherical outer shell of density $\rho_{Al} = 2900$ kg/m^3 (see Fig. 1.14). This simplified model accounts for the asymmetrical composition of the

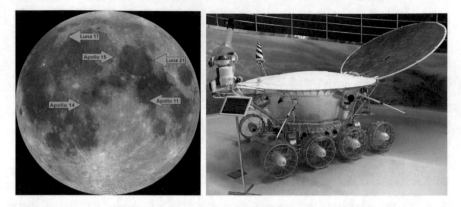

Fig. 1.13 Left: the position of the various retro-reflectors deployed on the Moon surface. Credits: NASA. Right: a model of the Soviet Moon rover Lunokhod used to deploy the French-built retro-reflectors. Photo courtesy of P. Milosević

Fig. 1.14 Model of a two components moon. C_{Al}, C_{Fe} are, respectively, the geometrical centres of the two components of aluminium and iron, while B is the centre of mass of the system. The distance values used in the model are: $t = C_{Al}C_{Fe} = 10$ km and $s = C_{Al}B = 1.98$ km

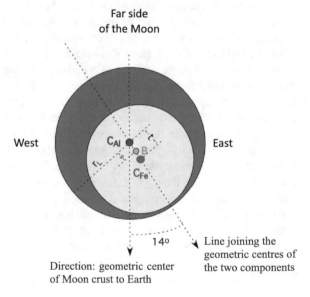

Far side of the Moon

West   East

14°

Direction: geometric center of Moon crust to Earth

Line joining the geometric centres of the two components

two faces of the lunar surface: the one facing the Earth which is rich of iron and the opposite one which is rich of aluminium. Additionally, the centre of mass of the moon turns out to be displaced from the spherical centre by $s = 1.98 \pm 0.06$ km along the direction $\theta_E = 14^o \pm 1^o$ to the East of the vector pointing towards the Earth.

If the action-reaction principle is violated, we should detect an effect due to the existence of a residual force $\vec{F}_{Al,Fe}$, which would depend on the two components. This force has the direction of the segment connecting the two geometrical centres

C_{Al}, C_{Fe} and it is applied in B:

$$\vec{F}_{Al,Fe} = \vec{F}_{Al} + \vec{F}_{Fe} \neq 0 \tag{1.38}$$

The most significant effect on the motion of the Moon would be determined by the component of the force tangent to the orbit,

$$F_t = F_{Al,Fe} \sin \theta_E \tag{1.39}$$

which would in turn produce a variation in the angular speed of the moon $\Delta\omega$. A simple calculation allows us to evaluate the magnitude of such an effect: compute the work done by the force F_t in a full rotation cycle (a lunar month of observations); this must be equal to the change in kinetic energy of the Earth-Moon system. Since the latter is one half of its potential energy, we have

$$2\pi r F_t = \frac{1}{2} M_\oplus \Delta\Phi = \frac{1}{2} G \frac{M_\oplus M_\mathbb{C}}{r^2} \Delta r = \frac{1}{2} F_{\oplus,\mathbb{C}} \Delta r \tag{1.40}$$

where $F_{\oplus\mathbb{C}}$ is the gravitational force between Earth and Moon. So we derive that

$$\frac{\Delta r}{r} = 4\pi \frac{F_t}{F_{\oplus,\mathbb{C}}} \tag{1.41}$$

Next, we differentiate Kepler's third law, $\omega^2 r^3 = constant$, to find the relation between $\frac{\Delta r}{r}$ and $\frac{\Delta\omega}{\omega}$. Then, from Eqs. 1.39 and 1.41, we conclude that

$$\frac{\Delta\omega}{\omega} = 6\pi \sin \theta_E \frac{F_{Al,Fe}}{F_{\oplus,\mathbb{C}}} \tag{1.42}$$

Finally, we need to evaluate $F_{Al,Fe}$ for this two-component model of the Moon. We introduce the factor $S(Al, Fe)$, the difference of the ratios between active and passive masses of the iron and aluminium components

$$S(Al, Fe) = \frac{M_{Al}^{(A)}}{M_{Al}^{(P)}} - \frac{M_{Fe}^{(A)}}{M_{Fe}^{(P)}}$$

that depends on the different nuclear composition of the two materials.

$$F_{Al,Fe} = S(Al, Fe) F_{Al} \tag{1.43}$$

$\vec{F}_{Al}$ ($\sim -\vec{F}_{Fe}$) is the internal force that the aluminium component exerts upon the iron component. We compute its modulus by integrating the field of gravitational force per unit mass f_{Al} of the aluminium body[12] on a mass element of the iron sphere:

$$F_{Al} = \int_{Fe} \rho_{Fe} f_{Al} dV \tag{1.44}$$

In order to deduce a_{Al} and calculate Eq. 1.44, we model the Moon as follows:

- The entire volume of the Moon is filled with material with density ρ_{Al}.
- The iron sphere is made of a sphere of density ρ_{Fe}.
- Superimposed on the latter, there is an identical sphere of density $-\rho_{Al}$.

Due to the spherical symmetry of the problem, it is straightforward to realize that the contribution to Eq. 1.44 from the sphere of density $-\rho_{Al}$ and centred at C_{Fe} is zero. We apply Gauss's theorem to the sphere of density ρ_{Al}. In a generic point, at a distance z from C_{Al} along the direction of the generic unit vector $\hat{k}$, we have

$$\vec{a}_G = \frac{4}{3}\pi G \rho_{Al} z \hat{k} \tag{1.45}$$

Substituting with this field in Eq. 1.44, we obtain

$$F_{Al} = \frac{4}{3}\pi G \rho_{Al} \rho_{Fe} t V_{Fe} \tag{1.46}$$

where V_{Fe} is the volume of the iron sphere and t is the distance between the centres C_{Al} and C_{Fe}.

The product $t V_{Fe}$ is deduced using the information of the Moon mass and the position s of barycentre. This yields

$$s M_{\mathbb{C}} = (\rho_{Fe} - \rho_{Al}) t V_{Fe} = \Delta \rho t V_{Fe} \tag{1.47}$$

Therefore, using Eqs. 1.43, 1.46 and 1.47, we conclude

$$\frac{F_{Al,Fe}}{F_{\oplus,\mathbb{C}}} = \frac{M_{\mathbb{C}}}{M_\oplus} \left(\frac{d_{\oplus,\mathbb{C}}}{R_{\mathbb{C}}}\right)^2 \frac{s}{R_{\mathbb{C}}} \frac{\rho_{\mathbb{C}}}{\Delta\rho} S(Al, Fe) \tag{1.48}$$

where we have introduced the distance $d_{\oplus,\mathbb{C}}$ between the Earth and the Moon, the radius of the Moon $R_{\mathbb{C}}$ and its mean density $\rho_{\mathbb{C}}$.

---

[12] The gravitational force per unit mass f coincides with the gravitational acceleration g if the Principle of Equivalence (see Chap. 3) holds.

Comparing this last expression with Eq. 1.42 we derive an upper limit for $S(Al, Fe)$, which depends on the accuracy of our knowledge of $\frac{\Delta\omega}{\omega}$, i.e. less than $1 \cdot 10^{-12}$.

This value is based on the monthly LLR measurements of $\dot{\omega} = -25.3\ arcsec/century$, where the known effects that distorts the lunar orbit due to the Earth's ocean tides have already been subtracted. The upper bound reads

$$S(Al, Fe) < \frac{1}{5}\frac{1}{6\pi}\frac{1}{\sin\theta_E}\frac{\Delta\omega}{\omega} \simeq 5 \cdot 10^{-14} \tag{1.49}$$

Its validity rests on the assumptions made on Moon composition and its simplified model. However, in the second part of the same paper, the authors (Van Buren 1986) generalize the Moon structure by assuming a multi-component composition, thus providing a more general bound for S.

In conclusion, the experiment supports the assumption that the passive mass of a body is proportional to its inertial one.

$$\frac{m_p}{m_i} = \alpha \tag{1.50}$$

with α a constant independent of material and composition.

There is still, in these days, an ongoing theoretical debate about this assumption within the framework of the principle of Universality of Free Falling (UFF). Delaying to Chap. 3 the verification of the equivalence principle ($m_G = m_I$), we infer, by a suitable choice of the measurement unit of these quantities:

$$m_p = m_a = m_i.$$

As we mentioned above, the proportionality (1.50) between active and passive masses is a manifestation of the action-reaction principle, or, in other words, a consequence of the momentum conservation in isolated **classical** systems.

We note also that the general relativity theory of A. Einstein *postulates* the identity among the various masses here discussed. Therefore, a violation of such an assumption would represent a deep crack in the otherwise robust theoretical framework of Einstein.

All the considerations made up to this point are based on body motions described in the framework of the classical mechanics. However, by imposing the equality $\vec{F}_1 = -\vec{F}_2$ we do not take into account the delays due to the propagation time of the gravitational interaction. They become important when the mass velocity v is no longer small to the respect of the light velocity c. Indeed, in the relativistic regime, the conservation of momentum is restated as conservation of the four-momentum.

In a two-body system, the gravitational interaction propagates in a finite time from one particle to the other, and we must take into account also the momentum (and energy) associated with the interaction field.[13]

References

Bartlett, D.F., Van Buren, D.: Equivalence of active and passive gravitational mass using the moon. Phys. Rev. Lett. **57** (1986)

Cavendish, H.: XXI. Experiments to determine the density of the Earth. Philos. Trans. R. Soc. **88**, 469 (1798)

Dicke, R.H.: The solar oblateness and the gravitational quadrupole moment. Ap. J. **159**, 1 (1970)

Dicke, R.H., Goldenberg, H.: Mark solar oblateness and general relativity. Phys. Rev. Lett. **18**, 313 (1967)

Kreuzer, L.B.: Experimental evidence of the equivalence of active and passive gravitational mass. Phys. Rev. **169**, 1007 (1968)

Duval, T.L., Harvey, J.W., et al.: Rotational frequency splitting of solar oscillations and Internal rotation of the Sun. Nature **310**(19–22), 22–25 (1984)

International Earth Rotation and Reference System Service: IERS Conventions (2020). IERS Technical note n. 36

Mecheri, R., et al.: New values of gravitational moments J_2 and J_4 deduced from helioseismology. Solar Phys. **222**, 191–197 (2004)

Rothleitner, S., Schlamminger, S.: Measurements of the Newtonian constant of gravitation. Rev. Sci. Inst. **88**, 111101 (2017)

Tiesinga, E., et al.: CODATA recommended values of the fundamental physical constants: 2018. Rev. Mod. Phys. **93**, 025010 (2021)

Further Reading

Curtis, H.D.: The two body problem. In: Orbital Mechanics for Engineering Students, 2nd ed. Butterworth Heinemann, (2010)

George T Gillies.: The Newtonian gravitational constant: recent measurements and related studies Rep. Prog. Phys. **60**, 151–225 (1997)

---

[13] A rigorous treatment of the fundamental laws of gravitation and of the complex problem of the four-momentum conservation in the framework of general relativity is given in Chap. 11 of *C. Moller, The Theory of Relativity, Claredon Press Oxford 1972.*

The Torsion Pendulum

2

2.1 The Torsion Balance

The torsion pendulum is the main tool of gravitational physics, used in a variety of configurations and operating modes. Cavendish was the first to use it for exploring the gravitational field, but the merit of its conception and development must be attributed to Coulomb and Michell, whose main goal was to measure electric forces: they realized, in 1777, what is considered the first experiment of modern physics. Indeed, Coulomb's law dates to 1785, some 13 years before Cavendish's claim of "having weighted the Earth" (Figs. 2.1 and 2.2).

They realized that the <u>horizontal</u> equilibrium condition allowed to remove almost completely the dominant pull of the Earth, even that associated to the centrifugal force coming from the Earth's rotation, but not the background noise of gravitational (or, for Coulomb, electrostatic) origin. Additionally, they became aware that to increase the sensitivity of the measurement, long observation (i.e. integration) times were needed, and that the pendulum should have a small damping coefficient, i.e. a long time decay constant.

At that time, the thermal origin of the intrinsic noise was unknown and therefore the experimentalists were unable to estimate the magnitude of such noise, once isolated from the seismic and environmental contributions. This instrument was extensively used during the whole nineteenth and twentieth centuries, in all kind of measurements requiring detection of small forces; just to mention a few investigations: to determine the value of G (more than 200 experiments since Cavendish, almost all of them using a pendulum), to test the equivalence principle (see Chap. 3), to verify several properties of the gravitational field such as the superposition prin-

---

The original version of this chapter was revised: Chapter have been updated with the correction. The correction to this chapter can be found at https://doi.org/10.1007/978-3-030-95596-0_15

F. Ricci and M. Bassan, *Experimental Gravitation*, Lecture Notes in Physics 998, https://doi.org/10.1007/978-3-030-95596-0_2

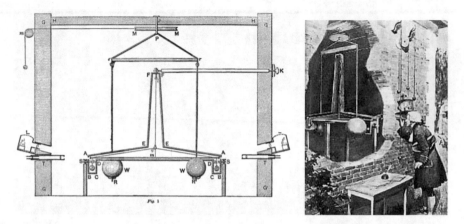

Fig. 2.1 The torsion balance used by Cavendish for "weighting the Earth". The lettering on the drawing refers to the description given in Cavendish (1798). On the right: an old print shows Cavendish reading the torsion angles through the optical lever, from outside the laboratory

Fig. 2.2 The first modern pendulum: the apparatus of Coulomb-Michell, for the study of electrostatics. Credits: Wikimedia Commons

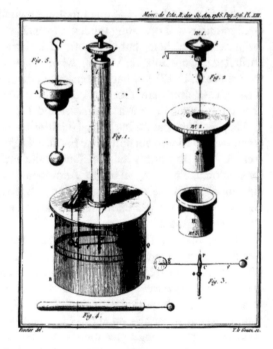

ciple and inverse square law, to evince gravitational anomalies, to search for a rest mass of the photon, to measure the Casimir force and to search for postulated weak forces of either long or short range, the so-called *fifth force*.

The pendulum still today plays a central role in the physics of experimental gravitation; besides, it is at the basis of the working principle of several versions of the gravimeters, which are widely employed in geophysics and geodesy.

The torsion pendulum has been built in uncountable versions: it can have the suspension wire made out of quartz fibre or tungsten wire, diameter of just a few microns or up to millimetres, it can have the inertial member, or *test mass* ranging from 10 *kg* down to a fraction of a gram. It has been implemented in vacuum chambers, inside tanks for submersion in water, placed in mountains tops or in deep caves, it has been cooled to cryogenic temperatures and heated up to incandescence. It is a robust and very versatile instrument assembled in many different versions; it can detect very small forces over a dynamic range of several orders of magnitude (Fig. 2.3).

In all these versions, its intrinsic noise sources have been carefully studied: thermal motion and the perturbations due to the transducer transforming mechanical signals in electromagnetic ones, both in free-running and feedback set-ups. Clearly, the intrinsic noise is only observable if the apparatus is effectively isolated from external noise sources, like ground vibrations. Few other scientific instruments, based on a mechanical device so simple in appearance, have been subject of studies for such a long time. Its evolution still continues (Gillies and Ritter 1993). The persisting discrepancies among the different measurements of *G*, still observed in our days, promote an improvement on the performance of this instrument, thus increasing the comprehension of its physical limits. In order to reduce the contributions from thermal noise and to increase the isolation from vibrations, the experimentalists have followed different strategies, like cooling the pendulum down to cryogenic

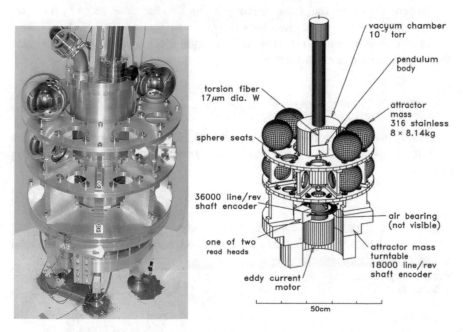

Fig. 2.3 A sophisticated, recent torsion pendulum, built and operated in the Eöt-Wash group at the University of Washington in Seattle for a measurement of *G* (Gundlach andMerkowitz 2000). Courtesy of Eöt-Wash group. On the right a diagram describing its components. Credits: reprinted figure with permission from (Gundlach and Merkowitz 2000). Copyright 2000 by the American Physical Society

temperatures or loading the suspension fibre, by reducing its diameter, until it reaches values close to the maximum safe load. Under these extreme conditions, non-linear effects start to be significant so that it is crucial to understand and optimize the anelastic properties of the system, which is a topic at the frontiers of current research on the dissipative properties of materials.

2.2   The Torsion of the Wire

A dumbbell, i.e. a rigid rod with two masses set at its end, is suspended by a long thin wire keeping the rod horizontally balanced. Applying a torque to the dumbbell, the rod rotates by an angle θ and a new equilibrium condition is reached, where the applied torque is balanced by the elastic one coming from the torsion wire. For simplicity, we will treat the wire as a cylinder of constant cross section: its torsion is a typical non-homogeneous shear deformation. We recall here that a generic volume deformation in an elastic body can be decomposed as the sum of

(a) a simple volume deformation, that is, *a deformation which changes the volume of the body but not is shape*;

(b) a simple slip deformation, that is, *a deformation which leaves the body's volume unchanged, but changes its shape*.

We consider the second case: referring to Fig. 2.4, the shear angle γ, for small deformations ($\tan \gamma \simeq \gamma$), is given by $\gamma = \frac{CC'}{AC}$.

Under elastic deformations condition, the angle γ is proportional to the applied shear stress (torque per unit area) Σ_τ:

$$\gamma = \frac{CC'}{AC} = \frac{\Sigma_\tau}{\mu} \tag{2.1}$$

It can be proven (Love 1994) that the proportionality coefficient μ, called shear modulus, is related to the Young's and Poisson's elasticity coefficients, E and σ, by the relation

$$\mu = \frac{E}{2(1 + \sigma)} \tag{2.2}$$

Considering the torsion of a cylinder, each section perpendicular to its axis (*slice*) remains on a plane, when subject to a twisting horizontal torque, due to its circular

Fig. 2.4 Simple shear
deformation

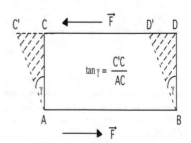

Fig. 2.5 Torsional
deformation of a cylinder of
infinitesimal height dz

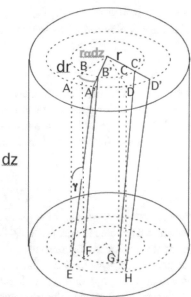

form; however, it ends up rotated with respect to the adjacent slices. We now consider
a slice of elementary height dz, and we focus on a circular sector of length dr and
width $r\, d\phi$ placed on either of the two bases of the cylinder (see Fig. 2.5).

Next, we assume that the corresponding sector on the other base is rotated with
respect to the first one by an angle which linearly grows with the cylinder height.
For a height dz of the elementary cylinder, such an angle is then αdz, where α is
the angle of rotation *per unit length*. For small deformations in the radial position r
of the reference section, the shear angle γ is obtained from the ratio of the arc $r\alpha dz$
and the height of the cylinder dz. Equation 2.1 gives

$$\gamma = r\alpha = \frac{\Sigma_\tau}{\mu} \tag{2.3}$$

We now relate this to the applied torque M_τ:

$$M_\tau = \int_0^R \Sigma_\tau(2\pi r)r\,dr = 2\pi\mu \int_0^R r^3\alpha\,dr = \frac{1}{2}\pi\mu R^4\alpha$$

Finally, we obtain the total torsion angle, that is, our measurable quantity, by multiplying α, the rotation angle per unit length, times the wire length L:

$$M_\tau = \mu\frac{\pi}{2}\frac{R^4}{L}\theta = k_\tau\theta \tag{2.4}$$

This equation, clearly valid for small angles θ, defines the torsion constant k_τ. In a
sensitive instrument, where a small torque should produce a large response, we want
k_τ to be as small as possible: the factor R^4/L is the reason why advanced torsion
pendulums have very thin and long suspension fibres. Besides the static deflection

measurement shown in Eq. (2.4), the pendulum can be operated in a dynamic fashion; if we apply Newton's law for rigid bodies, neglecting for now damping:

$$M_\tau = k_\tau \theta = I\ddot{\theta} \tag{2.5}$$

where I is the moment of inertia of the pendulum load about the rotation axis. The corresponding moment of inertia of the fibre itself is assumed negligible. From Eq. 2.5, a simple harmonic oscillator, we drive the resonance frequency ν_τ of free oscillations of the torsion pendulum:

$$\nu_\tau = \frac{1}{2\pi}\sqrt{\frac{k_\tau}{I}} = \frac{\sqrt{\pi}}{4}\frac{R^2}{\sqrt{IL}}\sqrt{\frac{E}{(1+\sigma)}} \tag{2.6}$$

where we have used eqs. (2.2), (2.4) and (2.6) In Eq. 2.6, the term $R^2/\sqrt{IL}$ summarizes the geometric properties of the pendulum. In order to have a very low natural frequency, approaching the *free motion* (where no restoring force or torque exists), we need a large I and a small fibre diameter $2R$. However, these are competing requirements, because the moment of inertia is typically proportional to the load mass M, and the fibre size must be large enough to hold the weight Mg. The safe load of a fibre scales with its cross section, i.e. with R^2, while the torsion constant is proportional to R^4; therefore, it is beneficial to have a lightweight inertial member and a very thin fibre. State-of-the-art torsion pendulums have natural frequencies in the mHz range.

2.3   Measurement Strategies

Despite its conceptual simplicity the torsion balance turns out to be, upon a detailed analysis, a complex device. Here, we will try to summarize some basic concepts, assuming that the system is characterized by a single fundamental mode of oscillation. More thorough treatments, which also consider the presence of several resonances, in both the free and forced regimes, as well as the non-linear effects, are given in Gillies and Ritter (1993) and in Metherell and Speake (1989).

Strategies for measuring the torsion angle of the balance can be grouped in two big categories:

(a) Direct measurements of the torsion angle and
(b) Measurements of the oscillation period.

Moreover, in the former case it possible to distinguish two operation modes:

(1) Angle measurement in *open loop* (free system) and
(2) Measurement in *closed-loop* configuration (system with feedback).

In case *(1)*, the balance is free to rotate around its own axis, which coincides with the fibre, and the torsion angle is deduced by observing the mean square displacement from the equilibrium position of the system. The sensitivity of the

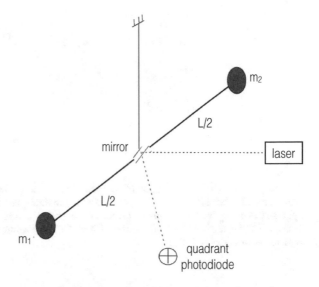

Fig. 2.6 Classical open-loop configuration with optical lever

instrument is defined as $S = d\theta/dM_\tau$. From Eq. (2.4), we simply have $S = 1/k_\tau$. High-performance pendulums have sensitivity of the order of $10^8 \, rad/Nm$.

To derive the order of magnitude of the smallest measurable signal M_τ, we need to specify the method for measuring the torsion angle θ. The measurement of $\Delta\theta$ must be performed using a system without mechanical contact, which in most devices is done applying either an optical lever method or an electrostatic read-out (Fig. 2.6).

2.4 Read-Out for the Torsion Balance

A light beam, emitted (in modern times) by a laser source, is directed towards a mirror placed at the centre of the rod and the reflected beam is aimed at a receiver: the light intensity is detected using a photodiode divided into four sensitive zones (*quadrant photodiode*). The cross section of the light beam falls differently upon the four sectors of the photodiode, whose electric signals can be easily processed to localize the centre of the beam. A quadrant photodiode can measure beam changes in two directions; for planar motion, as is the case of a torsion balance, we are only concerned with one angular motion (Fig. 2.7).

A small change $\delta\theta$ in the equilibrium position of the balance causes an equal change $\delta i = \delta\theta$ in the incidence angle i of the light beam on the mirror. As shown in Fig. 2.8, for small oscillations the position x of the centre of the reflected beam on the receiver depends on the distance travelled by the reflected beam s, the angle ϕ between the normal to the photodiode surface and the incident beam and i, the incidence angle of the beam on the mirror (Cook 1979).

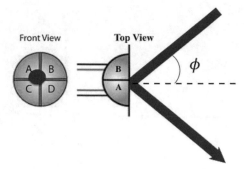

Fig. 2.7 The principle of operation of an optical lever with quadrant photodiode detector: if the beam (red) hits the surface not exactly at its centre, the four quadrants detect different amounts of light. In this example, the horizontal position is deduced by combining the readings $(A + C) - (B + D)$, while the vertical one by $(A + B) - (C + D)$. The reading is then normalized by the total detected power $(A+B+C+D)$.

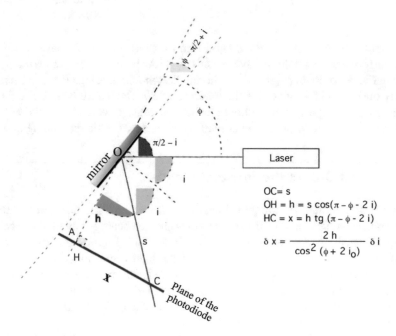

Fig. 2.8 The method of the optical lever

As the torsion angle changes, a small variation in the incidence angle δ_i causes the beam to move by δx on the photodiode; the sensitivity of the method depends on the minimum value of the displacement δx_{min} detected by the photodiode.

$$\delta x = \frac{2\,s}{cos(\phi + 2i_o)}\delta i \qquad (2.7)$$

Using a quadrant photodiode, changes in the position of the light beam of order $\delta x_{min} \sim 10^{-10}$ m can be observed. Assuming for laboratory experiments $h \sim 1\ m$, the minimum measurable value of the torsion angle is $\delta\theta_{min} \sim 10^{-8}$ rad. This value corresponds, from previous considerations (Eq. 2.4), to a minimum detectable torque: $M_{\tau,\,min} = k_t\delta\theta_{min} \sim 10^{-16}$ N m

In practice, other effects, like thermal noise or integration time, limit the sensitivity in the order of $f\,Nm$.

With an adequate optical configuration for recording the light shining on the photodiode, it is also possible to distinguish the translational degrees of freedom of the mirror from the rotational ones. We refer to Fig. 2.9 for the calculation of the displacement x of the laser spot on a plane placed at a distance D from a convergent lens of focal distance f, followed by a rotation δ_i and a translation z of the mirror placed at a distance L from the lens. We shall assume that, at rest ($\delta_i = 0$), the reflected beam travels along the optical axis of the lens.

To first order, it is sufficient to linearly add the two effects. We assume that, initially, the light beam falls on the mirror with an incidence angle equal to i_o. Recalling the equation that relates source (p=L) and image (q) positions in a thin lens:

$$\frac{1}{L} + \frac{1}{q} = \frac{1}{f} \qquad (2.8)$$

where q is the coordinate of the mirror image plane. By simple application of elementary geometry to the diagram of Fig. 2.9, we deduce that

$$x \simeq 2z \sin i_o\left(1 - \frac{D}{f}\right) + 2\tan(\delta_i)\left[L - D\left(\frac{L}{f} - 1\right)\right] \qquad (2.9)$$

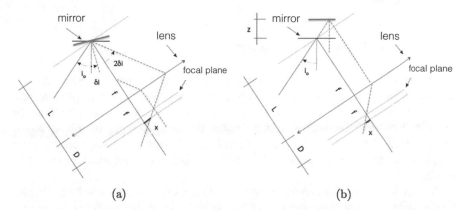

(a)   (b)

Fig. 2.9 Diagram for the calculation of the transfer function between either a rotation (**a**) or a translation (**b**), of a mirror placed at a distance L from a convergent lens of focal distance f and the corresponding displacement of a light beam on a plane orthogonal to the optical axis at a distance D behind the lens

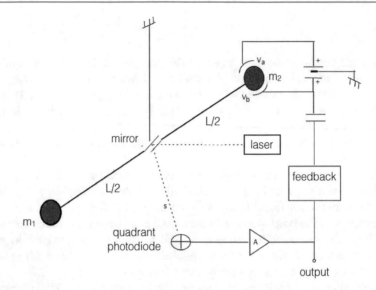

Fig. 2.10 Electrostatic feedback applied to the torsion pendulum

It follows that, if $D = f$, the displacement x only depends on the value δ_i, the mirror rotation angle. Therefore, by placing the photodiode in the focal point of the lens, the device will only be sensitive to the system's rotations, but not its translations.

On the other hand, note that by placing the photodiode in the image plane of L, i.e. at a distance D from the lens equal to

$$D = \frac{Lf}{L - f}$$

the term related to the rotation vanishes and the device is only sensitive to the translations of the mirror.

2.4.1 Pendulums with Feedback

The alternative approach to measure the torsion angle consists in introducing a feedback[1] in the torsion balance, which is done by applying a force proportional to the torsion signal from one of the suspended masses. In this case, the balance remains at rest and the external torque applied on the system is proportional to the output signal of the feedback system. To maintain the system in this working point a typical solution is to apply an electrostatic force to one of the masses, as shown in Fig. 2.10.

A capacitor pair is formed by two polarized plates with the same voltage V_0, and by the mass m_2, that plays the role of a grounded object, placed between the two

---

[1] See Appendix B for feedback and controls.

plates. The error signal from the feedback circuit V is added to the bias voltage in one capacitor while it is subtracted from the other one. If A is the surface of each plate, in the case of a balanced capacitance transducer with $C_a = C_b = C$, we have a total force acting on the mass m_2, which is proportional to the error signal V:

$$F_{fb} = \frac{1}{2\epsilon_o A} C^2 [(V_0 + V)^2 - (V_0 - V)^2] = 2\frac{C^2 V_0}{\epsilon_o A} V \qquad (2.10)$$

Thus we have a dynamical system that, when excited by an external perturbation $M(t)$ with frequency components in the bandwidth of the control system, remains locked at the operating point. This is due to the compensation produced by the electrostatic force, which in turns is proportional to the error signal V used as the output signal of the instrument. This method of measuring the response of the torsion balance allows to increase, with respect to the method of free oscillations, the dynamical range of the instrument, i.e. it allows to detect external forces or torques in a wider range of strength, while keeping constant the sensitivity of the quadrant photodiode. Besides, it overcomes a problem, suggested in the literature (Kuroda 1995), of a non-linear response of the fibre under torsion, that is, a breakdown of the simple proportionality law of Eq. (2.4), with a more complex, and not easily manageable coefficient $k_t = k_t(\theta)$. In feedback operation, the torsion angle remains "*nailed*" to zero, and k_t is bound to remain a constant value.

2.5   A Detailed Model for the Torsion Balance

A deeper analysis of the sensitivity limits of this device would require a much more extended discussion than what just provided in the previous sections, since there are many possible causes that could affect its sensitivity.

Two are the unavoidable noise sources intrinsic to the apparatus: the noise of the read-out, i.e. the system transducing the variations of the torsion angle $\delta\theta$ in electric signals, and the thermal noise of the balance. Additional noise from the detection amplifier circuit can be accounted for in the overall transducer noise while, in the case of optical transducers, we will need to consider the *shot* noise, the noise associated to radiation pressure, the amplitude and phase noises of the laser source, in addition to those related to the current and voltage fluctuations of the photodiode. We will discuss all these sources of noise when we study the interferometers for the detection of gravitational waves. In a system with feedback, also the possible contributions from the feedback circuit must be considered.

The thermal noise of the torsional oscillator plays a central role in defining the sensitivity limit of the instrument, as it is the case for a large part of gravitational experiments: Chap. 9 later on in this book discusses thermal noise in detail. Here, we will just recall that the power spectrum of the acceleration thermal noise is proportional to the temperature and decreases by reducing the anelastic dissipation of the system. For this reason, a careful choice of the material of the torsion wire and of the method of fastening it at both ends is required to reduce this noise contribution.

Here, we will try to analyse a particular interaction of the system with its environment (anthropic noise), a cause limiting the sensitivity different from the above-mentioned intrinsic noise sources (Fan et al. 2008).

In principle, the seismic motion of the ground and hence of the connection point of the torsion wire is filtered by the torsion pendulum. Such filtering action would have an optimal efficiency by carefully designing the fastening system of the wire. Assume, as a limiting case, that the connection region of the suspension wire on the rod with its end masses m_i is reduced to a perfectly centred, geometric point: there will be no spurious couplings between translational and rotational degrees of freedom. In a real apparatus this is not the case. As consequence, a horizontal acceleration of the connection zone entails a pendular motion which, in turn, introduces a torsional motion of the rod. This is why the seismic noise places a limit on the sensitivity of the experiment. To express this argument quantitatively, we report here a detailed model for the torsion balance and to derive its Lagrangian equations of motion (Ying 2004). A simpler but still realistic and comprehensive approach is developed in (Bassan 2013).

We refer to Fig. 2.11 and introduce two Cartesian reference frames including the rotation angles with respect to their respective axes: the first one $O_0, X, Y, Z, \alpha_o, \beta_o$ follows the connection point of the wire and is fixed to the laboratory, while the second $O_1, X_1, Y_1, Z_1, \alpha_1, \beta_1, \gamma_1$ follows the rod and its origin coincides with the connection point of the wire to the rod. $2\,h$, $2\,w$ and $2\,l$ are the height, length and

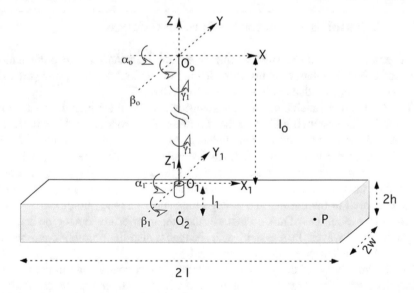

Fig. 2.11 Diagram of a more realistic torsion pendulum. O_o is the upper attachment point of the torsion wire, O_1 the attachment point from the wire to the pendulum bar, O_2 is the system centre of mass. We have introduced two Cartesian reference frames which include the rotation angles with respect to their respective axes: the first one $O_0, X, Y, Z, \alpha_o, \beta_o$ moves with the attachment point of the wire, while the second $O_1, X_1, Y_1, Z_1, \alpha_1, \beta_1, \gamma_1$ moves along with the rod

width of the rod; l_o is the length of the torsion wire and l_l is the distance between the centre of mass of the system O_2 and the attachment point O_1.

Under the hypothesis that the point O_o is fixed, the degrees of freedom of this system are 5, namely, the angles α_o, β_o, α_1, β_1, γ_1. If we want to include in the analysis the effects of the seismic noise, we should consider the (stochastic) variables X, Y, Z, describing the motion of the suspension attachment point.

Let us express x, y, z, the coordinates of a generic point P on the rod, as functions of the degrees of freedom. To this end, we write the equations that transform the variables in the reference system O_1, X_1, Y_1, Z_1 into the ones corresponding to the system O_0, X, Y, Z:

$$
\begin{aligned}
x = &-l_o sin\beta_o + x_1 cos\beta_1 cos\gamma_1 \\
&- y_1 cos\beta_1 sin\gamma_1 \\
&+ z_1 sin\beta_1
\end{aligned}
\tag{2.11}
$$

$$
\begin{aligned}
y = &\, l_o cos\beta_o sin\alpha_o \\
&+ x_1 (cos\alpha_1 sin\gamma_1 - cos\gamma_1 sin\beta_1 sin\alpha_1) \\
&+ y_1 (cos\alpha_1 cos\gamma_1 - sin\gamma_1 sin\beta_1 sin\alpha_1) \\
&- z_1 cos\beta_1 sin\alpha_1
\end{aligned}
\tag{2.12}
$$

$$
\begin{aligned}
z = &-l_o cos\beta_o cos\alpha_o \\
&+ x_1 (sin\alpha_1 sin\gamma_1 - cos\gamma_1 sin\beta_1 cos\alpha_1) \\
&+ y_1 (sin\alpha_1 cos\gamma_1 + sin\gamma_1 sin\beta_1 cos\alpha_1) \\
&+ z_1 cos\beta_1 cos\alpha_1
\end{aligned}
\tag{2.13}
$$

From these equations, we derive the expressions of $\dot{x}$, $\dot{y}$, $\dot{z}$ (being $\dot{x}_1 = \dot{y}_1 = \dot{z}_1 = 0$). We then integrate the kinetic and potential energies of the mass element $\rho\, dx_1 dy_1 dz_1$ over the volume of the rod ($l \leq x_1 \leq l$, $-w \leq y_1 \leq w$, $l_1 - h \leq z_1 \leq l_1 + h$), thus obtaining the expression for the Lagrangian of the system:

$$
\begin{aligned}
L = &\frac{1}{2}ml_o{}^2 cos^2\beta_o \dot{\alpha}_o{}^2 + \frac{1}{2}ml_o{}^2\dot{\beta}_o{}^2 \\
&\left[\frac{1}{6}m(w^2 + l^2) + \frac{1}{6}m(3l_1{}^2 \right. \\
&\left. + h^2 - w^2)cos\beta_1 + \frac{1}{6}m(w^2 - l^2)cos^2\beta_1 cos^2\gamma_1 \right]\dot{\alpha}_1{}^2 \\
&+ \left[\frac{1}{6}m(h^2 + w^2 + 3l_1{}^2) + \frac{1}{6}m(l^2 - w^2)cos^2\gamma_1 \right]\dot{\beta}_1{}^2 \\
&+ \frac{1}{6}m(w^2 + l^2)\dot{\gamma}_1{}^2 \\
&+ ml_o l_1 cos\beta_o cos\beta_1 cos(\alpha_o - \alpha_1)\dot{\alpha}_o\dot{\alpha}_1 \\
&+ ml_o l_1 cos\beta_o sin\beta_1 sin(\alpha_o - \alpha_1)\dot{\alpha}_o\dot{\beta}_1
\end{aligned}
$$

$$+ ml_ol_1 sin\beta_o cos\beta_1 sin(\alpha_o - \alpha_1)\dot{\alpha}_1\dot{\beta}_o$$

$$+ ml_ol_1[cos\beta_o cos\beta_1 + sin\beta_o sin\beta_1 cos(\alpha_o - \alpha_1)]\dot{\beta}_o\dot{\beta}_1$$

$$+ \frac{1}{3}m(w^2 - l^2)cos\beta_1 sin\gamma_1\dot{\alpha}_1\dot{\beta}_1$$

$$+ \frac{1}{3}m(w^2 + l^2)sin\beta_1\dot{\alpha}_1\dot{\gamma}_1$$

$$+ mg(l_o cos\alpha_o cos\beta_o + l_1 cos\alpha_1 cos\beta_1)$$

$$- \frac{1}{2}k\gamma_1{}^2 \tag{2.14}$$

where m is the mass of the rod, k the torsion constant of the wire and g the acceleration of gravity.

In the approximation of small angles, a simplified expression results in

$$L = \frac{1}{2}ml_o{}^2\dot{\alpha}_o{}^2 + \frac{1}{2}ml_o{}^2\dot{\beta}_o{}^2$$

$$+ \frac{1}{6}m(h^2 + w^2 + 3l_1{}^2)\dot{\alpha}_1{}^2$$

$$+ \frac{1}{6}m(h^2 + l^2 + 3l_1{}^2)\dot{\beta}_1{}^2$$

$$+ \frac{1}{6}m(w^2 + l^2)\dot{\gamma}_1{}^2$$

$$+ ml_ol_1\dot{\alpha}_o\dot{\alpha}_1$$

$$+ ml_ol_1\dot{\beta}_o\dot{\beta}_1$$

$$+ mg\left[l_0\left(1 - \frac{1}{2}\alpha_o{}^2 - \frac{1}{2}\beta_o{}^2\right)\right.$$

$$\left. + l_1\left(1 - \frac{1}{2}\alpha_1{}^2 - \frac{1}{2}\beta_1{}^2\right)\right]$$

$$- \frac{1}{2}k\gamma_1{}^2 \tag{2.15}$$

Comparing Eq. (2.14) with (2.15) we can see that, under our assumption of small angles approximation, every significant coupling between the different degrees of freedom cancels out. Beginning instead with the complete Lagrangian of Eq. (2.14) we now derive the equations of motion using the well-known Lagrange relation of analytical mechanics:

$$\frac{\partial L}{\partial q_i} - \frac{d}{dt}\left(\frac{\partial L}{\partial \dot{q}_i}\right) = 0 \tag{2.16}$$

where q_i is a generic degree of freedom. Applying it to the torsional degree of freedom γ_1, we obtain an equation that includes coupling terms

$$\frac{1}{3}m(w^2 + l^2)\ddot{\gamma}_1 - \frac{1}{3}m(l^2 - w^2)cos^2\gamma_1 sin2\gamma_1 \dot{\alpha}_1^2$$
$$+\frac{1}{6}m(l^2 - w^2)sin2\gamma_1\dot{\beta}_1^2 =$$
$$= -\frac{1}{3}m\big[(l^2 - w^2)cos2\gamma_1 + (l^2 + w^2)\big]\dot{\alpha}_1\dot{\beta}_1$$
$$+\frac{1}{3}m(l^2 + w^2)sin\beta_1\ddot{\alpha}_1 \qquad (2.17)$$

for small values of the angle α_1 and β_1, this equation reduces to the form

$$\ddot{\gamma}_1 + \omega_T{}^2\gamma_1 = -\frac{2l^2\dot{\alpha}_1\dot{\beta}_1}{l^2 + w^2} - \beta_1\ddot{\alpha}_1 = F \qquad (2.18)$$

where $\omega_T = \sqrt{k/I_z}$ and $I_z = \frac{1}{3}m(l^2 + w^2)$ is the moment of inertia with respect to the z-axis of the rod.

From this relation we infer that the swaying motion, caused, for instance, by seismic noise and described by the degrees of freedom α_1 and β_1 can act as a forcing term F of the twisting motion through the coupling terms. Analogously, we can see that α_1 and β_1 depend on the degrees of freedom α_0 and β_0 and the possible translational motion of the suspension point O.

We conclude that these coupling terms determine a serious sensitivity limitation for the torsion balance associated to the seismic motion of the attachment point of the system O.

References

Bassan, M., et al.: Torsion pendulum revisited. Phys. Lett. A **377**, 1555–1562 (2013)

Cavendish, H.: XXI. Experiments to determine the density of the earth. Phil Trans. R. Soc. **88**, 469 (1798)

Cook, R.O., Hamm, C.W.: Fiber optic lever displacement transducer. Appl. Optics **18**, 3230 (1979)

Fan, X.-D., et al.: Coupled modes of the torsion pendulum. Phys. Lett. A **372**, 547–552 (2008)

Gillies, G.T., Ritter, R.C.: Torsion balances, torsion pendulums, and related devices. Rev. Sci. Instrum. **64**, 283 (1993)

Gundlach, J.H., Merkowitz, S.M.: Measurement of Newton's constant using a torsion balance with angular acceleration feedback. Phys. Rev. Lett. **85**, 2869 (2000)

Kuroda, K.: Does the time-of-swing method give a correct value of the Newtonian gravitational constant? Phys. Rev. Lett. **75**, 2796 (1995)

Love, A.E.H.: A Treatise on the Mathematical Theory of Elasticity. Dover, New York

Metherell, A.J.F., Speake, C.C.: The dynamics of the double-pan beam balance. Metrologia **19**, 109 (1989)

Ying, Tu., et al.: An abnormal mode of torsion pendulum and its suppression. Phys. Lett. A **331**, 354–360 (2004)

Further Reading

Adelberger, E.G., Gundlach, J.H., Heckel, B.R., Hoedl, S., Schlamminger, S.: Torsion balance experiments: A low-energy frontier of particle physics. Progress in Particle and Nuclear Physics **62**, 102–134 (2009)

The Equivalence Principle

3

3.1 Statement of the Equivalence Principle

The equivalence principle is the basis of the Newtonian gravitational theory as well as Einstein's one. It can be stated as follows.

The property of a body that determines its response to any applied force (m_i = inertial mass) coincides with the property of the body that determines its response to gravitational attraction (m_g = gravitational mass)

$$m_i = m_g \tag{3.1}$$

Alternatively, it can be expressed as
All bodies, independently of its nature and internal structure, fall with the same acceleration when subject to a gravitational field.
Such a formulation of the equivalence principle is currently known as the *Weak Equivalence Principle* (WEP) by the scientific community.

Einstein extended the WEP formulation by including the statements regarding the Lorentz invariance. In other words, he assumed that *the outcome of any local non-gravitational experiment in a freely falling laboratory is independent of the velocity of the laboratory and of its location in space-time.* This is known as the Einstein Equivalence Principle. The consequence is that

- the physical phenomena can be described in terms of events in a four-dimensional differentiable space-time;

The original version of this chapter was revised: Chapter have been updated with the correction. The correction to this chapter can be found at https://doi.org/10.1007/978-3-030-95596-0_15

© The Author(s), under exclusive license to Springer Nature Switzerland AG 2022, corrected publication 2023
F. Ricci and M. Bassan, *Experimental Gravitation*, Lecture Notes in Physics 998, https://doi.org/10.1007/978-3-030-95596-0_3

- the general equations of gravitation can be expressed in a form independent of the choice of coordinates, termed *covariant*;
- the theory must be relativistic. In other words, in the limit of no gravitational field, neglecting possible contributions of gravitational self-energy that can be calculated in the Newtonian approximation, the laws of physics must reduce to those of special relativity.

This last consideration implies that the elementary space-time interval ds^2, in absence of gravity, can be written in terms of the symmetric tensor

$$\eta_{\mu\nu} = \begin{pmatrix} 1 & 0 & 0 & 0 \\ 0 & -1 & 0 & 0 \\ 0 & 0 & -1 & 0 \\ 0 & 0 & 0 & -1 \end{pmatrix} \tag{3.2}$$

$$ds^2 = \eta_{\mu\nu}dx^\mu dx^\nu \tag{3.3}$$

It also follows that the time intervals measured by a clock in motion and in absence of gravity only depend on its velocity and not on the acceleration.

The slowdown of clocks in motion is a phenomenon that is well demonstrated experimentally. One of the experiments devoted to this end was performed in 1977 at the European Centre of Nuclear Research (CERN) (Bailey et al. 1977). In this case, subatomic particles played the role of natural clocks, making use of the fact that their mean lifetime $\bar{\tau}$ is an intrinsic property of the particle. The mean lifetimes of positive and negative muons were measured. The muon is an unstable particle that disintegrates spontaneously in an electron and two different neutrinos after $2.2 \times 10^{-6}\,s$, on average. Muons with a speed of $v = 0.9994\,c$, corresponding to a Lorentz factor of $\gamma = \left(\sqrt{1 - (v^2/c^2)}\right)^{-1} = 30$, were accelerated around a ring of 14 m in diameter with a centripetal acceleration of 10^{18} times the gravitational acceleration g. The measured lifetime was found in excellent agreement with the well-known relation:

$$\bar{\tau} = \gamma\bar{\tau}_o \tag{3.4}$$

where $\bar{\tau}$ is the mean lifetime of the muons measured in the laboratory and $\bar{\tau}_o$ is the muon's mean lifetime at rest.

The information we obtain with this experiment is connected with the time transformation given by the previous equation: the time interval between two events Δt_o that occur in the same place in a moving system (events of injections and disintegration of the muon, in the example mentioned), if observed in the laboratory becomes dilated by the factor γ with respect to the corresponding time interval as measured by an observer in the moving (but at rest with the muons) reference system:

$$\Delta t = \gamma\Delta t_o \tag{3.5}$$

In addition to this important experiment, hundreds of measurements have been made, beginning with the famous, groundbreaking experiment of Rossi and Hall

(1941) on beams of unstable particles (muons, pions, hyperions, …), which have demonstrated that the mean lifetime, from creation until spontaneous disintegration, depends on the velocity just as predicted by Eq. (3.5). At present, in any high energy experiment, where elementary particles collide at velocities close to c, the events are analysed using kinematic laws predicted by special relativity.

Experimental evidence also shows that acceleration does not play any role in modifying the rate of clocks. In fact, at equal speeds, the mean lifetimes of rectilinear muon beams (with no acceleration) and those from the muons inside an accumulation ring (which withstand a huge acceleration) are stretched in the same way.

Another milestone experiment on the slowing down of moving clocks was carried on by Hafele and Keating (1972) using caesium atomic clocks. They were initially accurately synchronized, then loaded in two airplanes of commercial airlines to perform a complete travel around the world. One plane travelled to the East while the other flew to the West. After each flight, the clocks where compared with similar clocks that remained on the ground. It was found that, in comparison with these latter clocks, the clock travelling to the West suffered a delay of $59 \pm 10 \, ns$, while the one to the East was $273 \pm 7 \, ns$ in advance. Such a result agrees with the theoretical expectations. Note that the most significant variation came from the eastbound clocks. This is easily explained using a reference system still with respect to the distant stars: for clocks travelling to East, the airplane speed and the one of the Earth rotation add up, while we must subtract them for the westbound clocks.

In this experiment, we also need to account for the effect of the Earth's gravitational field, which depends on the flight altitude, thus changing the clock rate travelling with respect to the ones that rested on ground. In Chap. 4, we will revisit in detail this effect.

3.2 Tests of Einstein's Equivalence Principle

The formulation of Einstein's Equivalence Principle (EEP) incorporates the WEP statement. In fact, the latter states that *A neutral body, small enough not to be affected by the inhomogeneity effects of gravity, that is initially at a point of space-time (event) with a given velocity, will describe at later times a time line that is independent of its internal nature.*

The EEP formulation additionally implies that *the result of any **local** and **nongravitational** experiment is independent of the place and time at which it was performed.* This further requirement is usually referred to as Local Position Invariance (LPI).

Some of the first recorded tests of the equivalence principle were performed by Galileo (circa 1590) and by Newton (1686), observing the period of oscillation of different pendulums. They achieved an accuracy of $2 \cdot 10^{-2}$ and 10^{-3}, respectively. The experiments performed by Eötvös between 1885 and 1909 are considered a fundamental test of the equivalence principle.

The purpose of the experiments was to compare the acceleration of two bodies of different nature placed in the same gravitational field. It is a typical *null experiment*, since, should WEP be valid, there would be no difference in the acceleration. To discuss its implications, we now assume that the principle is violated and therefore

$$m_i \neq m_g \tag{3.6}$$

The equivalence of mass and energy, as stated by special relativity, tells us that there will be several contributions to the inertial mass of the body, which will depend on the nature of the body itself: electromagnetic energy, energy due to weak and strong interactions, ...

We shall therefore investigate, for each of these contributions, if an associated violation of the EEP of different type exists. We then write:

$$m_i = m_g + \sum_k \eta^k E_k / c^2 \tag{3.7}$$

where the index k identifies the type of contribution to the energy, E_k is the k-th type of internal energy of the body and η^k is a dimensionless parameter that measures the magnitude of the violation. Let us now consider two bodies that fall with different accelerations:

$$a_1 = \left(1 + \sum_k \eta^k E_k^{(1)} / m_i^{(1)} c^2\right) g \qquad a_2 = \left(1 + \sum_k \eta^k E_k^{(2)} / m_i^{(2)} c^2\right) g \tag{3.8}$$

The experimental parameter to be measured is

$$\eta = \frac{2|a_1 - a_2|}{|a_1 + a_2|} = \sum_k \eta^k \left(E_k^{(1)} / m_i^{(1)} c^2 - E_k^{(2)} / m_i^{(2)} c^2\right) \tag{3.9}$$

where η is known as the Eötvös coefficient, in honour of the Hungarian Baron Lorand von Eötvös who used the torsion balance to test the equivalence principle (Fig. 3.1). As mentioned in the previous chapter, to this day this instrument, redesigned in a modern way, still allows us to obtain the highest accuracy when verifying this principle.

A rod, with two masses of different composition attached to its extremes, is suspended by a quartz wire. In principle, the verification of WEP can be accomplished by assembling two masses of different nature at the ends of the transverse bar. The angle at which the restoring torque exerted by the wire balances the external torque is the observable that lets us estimate the effect under investigation (Fig. 3.2).

The experiment is done in a laboratory rotating with the Earth, and hence the force acting on each of the two masses is the vector sum of the centrifugal force due to Earth's rotation, directed away from the rotation axis and the gravitational force directed towards the Earth centre. The centrifugal force has a component parallel to the tangent plane of the Earth's surface (i.e. horizontal). We introduce a reference frame with spherical coordinates, with the origin at the Earth's centre and the polar

Fig. 3.1 Loránd Baron Eötvös de Vásárosnamény. Picture by Aladár Székely, 1912. Public domain image

Fig. 3.2 Principle of operation of torsion pendulum experiments aiming at verifying the equivalence principle

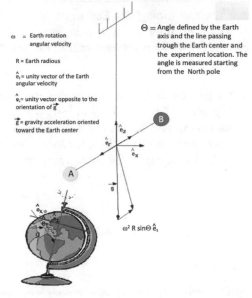

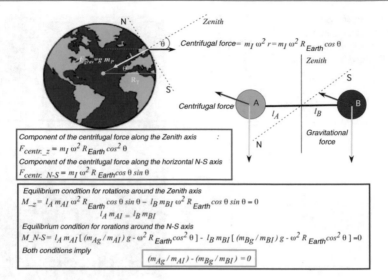

Fig. 3.3 Basics of Eötvös' experiment

axis (passing through the Earth's poles) coinciding with the rotation axis. The laboratory is positioned at a polar angle (colatitude) θ (its geographic latitude is then $\phi = \pi/2 - \theta$). In this system, each of the test masses of the torsion balance is subject to

- the gravitational force that depends, obviously, on the gravitational mass m_g and is purely radial, directed along $-\hat{z}$, where $\hat{z}$ denotes the local vertical;
- the centrifugal force that depends on m_i has both a radial and a polar (horizontal) component, oriented along the local N-S direction; both depend on the latitude of the laboratory on the Earth's surface:

$$F_z = -G\frac{m_g M_E}{R_E^2} + m_i \omega^2 R \sin^2 \theta \qquad = m_g g - m_i a_z \qquad (3.10)$$

$$F_{N-S} = \ m_i \omega^2 R \sin \theta \cos \theta \qquad\qquad = m_i a_x \qquad (3.11)$$

These two components of the force, acting on each of the test masses, located at distances r_A and r_B from the suspension fibre can induce rotation around two different axes, the horizontal component around the vertical axis and vice versa (see Fig. 3.3).

for the vertical components

$$M_z = m_i{}^A a_x r_A - m_i{}^B a_x r_B \qquad (3.12)$$

for the horizontal components

$$M_x = (m_g^A g - m_i{}^A a_z) r_A - (m_g^B g - m_i{}^B a_z) r_B \qquad (3.13)$$

By an adequate choice of r_B we can prevent the system from rotating around the x-axis. In other words, we choose r_B such that $M_x = 0$:

$$r_B = \frac{(m_g^A g - m_i{}^A a_z)}{(m_g^B g - m_i{}^B a_z)} r_A \tag{3.14}$$

Under these conditions, the residual component of the torque M_z results

$$M_z = m_i{}^A a_x r_A \frac{m_g^B / m_i^B - m_g^A / m_i^A}{m_g^B / m_i^B - a_x / g} \tag{3.15}$$

When $M_z = 0$, the system does not deviate from the equilibrium condition, that is,

$$m_g^B = m_i^B \qquad m_g^A = m_i^A$$

From formula (3.15), we can see that by rotating the pendulum bar by π, if and only if WEP is violated, the non-vanishing torque changes sign and the apparatus changes its equilibrium position.

In 1909 Lorand Baron von Eötvös, after devising an apparatus capable of detecting rotations angles of 10^{-11} rad (see Fig. 3.4), obtained an upper limit on the value of the η parameter, measuring WEP violations, of order 10^{-8}.

A similar experiment was performed by Roll et al. (1964) in Princeton, (USA) and by Braginsky and Panov (1972) in Moscow (Russia). In both cases, they made use of the acceleration produced by the Sun's gravity, rather than the Earth's, which

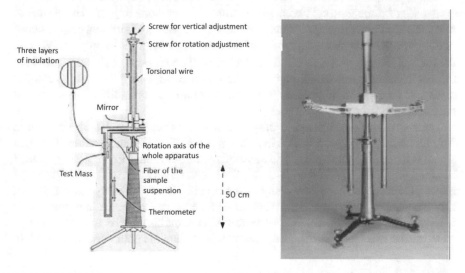

Fig. 3.4 The right panel shows a picture of one of the experimental devices used by Eötvös. On the left panel, a diagram of the working mechanism. Images from The Eötvös Lorand Virtual Museum

modulates the torsion signal with a period of 24 h. Despite the fact that the horizontal component of the solar acceleration is weaker than the Earth's one, $0.59 \ \mathrm{cm \, s^{-2}}$ versus $1.67 \ \mathrm{cm \, s^{-2}}$, the advantage of not having to manually rotate the apparatus, combined with other experimental advantages available with the technology of the '60s, allowed an improvement of three orders of magnitude with respect to Eötvös' result.

For simplicity sake, let us consider the system as simply composed of two masses A and B of different material. The resulting torque due to the solar acceleration $g_\odot$ is

$$M_S = (m_g^A g_\odot - m_i^A a) r_A + (m_g^B g_\odot - m_i^B a) \frac{m_i^A r_A}{m_i^B}$$

or after reducing terms:

$$M_S = m_i^A r_A \Big(\frac{m_g^A}{m_i^A} - \frac{m_g^B}{m_i^B} \Big) g_\odot$$

which vanishes when the ratio between inertial and gravitational mass is independent of the nature of the bodies.

Note that the direction of the solar acceleration changes with time, and therefore a modulation[1] with a period of 24 h is expected in the response of the system, in the case that WEP is violated. In general, the modulation technique allows to improve significantly the signal-to-noise ratio and avoids having to rotate the whole system by π rad.

The experimental device of Dicke, Roll and Kroktov is depicted in Fig. 3.5. It is composed of three masses: two made of aluminium and one of gold. The choice of materials is driven by the requirement of selecting materials of different composition; in this case, they considered the ratio of the number of neutrons to protons, N_n/N_p, the ratio of energy of the electron in level K to the rest mass, $E_e^{(K)}/m_e$, and the ratio of the nuclei electrostatic energy to the atomic mass, $E_{Nu}^{(elettr.)}/M_A$. Table 3.1 reports the corresponding values to illustrate the microscopic differences between the chosen materials.

As shown in Fig. 3.5 the gold mass is placed between the plates of a capacitor. This allows the device to function in a *closed feedback loop* configuration. The light signal is reflected on the mirror and detected by the photo-multiplier. A slit is placed between the mirror and the photo-multiplier in such a way that when the light beam moves, the recorded light's intensity also changes. However, the dynamic range of the apparatus depends on the transverse size of the beam and the width of the slit. To increase the dynamic range, the photo-multiplier signal is used to generate an error signal that drives the voltage across the capacitor where the gold mass is inserted. In

---

[1] See chapter *Modulation Techniques*.

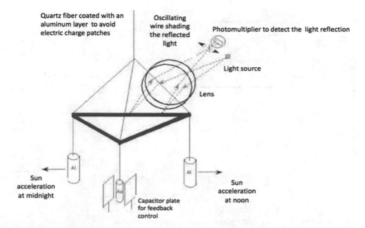

Fig. 3.5 The torsion pendulum of Dicke's experiment. Adapted from Roll et al. (1964)

Table 3.1 Differences in material parameters for the test masses (Roll-Krotkov-Dicke experiment)

Ratios	Aluminium	Gold
N_n / N_p	1.08	1.5
$E_e^{(K)} / m_e c^2$	0.03	0.16
$E_{Nu}^{(elettr.)} / M_A c^2$	0.03	0.16

this configuration, one can thus exert a force that maintains the mass at rest.[2] The detection is therefore derived from the measurements of the amplitude and phase of the error signal itself. In Fig. 3.6, we report the diagram for the detection published in the original paper by Roll, Dicke and Krotov.

With this apparatus, Roll, Dicke and Kroktov were able to obtain a limit $\eta \leq 0.96 \cdot 10^{-11}$ (Roll 1964).

At the Lomonosov University in Moscow, Braginsky and Panov (1972) calibrated a similar device but with a longer wire and a more complex configuration of suspended masses. The purpose was to increase the sensitivity (longer wire) and to reduce the spurious gravitational couplings with the surrounding environment, by making the system more symmetric, which means reducing the quadrupolar, sextupolar and octopolar mass moments of the system. In practice, compensation masses must be added appropriately. In the work they published, Braginsky and Panov did not provide many details about the experimental configuration and the reduction of systematic errors. This, in addition to the political climate due to the iron curtain of the Soviet world, generated considerable suspicion on the validity of the experiment

---

[2] This condition only holds for the Fourier components that lay within the characteristic frequency range of the control system.

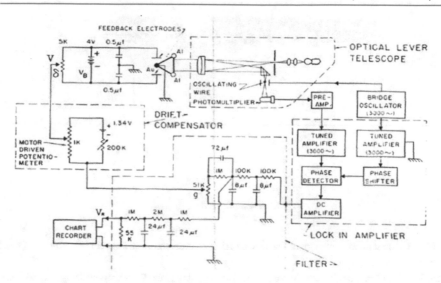

Fig. 3.6 Scheme of the implementation and detection system of the Roll, Dicke and Krotov experiment. Credits: (Roll 1964)

in the Western scientific circles. In any case, they concluded their paper with the
claim of having obtained a limit value for η of order $1 \cdot 10^{-12}$.

In a series of experiments since 1999, always using torsion pendulums, the
"Eot-Wash" gravitational group of E. Adelberger in the University of Washington
(Seattle—USA) obtained a value of $\eta \sim 1.4 \cdot 10^{-13}$ (Baeßler et al. 1999, Adelberger
2001). They used the Earth gravitational attraction, like Eötvös, and modulation of
the signal, like Dicke: this was achieved by operating the torsion pendulum on a
platform rotating at $\sim$ mHz frequency, much higher, and much less affected than
Dicke's by spurious effects with 1-day periodicity. By combining the turntable frequency with that of the Earth rotation and revolution, they have set upper limits on
η also for the attraction towards the Sun and towards the Galactic Centre: the latter
is of interest because nothing is known about WEP and Dark Matter (Wagner 2012).
Among the many ingenuous ideas they implemented in the apparatus, we mention a
careful arrangement of the compensation masses to balance the local gravity gradients. This type of apparatus is limited by the thermal noise of the suspension wire;
that is why there have been efforts to develop a torsion pendulum operating at low
temperatures.

3.3   Verification of EEP at the Atomic Level

On November 2004 Sebastian Fray and collaborators from the Max Planck Institute
for Quantum Optics at Garching and the Tubingen and Munich Universities have
compared the acceleration of two isotopes of rubidium in the Earth's gravitational
field (Fray et al. 2004), using an atomic interferometer. In agreement with the equivalence principle, the atoms are accelerated in the same way. This type of experiments

is part of the effort to verify to what extent the theory of gravity is compatible with the laws of physics in the microscopic world. According to some theories, when gravitational experiments are done on quantum objects like atoms, it might be possible to observe a violation of the equivalence principle, thus unveiling a new type of physics.

The German group based its experiment on the use of an atom interferometer, a technique already developed to measure the Earth gravity "g" with a precision of 10^{-9}.

In principle, an atom interferometer is similar to the optical counterpart, but there are matter waves interfering; in other words, it uses beams made of atoms instead of light; the role of the beam splitter is performed by stationary electromagnetic waves that are used both to divide and to combine the atom beams.

The experiment works by capturing in a magneto-optical trap[3] $\sim 2 \cdot 10^9$ atoms of the isotopes ^{85}Rb or ^{87}Rb. Using laser beams the atoms are accelerated upwards. When the beams are turned off, the atoms fall under the influence of gravity.

The interferometer has allowed to measure the acceleration of both types of atoms, g_{85} and g_{87}, reaching a result in agreement with the equivalence principle:

$$\frac{g_{85} - g_{87}}{g_{85}} = 1.2 \cdot 10^{-7} \pm 1.7 \cdot 10^{-7}$$

Fray and his collaborators have also compared the relative acceleration of Rb_{85} atoms in two different quantum states, finding that they coincide within the experimental error.

A similar experiment, carried out in 2014 by the group of G. Tino in Firenze (I) (Tarallo et al. 2014), has tested the UFF for two different isotopes of Strontium: bosonic ^{88}Sr isotope which has no spin versus the fermionic ^{87}Sr isotope which has a half-integer spin. This search is of interest because it can probe theories, recently proposed, predicting some spin-gravity coupling. The limit set on the η parameter for this particular form of coupling is $\eta < 2 \ 10^{-7}$.

3.4 Experiments in Microgravity Conditions

In 1972 P. Worden and F. Everitt proposed an ambitious experiment to verify the equivalence principle. It consists of placing two masses of different composition inside a satellite orbiting around the Earth with the purpose of obtaining free fall conditions over a long period of time. The device should be designed with a system monitoring the position of the masses with high sensitivity: this is obtained by cooling

---

[3] A magneto-optic trap (MOT) is a sophisticated apparatus where atoms are slowed down (to energies corresponding to μK) by elastic collisions with photons of laser beams; these "cold" atoms are then trapped by magnetic field, through the Zeeman effect, in a small region of space: localized clouds of up to 10^8 atoms can be generated and maintained indefinitely. The techniques of laser cooling were awarded a Nobel prize for physics in 1997.

the test masses down to $1.8\ K$, in order to detect their residual displacements between magnetic coils, via the use of SQUIDs[4]

The project, called STEP (Satellite Test of the Equivalence Principle) (Overduin et al. 2012), has been pursued for many years under the direction and supervision of both NASA and ESA: the goal was to measure the equivalence between inertial and gravitational mass with an accuracy of $1 \cdot 10^{-17} - 1 \cdot 10^{-18}$.

To this purpose, the accelerations of four pairs of test masses in orbit would be compared. The test masses in free fall would be placed in a ultra-high vacuum environment, isolated from external perturbations and within a cryostat equipped with superconducting magnetic shields. The differential accelerations of the test mass are measured by a superconducting circuit coupled to a SQUID with a sensitivity of $10^{-22}\ Wb$ in 1 Hz band. The material for the test masses was to be chosen among niobium, platinum-iridium and beryllium. The reasons of this limited choice are related to their physical properties, such as the ratio protons/neutrons and the nuclear binding energy, that need to be quite different, in order to maximize the contribution to a violation of the equivalence principle. Besides, the materials need being machinable with high mechanical accuracy. The test masses of platinum-iridium are placed at the centre, while the beryllium ones remain on the outer part, thus avoiding the condition of a position difference too large to be compensated by the actuation systems of the apparatus.

The test masses of different materials along with the positions transducers would be placed inside the satellite (see Fig. 3.7), whose attitude is controlled by micro-thrusters. These serve the purpose of cancelling the accelerations caused by the residual atmosphere, by radiation pressure and by solar wind, which could influence the motion of the masses. Such a set-up of the satellite is called "drag free".

This technique reduces the low-frequency noise in the acceleration produced by non-gravitational interactions. The perturbations associated with gravity gradients are eliminated via the precise positioning, in the same point, of the centres of mass of both test masses. Figure 3.8 shows a pair of masses in their operating configuration.

The orbit was designed as approximately circular and synchronized with the solar motion in order to minimize fluctuations in temperature, with an optimal height close to 550 km. The expected duration of the mission is 6 months. However, despite its potential and its elegant technology, the joint ESA-NASA mission is, to this day, far from being approved.

The MICROSCOPE experiment (Touboul et al. 2012) is a mainly French (CNES-DLR-ESA) mission with a less ambitious goal, $\eta \sim 1 \cdot 10^{-15}$, but it achieved its final scientific target. It was launched on 26 April 2016, operated for over 2 years, orbiting at a height of 712 km, and was deactivated on 18 October 2018, after completing all its scientific goals (Fig. 3.9).

The satellite payload is composed of two differential, almost identical micro-accelerometers made of concentric cylindrical test masses. The first instrument has

---

[4] The Superconducting Quantum Interference Device (SQUID) is a very sensitive magnetometer based on the Josephson effect. It is described in Sect. 8.2.5.

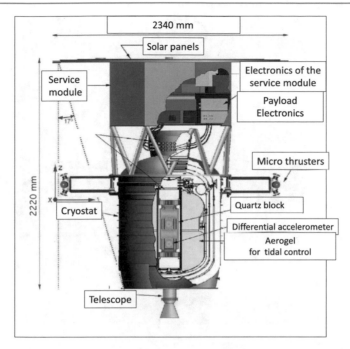

Fig. 3.7 Diagram of the working principle of the STEP satellite. Credits: Overduin et al. (2012)

Fig. 3.8 Test masses of the STEP experiment. Credits: courtesy of STEP team

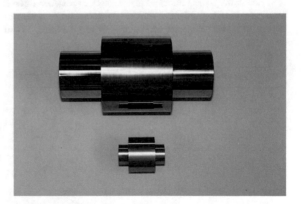

masses of the same material (Pt) and is dedicated to evaluate the experimental accuracy of the measurement (verification instrument). On the other hand, the second accelerometer has masses of different materials (Pt-Ti), and is thus suited to test the equivalence principle, looking for different behaviour in the field of the Earth (Fig. 3.10).

The attitude of the satellite and its thermal resistance are controlled in an active way, so that the satellite follows, using FEEP thrusters (field emission electric propulsion), the two test masses in their drag-free gravitational motion. Testing this micropropulsion system is the other technological target of the mission itself. One important technological achievement of MICROSCOPE is the proof that the differential

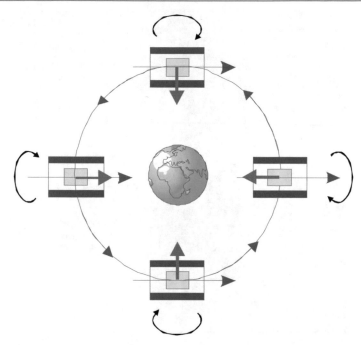

Fig. 3.9 The principle of operation of MICROSCOPE, and similar EP satellite experiments: if WEP is violated, the two cylindrical test masses experience different accelerations (red arrows) as the satellite orbits the Earth. The differential acceleration is measured by the feedback voltages applied to the test masses to keep them in equilibrium. Credits: Toubul and Rodrigues (2001)

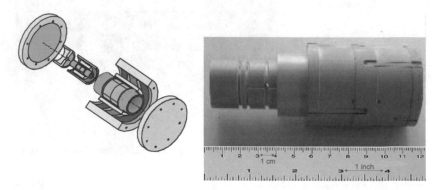

Fig. 3.10 On the left a sketch of the differential accelerometer of the MICROSCOPE experiment, with the two test masses surrounded by electrodes to measure their relative position, both in radial and axial directions. On the right a prototype of the accelerometer core with an inner gold-coated silica cylinder carrying the radial electrodes. Credits: (Toubul and Rodrigues 2001)

accelerometers, built by ONERA, can detect the effects of gravity gradients due to the force difference between the attraction exerted by the Earth on the internal and external masses.

While analysis of data is still underway, the French team has published (Touboul et al. 2017) a preliminary result confirming validity of the equivalence principle: the two test masses (Ti and Pl) fall, in the Earth field, with the same acceleration with a $1\,\sigma$ statistical confidence:

$$\eta \leq [-1 \pm 9(\text{statistics}) \pm 9(\text{systematic})] \times 10^{-15}$$

Last, we mention the Italian proposal for the satellite *Galileo Galilei* (GG) (Nobili et al. 2009), still far from being approved. GG is a proposal of a small satellite at low orbit dedicated to the verification of the equivalence principle. The preliminary study phase of GG, directed by A. Nobili from Pisa University, aims to investigate how it would be possible to test the equivalence principle with $\eta \sim 1 \cdot 10^{-17}$ using a device operating at room temperature. The main novelty of GG compared with STEP and MICROSCOPE is to modulate the possible differential signals of WEP violations at a relatively high frequency (~ 2 Hz), by forcing the test masses (concentric cylinders, just as in STEP and MICROSCOPE) to rotate. The modulation frequency is increased, with respect to other experiments, by a factor of more than 10^4, thus reducing the low-frequency noise, due to many sources associated to the electrical and mechanical systems, which has a typical $1/f$ dependence.

To conclude this chapter on an amusing note, we mention a qualitative experiment (i.e. with accuracy not assessable) performed in 1971 by NASA astronaut D. Scott, of the Apollo 15 mission: following Galileo's suggestion, he dropped a hammer and a feather on the Moon surface, observing how, in absence of atmosphere, the two bodies fell with the same acceleration. The video clip is available on YouTube ("Hammer vs Feather").

References

Bailey, J., et al.: Measurements of relativistic time dilatation for positive and negative muons in a circular orbit. Nature 268, 301–305 (1977).

Rossi, B., Hall, D.: Variation of the rate of decay of Mesotrons with momentum. Phys. Rev. 59, 223 (1941).

Hafele, J.C., Keating, R.E.: Around-the-world atomic clocks. Science 177, 166 (1972).

Roll, P.G., Kroktov, R., Dicke, R.H.: The equivalence of inertial and passive gravitational mass. Ann. Phys. 26, 442–517 (1964).

Braginsky, V.B., Panov, V.I.: Verification of equivalence of inertial and gravitational masses. Sov. Phys. JETP 34, 463–466 (1972).

Baeßler, S., et al.: Improved test of the equivalence principle for gravitational self-energy. Phys. Rev. Lett. 83, 3585 (1999).

Adelberger, E.: New tests of Einstein's equivalence principle and Newton's inverse-square law. Class. Quantum Grav. 18, 2397 (2001).

Wagner, T.A., et al.: Torsion-balance tests of the weak equivalence principle. Class. Quantum Grav. 29, 184002 (2012).

Fray, S., Alvarez, Diez C., Hansch, T.W., Weitz, M.: Atomic interferometer with amplitude gratings of light and its applications to atom based tests of the equivalence principle. Phys. Rev. Lett. **93**, 240404 (2004)

Tarallo, M.G., et al.: Test of Einstein equivalence principle for 0-spin and half-integer-spin atoms: search for spin-gravity coupling effects. Phys. Rev. Lett. 113, 023005 (2014).

Overduin, J., Everitt, F., Worden, P., Mester, J.: STEP and fundamental physics. Class. Quantum Grav. 29, 184012 (2012).

Touboul, P., Metris, G., Lebat, V., Robert, V.: The MICROSCOPE experiment, ready for the in-orbit test of the equivalence principle. Class. Quantum Grav. 29, 184010 (2012).

Toubul, P., Rodrigues, M.: The MICROSCOPE space mission. Class. Quantum Grav. 18, 2487 (2001).

Touboul, P., et al.: The MICROSCOPE mission: first results of a space test of the equivalence principle. Phys. Rev. Lett. 119, 231101 (2017).

Nobili, A., et al.: Galileo Galilei" (GG) a small satellite to test the equivalence principle of Galileo: Newton and Einstein. Exp. Astron. 23, 689–710 (2009).

Further Reading

Nordtvedt, K.: Equivalence principle for massive bodies I phenomenology and II theory. Phys. Rev. 169, 1014 (1968).

Principles of Metric Theories

4

4.1 Introduction

In the previous chapter, we discussed the equivalence principle (EP) and underlined how it constitutes the foundation of Newton's as well as Einstein's theory. We recall here its weak formulation (WEP), following Will (2014):

the trajectory of a freely falling test body in a given point of space-time (event) is independent of its internal structure and composition.

and we remark that this postulate is not sufficient to support Einstein's theory. Indeed, this is based on a more extended postulate that takes the name of Einstein Equivalence Principle (EEP):

(a) WEP is valid.
(b) The outcome of any local non-gravitational experiment is independent of the velocity of the freely-falling reference frame in which it is performed—Local Lorentz Invariance (LLI).
(c) The outcome of any local non-gravitational experiment is independent of where and when in the universe it is performed—Local Position Invariance (LPI).

As an example, we mention that, as a consequence of EEP, any measurement of the fine structure constant of electromagnetism, $\alpha = 1/137$, must give the same value independently on the reference frame (LLI), and the location and time of the measurement (LPI). This statement is at present challenged, both on theoretical and

---

The original version of this chapter was revised: Chapter have been updated with the correction. The correction to this chapter can be found at https://doi.org/10.1007/978-3-030-95596-0_15

© The Author(s), under exclusive license to Springer Nature Switzerland AG 2022, corrected publication 2023
F. Ricci and M. Bassan, *Experimental Gravitation*, Lecture Notes in Physics 998, https://doi.org/10.1007/978-3-030-95596-0_4

experimental ground, but no clear evidence of a time (Uzan 2003) or space (Webb 2011) variation of $\alpha = 1/137$ has yet emerged.

At this point, we need to introduce a further extension of the EP, called Strong Equivalence Principle (SEP). Its formulation is almost identical to the previous one (EEP), except for substituting, in the statement defining LLI and LPI, the words *non-gravitational* with *both gravitational and non-gravitational*. In other words, we extend the EP to all phenomena that include self-gravitating objects, or bodies that exert a gravitational interaction with themselves, like stars, planets, black holes....or torsion balances.

EEP implies that gravity must be described by a *metric* theory, i.e. a theory where:

(1) Space-time has a metric described by a symmetric tensor $g_{\mu\nu}$.
(2) Trajectories of test masses in free fall describe the geodetic lines of this metric.
(3) In local Lorentz frames, geodetics are straight lines and the laws of non-gravitational physics are the laws of special relativity.

Phenomena are described in terms of events of the space-time, whose structure is characterized by privileged trajectories that bodies in "free fall" describe. To these lines, we associate a local reference frame where laws of special relativity hold, the so called *Lorentz frames*. Validity of LLI implies that, in any point of space-time, the laws of physics must be the same, regardless of the local frame chosen.

Note that the choice of local Lorentz frame is not unique: we can always have reference frames associated with the same event, but in mutual relative motion.

The metric and tensor nature of the gravitational field follows, in particular, from the limit requirement that, in absence of gravity, the laws of physics must be those formulated in special relativity, i.e. in a space-time described by the Minkowski metric $\eta_{\mu\nu}$.

$$
\eta_{\mu\nu} = \begin{pmatrix} 1 & 0 & 0 & 0 \\ 0 & -1 & 0 & 0 \\ 0 & 0 & -1 & 0 \\ 0 & 0 & 0 & -1 \end{pmatrix}
$$

Therefore, there must be one or more tensor fields that interact with gravity: in principle, these interactions could be described by different formalisms, according to their different nature. However, the joint verification of WEP and LLI forces the set of possible tensor fields to locally reduce, in a generic point (event) $P \equiv x^\mu$ of space-time, to those fields that satisfy $\psi_{\mu\nu}(i) = \Phi^{(i)}(P)\eta_{\mu\nu}$. $\Phi^{(i)}(P)$ could, in principle, change from point to point and from field to field.

As LPI also holds, we demand that an experiment involving those fields provide the same results in all points of space-time. Formally, this implies having $\Phi^{(i)}(P) = C^{(i)}$, where C is a constant associated to each of the possible choices of fields $\psi^{(i)}$. Else, it might happen that a unique universal field $\Phi(P)$ exists, that all fields transform into: $\Phi^{(i)}(P) = \Phi(P)C^{(i)}$. In the first case, we can rescale all coupling constants in such a way as to set every $C^{(i)} = 1$. In the second case too, we can rescale the coupling constants and define a transformation such that $\Psi' = \Phi^{-1}(P)\Psi$.

If all fields assume the same local form, they can be reduced to a unique tensor field, symmetric and rank-2: $g_{\mu\nu}$.

Summarizing, $g_{\mu\nu}$ describes a space-time where it exists a family of privileged curves (trajectories), i.e. the geodesics: we define them as the curves that minimize the distance between two events in space-time.

On the other hand, we can assert, based on WEP, that in a free-falling reference frame, the effect of gravity is canceled, and test particles move of uniform linear motion. These reference frames in free-fall are the local Lorentzian frames, where the geodesics that free bodies follow are straight lines.

In any point-event P of space-time, there exist *local* reference frames where space-time is flat and corresponding geodesics are straight lines. Therefore, the tensor fields must reduce to $||\eta||$ in any free-falling frame, anywhere in space-time, where the laws of non-gravitational physics are those of special relativity.

4.2 Experiments to Verify LLI e LPI

In the previous section, we have outlined the reasons why LLI and LPI are the foundations of all metric theories of gravitation. We must now discuss the experimental basis that those principles rest on.

We have recalled, in Chap. 3, that particle physics has provided countless experimental verifications of Lorentz transforms for time intervals, proving they are determined by relative velocity and not by acceleration of bodies. Besides, LLI was experimentally verified in many different fields of physics. *In primis*, we recall the experiment of Brillet and Hall (1979): it is an improved version of the glorious experiment of Michelson and Morley that proved the isotropy of the speed of light. With sensitivity 4000 times better than several previous, Michelson-like experiments. Brillet and Hall measured the beat frequency between two single mode lasers, one positioned on a rotating platform and the other on the ground. They set a limit on the anisotropy of space: $3 \cdot 10^{-15}$. Actually, we should note that, in order to measure the frequency of the rotating laser, Hall and Brillet used a Fabry-Perot interferometer or, more exactly, an *etalon*: thus, light travels back and forth. The limit that can be set, with this experiment on the round trip anisotropy is $(v/c)^2 < 3 \cdot 10^{-15}$, to be compared with the value $\sim 10^{-8}$ of Michelson-like experiments, using the velocity v of the Earth.

Hughes (1960) and Drever (1961) independently set an upper limit of about 10^{-20}, with two very sophisticated experiments exploiting the magnetic resonance of the ^{7}Li nucleus in its fundamental state, characterized by a nuclear spin $I = 3/2$.

To get an insight to this experiment, we must mention a result obtained, 2 years before, by Cocconi and Salpeter (1958): assume an anisotropy of space that produce a change Δm in a particle of mass m; Cocconi and Salpeter proved that, in a Coulomb field, the binding energy E would change as:

$$\Delta E = \frac{\Delta m}{m} \bar{T} P_2(cos\theta) \tag{4.1}$$

where $\bar{T}$ is the mean kinetic energy of the particle, θ the angle between the direction of particle acceleration and the direction of the anisotropy, assumed to be toward the galactic center; P_2 is the Legendre polynomial of order 2. The same prediction also holds in the nuclear case, if the nucleus is modeled as a single particle in a quadratic potential with spherical symmetry. In an external magnetic field, the energetic fundamental state of the particle is split into four equispaced levels: there is then only one observable nuclear transition line. Moreover, the direction of acceleration of the particle depends on the direction of the external magnetic field. Therefore, if an anisotropy exists, such degeneracy should be removed and, with enough resolution, a triplet should be observed, with a separation changing as a function of *sidereal time* T_s, as the direction toward the galactic center is $\theta = \theta(T_s)$.

The limit 10^{-20} was set by Hughes and Drever by observing the broadening of the spectral line of the hyperfine transition versus the direction of the magnetic field with respect to the galactic reference frame centered on the Earth.

Note that, anyways, these experiments cannot provide indications on the nature of the metric tensor $g_{\mu\nu}$ of the space-time and cannot assure the existence of a global Lorentz reference frame where the metric tensor is $g_{\mu\nu} = \eta_{\mu\nu}$.

4.3   The Experiment of Pound–Rebka

An experimental proof of EP can be performed with a torsion balance with two end masses of different materials. It implies that bodies follow lines in the event space that can be identified as *geodesics* of the tensor metric $g_{\mu\nu}$. These lines can indeed be defined as geodesics if the motion of the bodies is not accelerated with respect to the origin of a local Lorentz frame. We know from special relativity that the choice of a Lorentz frame requires, beyond the three spatial coordinates, also a reference clock: therefore, this measurement of relative motion inevitably requires knowledge of if and how the flow of time can change from one point to another, when a gravitational field is present. The experiment of Pound and Rebka (PR) (Pound 1960) shed light on this issue: the idea is to compare the behaviour of two "reference clocks" in two different position in a uniform gravitational field (g on the Earth surface). They used a tower in the Harvard campus, with two measuring stations with height difference $h = 22.6$ m, and exploited the *Mössbauer effect*. In this case, the (atomic) clocks are provided by the energetic levels of the excited isospin state $I = 5/2$ in ^{57}Fe nuclei. The nuclei are prepared in this state starting from the source isotope ^{57}Co that, by electronic capture, transforms into ^{57}Fe. This nuclide has two possible decays (see the lower pane in Fig. 4.1): the main one provides γ radiation with energy $\mathcal{E} = 14.4$ keV.

The Lorentzian shape of this emission line, although extremely narrow ($\delta\nu_{Lor}/\nu = 1.13 \cdot 10^{-12}$), is still 200 times larger than the frequency shift expected by the gravitational red-shift, as shown below in this section: $\delta\nu_{GR}/\nu = gh/c^2 = 2.4 \cdot 10^{-15}$. For this reason, in order to compare the emission lines of isotopes in identical conditions, we need to change the energy of the γ rays emitted by the source inducing Doppler shifts.

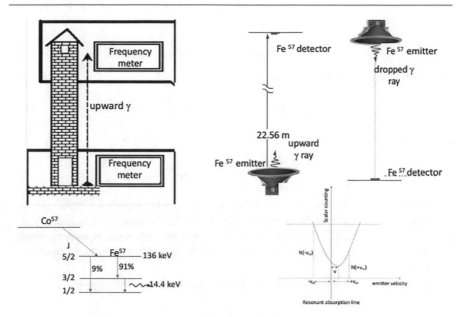

Fig. 4.1 Schematic of the Pound–Rebka experiment

So, the γ source is moved with respect to the detector till the point when absorpion takes place, and this happens when the emitted energy is exactly the required one. For this reason, the radioactive source is put into oscillation, on a loudspeaker core, with an oscillating speed v of few mm/s. In this way, one can compare the frequency of emitted γ downward with that of γ emitted upward, sweeping across the linewidth with a frequency span of the order of the ratio v/c, i.e. $\sim 3 \cdot 10^{-11}$.

The PR experiment was carried out at the Jefferson physics laboratory of Harvard University, with two pairs of Mössbauer emitter and detector, separated by a vertical distance $h = 74$ feet $\simeq 22.56$ m. They measured a relative frequency difference, between upward and downward photons, given by

$$\frac{\Delta \nu}{\nu} = 4.92 \cdot 10^{-15} \tag{4.2}$$

To understand this result, we need quantum theory, WEP and energy conservation. The particle emitting the photon at the top of the tower (*above*) loses a fraction of mass $\Delta m_a c^2 = -2\pi\hbar\nu_a$, where $2\pi\hbar$ is Planck's constant.[1] When the photon is absorbed at the bottom (*below*), a free falling observer will see his apparatus increase its inertia by $\Delta m_b c^2 = 2\pi\hbar\nu_b$. However, conserving the total energy of the above + below system, also including the potential energy of gravitational field $m\Phi$, we get:

---

[1] Although $2\pi\hbar$ might seem akward, we try and stay away from h, that in this book has very different meanings.

$$\Delta m_a c^2 + \Phi_a \Delta m_a + \Delta m_b c^2 + \Phi_b \Delta m_b = 0 \tag{4.3}$$

or, substituting the above relations for Δm:

$$- \nu_a - \Phi_a \nu_a / c^2 + \nu_b + \Phi_b \nu_b / c^2 = 0 \tag{4.4}$$

and finally:

$$\frac{\nu_a}{\nu_b} = \frac{1 + \Phi_a / c^2}{1 + \Phi_b / c^2} \simeq 1 + \frac{(\Phi_b - \Phi_a)}{c^2} = 1 + \frac{g\,h}{c^2} \tag{4.5}$$

The photon, in the gravitational field, undergoes a frequency (and energy) shift of opposite sign when it goes up (*red shifted*) with respect to the photon going down (*blue shifted*). We then expect the two receivers, above and below, measure γ radiation with a frequency difference due to the Doppler shift given by: $2\,(gh/c^2) \simeq 4.95 \cdot 10^{-15}$.

The agreement between theory and the Pound–Rebka experiment is

$$\frac{\Delta \nu_{spe}}{\Delta \nu_{teo}} = 1.05 \pm 0.10 \tag{4.6}$$

This experiment was repeated in 1965 by Pound and Snider (1965), achieving an accuracy of 1%. The formally rigorous interpretation of this experiment in the light of GR requires advanced notions, such as hyperbolic motion (see Misner et al. (1973) Sects. 6.2–6.6). One must describe the motion of an object subject to a constant acceleration with respect to either an inertial reference system *co-moving* with the object, or to inertial systems that must change at any instant. We try to provide here a simpler explanation:

In the PR experiment the two clocks at the top and bottom of the tower are at rest in the laboratory system $(t_s, \vec{x}_s)$, and their proper time flows according to:

$$cd\tau = \sqrt{g_{\mu\nu} dx^\mu dx^\nu} \tag{4.7}$$

As the clock is stationary in the lab frame, $(dx^i = 0)$, we have

$$dt = \frac{d\tau}{\sqrt{g_{oo}}} \tag{4.8}$$

The coordinate time interval Δt between two pulses is the same at the points of emission and reception: indeed the two e.m. pulses travel in world lines that, although not at 45° (space-time is not flat) are identical and just displaced by Δt, because the geometry does not depend on time. Thus, for the two clocks in positions where the value of g_{oo} is different, the proper time will flow in two different ways:

$$\Delta t = \frac{\Delta \tau_a}{\sqrt{g_{oo}(x_a)}} = \frac{\Delta \tau_b}{\sqrt{g_{oo}(x_b)}}$$

from this we deduce

$$\frac{\nu_a}{\nu_b} = \frac{\Delta\tau_b}{\Delta\tau_a} = \sqrt{\frac{g_{oo}(x_a)}{g_{oo}(x_b)}}$$

We shall see in the next chapter that, in the case of weak and uniform gravitational field, the approximation

$$g_{oo}(x) \simeq 1 - 2\Phi(x)/c^2$$

holds, and we can conclude that in a uniform gravitational field, as the two clocks are positioned at a height difference $x_a - x_b = h$, we have

$$\frac{\nu_a - \nu_b}{\nu_b} = \frac{g\,h}{c^2} \tag{4.9}$$

that is the same result previously found.

The accuracy of these measurements can be defined through a parameter α_{rs} expressing the measured deviation from the result (4.9):

$$\frac{\nu_a - \nu_b}{\nu_b} = \frac{g\,h}{c^2} \cdot (1 + \alpha_{rs}) \tag{4.10}$$

with α_{rs} vanishing when the EEP is valid.

The frequency shift of photons in a gravitational field has been verified in numerous other experiments; for example, by observing the spectrum of deuterium in the gravitational field of the Sun, with $\alpha_{rs} = 5 \cdot 10^{-2}$. In more recent times, following the development of the time standards with stability approaching $1/10^{16}$, the accuracy limit of 10^{-2} has been largely exceeded, in June 1976, by the experiment of Vessot and Levine (1980), named Gravity Probe A: they used a Scout missile to fly a hydrogen atomic clock at a height of $10000\,$km, comparing it with an identical clock on ground. One of the difficulties of the experiment was to correct the Doppler frequency shift, depending on position and velocity of the spacecraft payload. The accuracy achieved in this case was $4 \cdot 10^{-4}$. This experiment has been recently repeated using two satellites of the European GNSS Galileo (Delva 2018), which were unintentionally launched into eccentric orbits. The longer observation time allowed to reduce the uncertainty to $\alpha_{rs} = (4.5 \pm 3.1)10^{-5}$.

In recent years, experiments using atom interferometers have reached sensitivities so high as to achieve measurements of the gravitational red shift at mm height difference, with accuracy of $7 \cdot 10^{-9}$ (Muller 2010).

4.4 Schiff's Conjecture

The experimental consequences of the parts of the Einstein Equivalence Principle described above, i.e. WEP + LLI + LPI, are quite different. Any complete gravitational theory must be characterized by a mathematical structure capable of making predictions for experiments testing each aspect of the principle. Actually, there is a limited number of ways for a gravity theory to be compatible with special relativity; therefore it is not surprising that there might be theoretical links among the three parts of the EEP. As an example, the same mathematical formalism used to produce the equations describing the free fall of a hydrogen atom should also produce equations describing the energy levels of Hydrogen in a gravitational field, thus setting the typical periodicity of a Hydrogen maser clock. It follows that, in a WEP violating theory, an EEP violation could also appear as a violation of LPI: WEP is, therefore, sufficient to imply EEP.

Around 1960, L. Schiff hypothesized that this type of link could be a *needed* feature of any self-consistent theory of gravity. More precisely, the Schiff conjecture states that *any* self-consistent complete theory of gravity incorporating WEP necessarily embodies EEP.

In other words, the validity of WEP guarantees the validity of both Lorentz and Local Position Invariance, and therefore of EEP. Moreover, if Schiff conjecture is correct, the experiments of Eötvos could be interpreted as an experimental verification of EEP, and therefore be the foundation of the hypothesis that interprets gravity as a curvature phenomenon of space-time.

A rigorous proof of this conjecture seems impossible; it is supported only on the basis of a number of robust and plausible arguments.

An elegant example of these arguments, due to Dicke, Haugan and Nordtvedt, is based on energy considerations. It is a simple conceptual experiment (*gedanken experiment*), where the core assumption is that the energy of a closed system, at the end of a series of transformation, must invariably have the same value than that at the beginning. We now try to explain it.

Consider a system in position $\vec{x}$, moving at speed $|\vec{v}| << c$: the total energy of the system has the general form:

$$E_c = M_R c^2 - M_R \Phi(\vec{x}) + \frac{1}{2} M_R v^2 + \ldots\ldots\ldots\ldots \tag{4.11}$$

where M_R is the rest mass of the system, Φ the external potential and v the speed absolute value. The dots " $\ldots$ " indicate possible terms of higher order in Φ and v^2

Assume now that our system is composed of n bound elementary particles of mass m_o; there is a binding energy E_B associated with it. The rest energy can be written as the difference between the rest energy of the n isolated particles and the binding energy E_B: if we have a EEP violation, the binding energy can depend, in general, on position $\vec{x}$ (LPI violation) and/or on speed $\vec{v}$ (LLI violation):

$$M_R c^2 = n \, m_o c^2 - E_B(\vec{x}, \vec{v}) \tag{4.12}$$

In the reasonable assumption of a weak EEP violation, we can write the binding energy as a series expansion of the mass anomalies due to the violating terms:

$$E_B(\vec{x}, \vec{v}) = E_B{}^o + \delta m_p{}^{ij} U^{ij} + \sum_{i=1}^{3} \sum_{j=1}^{3} \delta m_I{}^{ij} v^i v^j + \ldots\ldots \qquad (4.13)$$

We have explicitly highlighted the passive gravitational mass variation, $\delta m_p{}^{ij}$, associated with LPI violation, and the inertial mass violation, associated with the LLI, $\delta m_I{}^{ij}$.

From now on we shall omit the explicit sum symbol, following Einstein's convention of summation on repeated indexes. Latin indexes (i, j, k ...) run over spatial coordinates 1, 2, 3.

We now consider two different quantum systems, e.g. two excited states of the ^{7}Li nucleus, in an external magnetic field. These states are characterized by different azimuth quantum number: if EEP is not violated, the energy spacing between the two levels is the same. Besides, the binding energy $E_B{}^o$ is, to zeroth order, the same for both states.

The systems at rest ($v^i = v^j = 0 \ \forall \ i, j$) in the gravitational potential Φ can perform transitions, emitting photons of frequency proportional to the energy change of the corresponding system. From (4.13) it follows that the frequencies emitted by the two systems will be different, due to the presence of the terms $\delta m_p{}^{ij} \Phi^{ij}$. This shows how violation of WEP can induce a violation of LPI.

We report a *gedanken-experiment* (Haugan 1979) for a quantum system subject to a cyclic transformation.

The quantum system of n identical, non-interacting particles, initially at rest at a height $z = h$, is endowed with an energy

$$E_{free}(z = h, v = 0) = n \, m_o[c^2 - \Phi(h)]$$

(for notation simplicity, we shall drop in the following the dependence on v, inessential to our purposes). This is the energy that must be conserved during all transformations that we now consider:

1. We allow the particles to interact, creating a new *composite* system. This will release a binding energy E_B, (see Eq. 4.13) and reduce its energy

$$E_{composite}(h) = [n \, m_o c^2 - E_B(h)] \left[1 - \frac{\Phi(h)}{c^2} \right]$$

We can convert E_B in a number of free particles and store them separately. We then have two systems: a composite body, with energy $E_{composite}(h)$ and a system of free particles with energy $E_B(h)$.

2. We now let both systems fall to the reference height $z = 0$: the composite system with an acceleration $\vec{a}$ (not necessarily equal to $\vec{g}$), while the free particles fall, by definition, with acceleration $\vec{g} = -\vec{\nabla}\Phi$.

3. We bring the two systems to rest, converting and storing their kinetic energy. We now have available the energies:

$$E_{composite}(0) + \left[n\, m_o - \frac{E_B(0)}{c^2} \right] \vec{a} \cdot \vec{h} + \delta m_I{}^{ij} g^i h^j$$

and, for system 2:

$$E_{released} = E_B(h) + \frac{E_B(0)}{c^2} \vec{g} \cdot \vec{h}$$

4. We now disassemble the composite system into the original n particles: this takes up the energy
$E_{free}(0) = E_{composite}(0) + E_B(0)$.

5. Finally, raise again the non-interacting system to height h, using an energy $nm_0 gh$. We have recomposed the initial system, with its energy $E_{free}(h)$, plus the following energy terms:

$$[E_B(h) - E_B(0)] - \left(nm_0 - \frac{E_B(0)}{c^2} \right) (\vec{a} - \vec{g}) \cdot \vec{h} + \delta m_I{}^{ij} g_i h_j \qquad (4.14)$$

If energy conservation holds, this amount must be zero. We rewrite the term in the square bracket with the help of Eq. 4.13 and define, for shorthand notation $M_c \equiv nm_0[1 - \frac{E_B(0)}{c^2}]$

$$\delta m_p{}^{ij}[\Phi_{ij}(h) - \Phi_{ij}(0)] - \delta m_I{}^{ij} g_i h_j - M_c(a^k - g^k)h_k \qquad (4.15)$$

We recall that $\vec{h}$ is small enough to consider $\vec{g}$ as constant. Therefore, we can rewrite $\Delta \Phi_{ij}/h_k$ as a derivative, and rearrange Eq. 4.15 as:

$$a^k = g^k + \frac{\delta m_p{}^{ij}}{M_c} \frac{\partial \Phi_{ij}}{\partial x_k} + \frac{\delta m_I{}^{kj}}{M_c} g_j \qquad (4.16)$$

This proves that a violation of LPI (second term on the r.h.s) or a violation of LLI (third term) will produce a violation of WEP ($\vec{a} \neq \vec{g}$).

Finally, we report one more argument, due to K. Nordvedt, supporting Schiff's conjecture.

Let A be a system falling with acceleration g_A: it can decay emitting a quantum of radiation of frequency ν. Assume now that such emission takes place when the system is at a height H from ground in an external field. Due to the gravitational field, the quantum is shifted to frequency ν'. Once the photon is emitted, the system, in the new quantum state B, will fall, if the equivalence principle is violated, with a different acceleration g_B. We can now identify, thanks to the LLI hypothesis, the inertial mass with the total energy of each body.

We can write in a formally simple way the concept that g_A and g_B depend on the internal energy of the quantum states involved in the transition:

Fig. 4.2 Experimental foundations of Einstein Equivalence Principle (Haugan 1979)

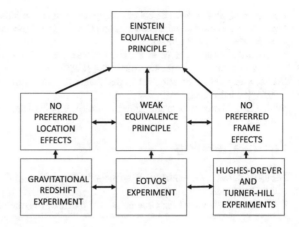

$$g_A = g\Big(1 + \alpha \frac{E_B}{m_A c^2}\Big) \quad g_B = g\Big(1 + \alpha \frac{E_B}{m_B c^2}\Big) \quad E_B - E_A = h\nu \qquad (4.17)$$

If we want to bring our system back to its initial condition, before the emission and at height H, we must use the kinetic energy of the system on ground, converted from the potential $m_B g_b H$, and the energy of the photon $h\nu'$. In other words, we must recover $m_A g_A H$ and the frequency shift $\nu - \nu'$. If we impose this condition, we obtain, to lowest order in $h\nu/mc^2$, a relative frequency shift given by:

$$Z = \frac{\nu - \nu'}{\nu'} = g(1 + \alpha)\frac{H}{c^2} = (1 + \alpha)\frac{\Delta\Phi}{c^2} \qquad (4.18)$$

This proves that violation of LPI must follow to violation of WEP. Figure 4.2 shows a diagrammatic representation of the relations underlying EEP.

References

Brillet, A., Hall, J.L.: Improved laser test of the isotropy of space. Phys. Rev. Lett. **42**, 549–552 (1979)

Cocconi, G., Salpeter, E.E.: A search for anisotropy of inertia. Nuovo Cimento **10**, 646 (1958)

Delva, P., et al.: Gravitational redshift test using eccentric Galileo satellites. Phys. Rev. Lett. **121**, 231101 (2018)

Drever, R.W.P.: A search for anisotropy of inertial mass using a free precession technique. Philos. Mag. **6**, 683 (1961)

Haugan, M.P.: Energy conservation and the principle of equivalence. Ann. Phys. **118**, 156 (1979)

Hughes, V.W., et al.: Upper limit for the anisotropy of inertial mass from nuclear resonance experiments. Phys. Rev. Lett. **4**, 342 (1960)

Misner, C.W., Thorne, K.S., Wheeler, J.A.: Gravitation. Freeman, San Francisco, USA (1973)

Muller, H., Peters, A., Chu, S.: A precision measurement of the gravitational redshift by the interference of matter waves. Nature **463**, 926 (2010)

Pound, R.V., Rebka, G.A.: Apparent weight of photons. Phys. Rev. Lett. **4**, 337 (1960)

Pound, R.V., Snider, J.L.: Effect of gravity on gamma radiation. Phys. Rev. B **140**, 788 (1965)
Uzan, J.P.: The fundamental constants and their variation: observational and theoretical status. Rev. Mod. Phys. **75**, 403 (2003)
Vessot R.F.C., LevineM.W. et al. *Test of Relativistic Gravitation with a Space-Borne Hydrogen Maser* - Phys. Rev. Lett., **45**, 2081-2084 (1980)
Webb, J.K., et al.: Indications of a spatial variation of the fine structure constant. Phys. Rev. Lett. **107**, 191101 (2011)
Will, C.M.: The confrontation between general relativity and experiment. Living Rev. Relativ. **17**, 4 (2014)

Further Reading

Bertotti, B. (ed.): Gravitazione Sperimentale–Experimental Gravitation; Atti dei convegni Lincei **34**, Roma (1977)
Ciufolini, I., Wheeler, J.A.: Gravitation and Inertia. Princeton University Press, Princeton (1995)

Tests of Gravity at First Post-Newtonian Order

5

5.1 Introduction

The criteria to assess the reliability of a theory are its completeness and self-consistency. A theory is *complete* when all the equations needed to describe how a system behaves in given conditions can be derived from first principles. A theory is *self-consistent* when the prediction of the outcome of a given experiment, obtained with different methods, is unique: as an example that we will discuss further on, regardless of whether the light is considered a wave or massless particle, both predictions about the light deflection due to the gravitational field should coincide.

Furthermore, its reliability is corroborated if it can make correct prediction about the Newtonian and the relativistic limits. In other words, in the so-called Weak Field and Slow Motion (WFSM) limit, when the gravitational field is weak (see the next section for a quantitative assessment of what we intend for *weak*) and the particles move with velocity $v \ll c$, the theory must recover the laws of Newtonian physics. On the other hand, in the absence of a gravitational field, the laws of the theory should lead to those of Special Relativity (SR). In previous chapters, we have discusses the essential pillars (such as WEP-LLI-LPI, that together yield EEP) that support the hypothesis that gravity is a metric theory. In such a theory, matter creates fields that, along with matter itself, generate the metric, but in turn, matter motion is determined by the metric.

Based on all the considerations made so far about the EEP, the laws of physics in covariant form can be formulated using SR, and generalized when accounting for space curvature. The idea is therefore to derive the laws of SR from an action, containing the pseudo-Euclidean (Minkowskian) metric tensor $\eta_{\mu\nu}$. The generalization will take place through a generic transformation of the coordinates $x'^{\mu} = x'^{\mu}(x^{\alpha})$.

The original version of this chapter was revised: Chapter have been updated with the correction. The correction to this chapter can be found at https://doi.org/10.1007/978-3-030-95596-0_15

F. Ricci and M. Bassan, *Experimental Gravitation*, Lecture Notes in Physics 998, https://doi.org/10.1007/978-3-030-95596-0_5

By transforming in a general form the vectors, tensors, differentials and integration elements, we end up rewriting the action in the new reference system of curved coordinates. It is interesting to note that at the end of such a transformation, the form of the action remains the same, provided that the pseudo-Euclidean metric tensor is replaced by the metric tensor $g_{\mu\nu}$, and the derivatives by covariant derivatives. These simple mnemonic rules are the formal statement of Einstein's Equivalence Principle.

Several theories have been formulated, and they differ only in how the metric is generated: that is because, in any case, if EEP is valid, matter only couples to the metric. In Einstein's General Theory of Relativity (this is the name he gave), there exists a unique gravitational field generated by the stress-energy tensor which contains contributions of both matter and other fields. We just mention, as examples, two other theories. In the Brans-Dicke-Jordan theory (Brans and Dicke 1961), matter and fields generate a scalar field Φ that, along with matter and fields, generates the metric. In Ni's theory (Ni 1973), it is assumed the existence of a flat metric in all the Universe where a proper time exists. This flat metric contributes, with matter and non-gravitational fields, to generate a scalar field Φ. All these fields then combine to generate *physical metric* $g_{\mu\nu}$, the metric that enters into the equivalence principle.

Einstein's General Relativity (GR) has emerged in the last century as the best and most robust theory of gravity, also due to the precise experimental verifications of a few, precisely predicted phenomena, observed in the Solar System, that do not have Newtonian counterparts. Three of these, the so-called *classical tests*, were proposed by Einstein himself (1916):

(a) The red-shift of electromagnetic (e.m.) radiation, discussed in Chap. 4,
(b) the precession of the perihelion of Mercury and
(c) the deflection of light rays.
     Two additional tests, also observable in the Solar System, were later proposed:
(d) the Shapiro time delay of e.m. radiation and
(e) the Lense-Thirring dragging of inertial frames.

All these phenomena are conveniently described using GR in the Weak Field, Slow Motion (WFSM) approximation, that we introduce in the following sections. Other effects, like generation of gravitational waves that takes place in regions of strong gravity, require instead a more complete GR approach. In the next chapter, we will expand the focus to other theories, and recall the Post-Newtonian Parametrization (PPN): a formalism general enough to represent in a comprehensive form the different predictions of a broad class of metric theories in a weak field limit, in order to compare them with the experimental observations.

5.2   Recollection of GR Formalism and Equations

In this section, we briefly review the main relations of Einstein's GR, taking a top-down approach, i.e. starting with the field equations that are often the arrival point of an introduction to GR.[1]

The synthesis of Einstein's Theory of Gravitation is enclosed in the field equations of GR (we neglect here the cosmological constant Λ)

$$R_{\mu\nu} - \frac{1}{2}g_{\mu\nu}R = \frac{8\pi G}{c^4}T_{\mu\nu} \tag{5.1}$$

$$T_{\mu}{}^{\nu}{}_{;\nu} = 0 \tag{5.2}$$

with μ, $\nu = 0$, 1, 2, 3. We recall here a few of the math tools and concepts of GR, in order to get an insight into the meaning of these equations.

- All formulas are written by applying Einstein's convention of summing on repeated indexes; for example: $v^{\alpha}{}_{\alpha} = \sum_{\alpha} v^{\alpha}{}_{\alpha}$ is the trace of the tensor $v_{\alpha\beta}$. The operation of summing over a repeated index is called "contraction".
- Indexes are raised or lowered (passing from covariant to contravariant) by contracting them with the metric tensor: $v_{\alpha} = g_{\alpha\beta}v^{\alpha}$
- The symbol appearing in Eq. 5.2: ";ν" (subscript: semicolon—ν), represents the covariant derivative (defined below) with respect to the coordinate x^{ν}. Not to be confused with the ordinary derivative, indicated by a comma: ",$\nu = \frac{\partial}{\partial x^{\nu}}$". In Eq. 5.2, the covariant derivative is applied to the energy-momentum tensor $T_{\mu\nu}$, representing matter properties. Equation 5.2 expresses the conservation law for mass-energy-momentum
- The basic object that describes the space-time geometry is the metric tensor $g_{\mu\nu}$, defined by the metric:

$$ds^2 = g_{\mu\nu}dx^{\mu}dx^{\nu} \tag{5.3}$$

  $g_{\mu\nu}$ takes the role of a *tensor* potential in the field theory of gravity, or geometrodynamics.
- We recall the definition of covariant derivative of a vector field v_j:

$$v_{\mu;\nu} = \frac{\partial v_{\mu}}{\partial x^{\nu}} - \Gamma_{\mu\nu}{}^{\rho}v_{\rho} \tag{5.4}$$

- where $\Gamma_{ij}{}^{r}$ are the Christoffel symbols (or Affine Connections) of second type:

$$\Gamma_{\mu\nu}{}^{\rho} = \frac{1}{2}g^{\rho\alpha}\left(\frac{\partial g_{\nu\alpha}}{\partial x^{\mu}} + \frac{\partial g_{\alpha\mu}}{\partial x^{\nu}} - \frac{\partial g_{\mu\nu}}{\partial x^{\alpha}}\right) \tag{5.5}$$

---

[1] It is assumed that the reader is already familiar with GR: the list of equations we report here cannot substitute a specific study of the foundations and formalism of GR.

The Christoffel symbols, made out of derivatives of the metric tensor, are often considered as the *fields* of gravitational theory. Just as the fields of electrodynamics, they are not invariant (are not tensors) under coordinate transformation.

- The Riemann tensor $R^\alpha{}_{\beta\mu\nu}$ is also called the curvature tensor in a four-dimensional space:

$$R^\alpha{}_{\beta\mu\nu} = \Gamma^\alpha{}_{\beta\nu,\mu} - \Gamma^\alpha{}_{\beta\mu,\nu} + \Gamma^\alpha{}_{\mu\rho}\Gamma^\rho{}_{\beta\nu} - \Gamma^\alpha{}_{\nu\rho}\Gamma^\rho{}_{\beta\mu} \tag{5.6}$$

- $R_{\mu\nu}$ is the Ricci tensor, obtained by contracting two indexes of the Riemann tensor:

$$R_{\mu\nu} = \Gamma^\alpha{}_{\mu\nu,\alpha} - \Gamma^\alpha{}_{\mu\alpha,\nu} + \Gamma^\alpha{}_{\mu\nu}\Gamma^\beta{}_{\alpha\beta} - \Gamma^\beta{}_{\mu\alpha}\Gamma^\alpha{}_{\nu\beta} \tag{5.7}$$

R is the curvature scalar: it is derived from Ricci tensor by further contracting the two indexes. Thus, the field equations 5.1 are a set of ten coupled, non-linear, second-order differential equations in the potentials $g_{\mu\nu}$.

A gravitational theory is valid and verified if we can make predictions on the motion of matter and compare them with measurements. We must therefore measure $g_{\mu\nu}$, i.e. determine the motion of a matter element with respect to another, and not just to a reference system.

This is a fundamental peculiarity of gravitation and of GR.

To state it differently, it is not sufficient to consider the motion of a freely gravitating particle with respect to a given reference frame: such motion is described by the *geodesic equation*, that is simply the generalization of the inertia principle to a curved space-time.

$$\frac{d^2x^\mu}{ds^2} + \Gamma^\mu{}_{\beta\gamma}\frac{dx^\beta}{ds}\frac{dx^\gamma}{ds} = 0 \tag{5.8}$$

Indeed, this is the explicit expression of the covariant derivative of the 4-vector velocity for a particle freely moving along a trajectory described by the curvilinear coordinate s (else, it can be considered as Newton's law for a particle in flat space-time, with the force field given by the second term, i.e. by the curvature).

However, as mentioned above, this equation does not allow us to deduce the fundamental nature of the phenomena in exam: we need to describe the motion of a particle with respect to another massive body. To this purpose, we now consider two particles moving along two geodesic lines, described by coordinates s and $s + ds$; let ξ^μ be the vector connecting the two geodesics. Choosing a locally Galilean reference frame, we compute the variation of the geodesic equation with respect to this vector, obtaining the *equation of geodesic deviation*:

$$\frac{\partial^2\xi^\gamma}{\partial s^2} + u^\mu u^\nu \xi^\rho R^\gamma{}_{\mu\nu\rho} = 0 \tag{5.9}$$

where u^α is the observer's 4-velocity. The corresponding equation in Newtonian gravity was

$$\frac{d^2\xi^i}{dt^2} = -\delta^{ij}\frac{\partial^2\Phi}{\partial x^j \partial x^k}\xi^k = -\delta^{ij}(\Phi,_{j,k})\xi^k \tag{5.10}$$

that defines the *gravitational gradient* or *tidal* tensor $\Phi_{,j,k}$.

The equation of geodesic deviation is the starting point to understand the working principle of any detector of gravitational waves.

5.3   The Weak Field, Slow Motion Approximation

We now return to the field Eqs. 5.1 and look for a solution of Einstein's equations in the case where we have matter, possibly in a stationary state of rotation. Assume we are in conditions of small perturbation of the flat (or Minkowskian) space-time:

$$g_{\mu\nu} \simeq \eta_{\mu\nu} + h_{\mu\nu} \tag{5.11}$$

where $h_{\mu\nu} \ll 1$ is the definition of Weak Field and we compute the perturbed metric in the linear approximation. We insert the perturbed metric tensor in the definitions (Eq. 5.5) of the affine connections and of the Ricci tensor, limiting our expansion to first order in h:

$$\Gamma^{\lambda}_{\mu\nu} \simeq \frac{1}{2}\eta^{\lambda\rho}[\frac{\partial}{\partial x^{\mu}}h_{\rho\nu} + \frac{\partial}{\partial x^{\nu}}h_{\rho\mu} - \frac{\partial}{\partial x^{\rho}}h_{\mu\nu}] \tag{5.12}$$

$$R_{\mu\nu} \simeq \frac{\partial}{\partial x^{\nu}}\Gamma^{\lambda}_{\lambda\mu} - \frac{\partial}{\partial x^{\lambda}}\Gamma^{\lambda}_{\mu\nu} \tag{5.13}$$

Note that, in agreement with the assumption of linear approximation, all tensor indexes are raised or lowered by the flat space-time metric $\eta_{\mu\nu}$ rather than $g_{\mu\nu}$. We have

$$\eta^{\lambda\rho}h_{\rho\nu} = h^{\lambda}{}_{\nu} \qquad\qquad \eta^{\lambda\rho}\frac{\partial}{\partial x^{\rho}} = \frac{\partial}{\partial x_{\lambda}}$$

We now combine the two expressions Eqs. 5.12, 5.13 to obtain

$$R_{\mu\nu} \simeq \frac{1}{2}\left(\frac{\partial^2}{\partial x^{\lambda}\partial x_{\lambda}}h_{\mu\nu} - \frac{\partial^2}{\partial x^{\lambda}\partial x^{\mu}}h^{\lambda}{}_{\nu} - \frac{\partial^2}{\partial x^{\lambda}\partial x^{\nu}}h^{\lambda}{}_{\mu} + \frac{\partial^2}{\partial x^{\mu}\partial x^{\nu}}h^{\lambda}{}_{\lambda}\right) \tag{5.14}$$

And finally, we derive the simplified form of Einstein's equations:

$$\frac{\partial^2}{\partial x^{\lambda}\partial x_{\lambda}}h_{\mu\nu} - \frac{\partial^2}{\partial x^{\lambda}\partial x^{\mu}}h^{\lambda}{}_{\nu} - \frac{\partial^2}{\partial x^{\lambda}\partial x^{\nu}}h^{\lambda}{}_{\mu} + \frac{\partial^2}{\partial x^{\mu}\partial x^{\nu}}h^{\lambda}{}_{\lambda} =$$
$$= -\frac{16\pi G}{c^4}\mathcal{T}_{\mu\nu} \tag{5.15}$$

The right-hand side of Einstein equations 5.1, includes the energy and momentum of the gravitational field: $\mathcal{T}_{\mu\nu} = T_{\mu\nu}(\text{matter, fields}) + t_{\mu\nu}(h)$. In this non-linearity lies the beauty and the complexity of GR ! These contributions are, as in all field theories, quadratic (at least) in the field amplitude $h_{\mu\nu}$ (see e.g. in Chap. 7, the energy and momentum carried by gravitational waves). In the WFSM approximation, however, we only retain first-order corrections to $\eta_{\mu\nu}$ and therefore neglect the gravitational *self-energy* contribution.

Note that the tensor $T_{\mu\nu}$, defined to first order in $h_{\mu\nu}$, satisfies the ordinary (non-covariant derivative) conservation law:

$$\frac{\partial}{\partial x^\mu} T^\mu_{\ \nu} = 0 \qquad (5.16)$$

Also note that the first term of the left-hand side of Eq. 5.15 is the D'Alembert operator, applied to the metric perturbation. We now try, via a suitable coordinate transformation, to set the other terms to zero: this is equivalent to what we do to get the Lorentz gauge transformation in e.m.

First, we change variable, defining the *trace-reversed* metric:

$$\tilde{h}_{\mu\nu} = h_{\mu\nu} - \frac{1}{2}\eta_{\mu\nu}h \qquad (5.17)$$

h is the trace, i.e. the scalar obtained by contracting the two indexes: $h = h^\mu_{\ \mu} = \eta^{\mu\nu}h_{\mu\nu}$.
The inverse transformation is

$$h_{\mu\nu} = \tilde{h}_{\mu\nu} - \frac{1}{2}\eta_{\mu\nu}\tilde{h} \qquad (5.18)$$

with $\tilde{h}$ defined in a similar way. It is easy to verify that $h = -\tilde{h}$.

Moreover, we apply a general coordinate transform preserving the weak field approximation:

$$x^\mu \to x^{'\mu} = x^\mu + \epsilon^\mu(x)$$

with ϵ^μ an arbitrary function of coordinates. These new coordinates imply a further change of the definition of metric, given by

$$g^{'\mu\nu} = \frac{\partial x^{'\mu}}{\partial x^\lambda} \frac{\partial x^{'\nu}}{\partial x^\rho} g^{\lambda\rho}$$

So, we now introduce a new $\tilde{h}'_{\mu\nu}$, defined as

$$\tilde{h}'_{\mu\nu} = \tilde{h}_{\mu\nu} - \frac{\partial \epsilon_\mu}{\partial x^\nu} - \frac{\partial \epsilon_\nu}{\partial x^\mu}$$

It can be shown that, due to *gauge invariance*, $\tilde{h}'_{\mu\nu}$ is still a solution of the linearized field equations. By choosing ϵ^μ such as to verify the four *Lorentz gauge* conditions,

$$\frac{\partial \tilde{h}^\mu_\nu}{\partial x^\mu} = 0 \qquad (5.19)$$

we can cast the field equations in the final form[2] :

$$-\frac{\partial^2 \tilde{h}_{\mu\nu}}{\partial x^\lambda \partial x_\lambda} = \Box \tilde{h}_{\mu\nu} = \frac{16\pi G}{c^4} T_{\mu\nu} \tag{5.20}$$

There is an almost perfect formal analogy with the corresponding e.m. equation for the 4-potential $A_\mu{}^{(em)} \equiv (\Phi^{(em)}, \vec{A}^{(em)})$ and the 4-current $j_\mu{}^{(em)} \equiv (c\rho^{(em)}, \vec{j}^{(em)})$, expressed in Gaussian units[3] :

$$\Box A_\mu{}^{(em)} = \frac{4\pi}{c} j_\mu{}^{(em)} \tag{5.21}$$

The tensor potential $\tilde{h}_{\lambda\mu}$ plays the role of the e.m. vector potential A_μ, while the energy-momentum tensor $T_{\mu\nu}$ plays the role of the sources, that in e.m. is taken by the 4-current $j_\mu{}^{(em)}$. Using this analogy, the set of solutions of Eq. 5.20 can be immediately derived as

$$\tilde{h}_{\mu\nu}(\vec{r}, t) = -\frac{4G}{c^4} \int \frac{T_{\mu\nu}(\vec{r}', t)}{|\vec{r} - \vec{r}'|} d^3 r' \tag{5.22}$$

We note that the stress energy tensor should be computed at an advanced time: $t^{adv} = t - |\vec{r} - \vec{r}'|/c$. This requirement is relaxed here, as we assume stationary motion of the sources.

We now need to identify the various components of $T_{\mu\nu}$: to this purpose, we apply analogous approximations to the continuity equation of the stress-energy tensor:

$$T^{\mu\nu}{}_{,\nu} = \frac{\partial}{\partial x^0} T^{\mu 0} + \frac{\partial}{\partial x^k} T^{\mu k} = 0$$

and compare these equations to the Euler equations of fluid dynamics for the motion of a perfect fluid (an ensemble of non-interacting particles) in absence of gravity:

$$\frac{\partial \rho}{\partial t} + \vec{\nabla} \cdot (\rho \vec{v}) = 0$$

$$\rho \frac{d\vec{v}}{dt} = -\vec{\nabla} p$$

where ρ is the fluid density, p its pressure and $\vec{v}$ its velocity, and $\frac{d}{dt} = \frac{\partial}{\partial t} + \vec{v} \cdot \vec{\nabla}$. The components of the matter tensor $T^{\mu\nu}$ are therefore identified in the following way

$$T^{00} = \rho c^2, \qquad T^{0j} = c\rho v^j, \qquad T^{jk} = \rho v^j v^k + p\delta^{jk} \tag{5.23}$$

---

[2] Note that, with our choice of metric signature $(+ - - -)$, we have $\frac{\partial^2}{\partial x^\lambda \partial x_\lambda} = \frac{\partial^2}{\partial x_0^2} - \nabla^2 = -\Box$, with opposite signs w.r.t the conventional definition of the D'Alembert operator.

[3] The Gauss unit system of e.m., (Jackson 1975) although unfamiliar in these days, best holds the analogy with GR. In this system, $\epsilon_0 = 1$; $\mu_0 = 1$ and the fields $\vec{E}$ and $\vec{B}$ have the same units.

where v^j are the components of the velocity $\vec{v}$.

It is now sufficient to plug the $T^{\mu\nu}$ values as given by Eq. 5.23 into the general solution Eq. 5.22 to obtain the components of the metric perturbation $\tilde{h}_{\mu\nu}$

$$\tilde{h}_{00}(\vec{r}) = -\frac{4G}{c^2} \int \frac{\rho(\vec{r}', t)}{|\vec{r} - \vec{r}'|} d^3 r' \tag{5.24}$$

$\tilde{h}_{00}$ is the so-called *gravitoelectric* potential while

$$\tilde{h}_{0j}(\vec{r}) = -\frac{4G}{c^3} \int \frac{\rho(\vec{r}', t)v'_j}{|\vec{r} - \vec{r}'|} d^3 r' \tag{5.25}$$

$\tilde{h}_{0j}$ is the so-called *gravitomagnetic* potential that depends on the *mass currents* $\vec{j} = \rho \vec{v}'$.

The other components,

$$\tilde{h}_{ij}(\vec{r}) = -\frac{4G}{c^4} \int \frac{\rho(\vec{r}', t)v'_i v'_j}{|\vec{r} - \vec{r}'|} d^3 r' \tag{5.26}$$

are higher order terms in the small quantity v/c.

We now assume the metric to be perturbed by a single isolated body of mass M, axisymmetric and possibly rotating with constant angular velocity Ω. At large distance r from the source, we expand the denominator in Taylor series around r:

$$\frac{1}{|\vec{r} - \vec{r}'|} = \frac{1}{r} + \frac{\sum_i x^i x'^i}{r^3} + O(\frac{1}{r^3})$$

the expressions for $\tilde{h}_{\mu\nu}$ simplify to

$$\begin{aligned} \tilde{h}_{00} &= -\frac{4G}{c^2}\frac{M}{r} \;; \\ \tilde{h}_{0j} &= -\frac{2G}{c^3}\frac{(\vec{r} \times \vec{J})_j}{r^3} \;; \qquad \tilde{h}_{ij} = O(v^2/c^2, \frac{1}{r^3}) \end{aligned} \tag{5.27}$$

where $\vec{J}$ is the angular momentum[4] of the rotating mass M.

We recall that $\tilde{h}_{\mu\nu}$ is related to the metric tensor $g_{\mu\nu}$ by

$$g_{\mu\nu} = \eta_{\mu\nu} + \tilde{h}_{\mu\nu} - \frac{1}{2}\eta_{\mu\nu}\tilde{h} \tag{5.28}$$

---

[4] Derivation of the second Eq. 5.27 from Eq. 5.25 is not obvious, but is an interesting exercise: use local cartesian coordinates, assume rotation around the z axis: $v'_j = (\Omega\hat{z} \times \vec{r}')_j$. Then integrate using the axial symmetry of the source: $\int d^3r' \; \rho(\vec{r}')x'_k = 0 \; \forall k$ and $\int d^3r' \; \rho(\vec{r}')x'^2 = \int d^3r' \; \rho(\vec{r}')y'^2 = \frac{1}{2}I_{zz}$.

and, noting that the trace is $\tilde{h} = -4\frac{GM}{rc^2}$, we convert back to the metric $g_{\mu\nu}$ using Eq. 5.18, and finally obtain, to the leading order in $\frac{GM}{rc^2}$

$$
h_{00} = -\frac{2GM}{rc^2}; \qquad h_{ij} = -\frac{2GM}{rc^2}\delta_{ij}; \qquad h_{0j} = -\frac{2G}{c^3}\frac{(\vec{r} \times \vec{J})_j}{r^3} \qquad (5.29)
$$

The GR metric, to first order in the WFSM approximation, can now be expressed in spherical coordinates ($x^0 = ct$, $x^1 = r$, $x^2 = \theta$ and $x^3 = \varphi$):

$$
ds^2 = (1 - \frac{2GM}{rc^2})c^2dt^2 - (1 + \frac{2GM}{rc^2})dr^2 - r^2(d\theta^2 + \sin^2\theta d\varphi^2) +
$$
$$
- \frac{4GJ}{c^2r}\sin^2\theta dt d\varphi \qquad (5.30)
$$

We remark that this metric, derived in the WFSM approximation, is indeed the weak field limit of the more general Kerr metric, that describes space-time around a rotating, axisymmetric mass.

5.4   Effects of Gravity on Photon Propagation

For a non-rotating source mass, we can drop from Eq. 5.30 the last term $\frac{4J}{cr^3}\sin^2\theta\, dt\, d\varphi$ (we shall use in Sect. 5.5 to describe the Lense-Thirring effect):

$$
ds^2 = (1 - \frac{2GM}{rc^2})c^2dt^2 - (1 + \frac{2GM}{rc^2})dr^2 -
$$
$$
r^2(d\theta^2 + \sin^2\theta d\varphi^2) \qquad (5.31)
$$

This is just an approximated version of the Schwarzschild metric:

$$
ds^2 = (1 - \frac{2MG}{rc^2})c^2dt^2 - \frac{dr^2}{(1 - \frac{2MG}{rc^2})} - r^2(d\theta^2 + \sin^2\theta d\varphi^2) \qquad (5.32)
$$

where ds^2 defines the proper time $(cd\tau)^2$ for massive particles, while $ds^2 = 0$ for photons.

In the following, we shall alternatively use one metric or the other.

Although we stay with our SI measurement units (i.e. do not use $c = G = 1$, as many theorists do), we find convenient, in the following sections, use the shorthand notation:

$$
\mathcal{R} = \frac{2GM}{c^2} \qquad (5.33)
$$

With this notation, we measure mass in metres: the Earth has then $\mathcal{R}_\oplus = 8.8$ mm and the sun $\mathcal{R}_\odot = 2.95$ km. $\mathcal{R}$ is known as the *Schwarzschild radius* of a celestial

Table 5.1 Schwarzschild radius and normalized gravitational potential $\Phi^{(g)}/c^2 = G\,M/Rc^2$ on the surface of some spherical bodies

	Mass (kg)	Radius R (m)	$\mathcal{R} = \frac{2GM}{c^2}$ (m)	$\Phi^{(g)}(R)/c^2$ at surface
Proton	$1.7\ 10^{-27}$	10^{-15}	$2\ 10^{-54}$	10^{-39}
Bowling ball	7.3	0.22	10^{-26}	$2\ 10^{-26}$
Earth	$6\ 10^{24}$	$6\ 10^{6}$	$9\ 10^{-3}$	$7\ 10^{-10}$
Sun	$2\ 10^{30}$	$7\ 10^{8}$	$3\ 10^{3}$	$2\ 10^{-6}$
White dwarf	10^{30}	10^{6}	$1\ 10^{3}$	$7\ 10^{-4}$
Neutron star	$3\ 10^{30}$	$4\ 10^{3}$	10^{4}	0.2

body. Table 5.1 shows, for a few familiar quasi-spherical bodies, the gravitational potential on the surface and the Schwarzschild radius related to their mass.

We now have the appropriate tools to address the Solar System tests of GR.

The classic experimental tests of the general relativity are based on observations performed in the weak gravitational field regime. In this approximation, we can predict the perturbation effects on the propagation of massless particles as photons in order to perform a comparison with the experimental data. There are two main effect to be considered:

(a) the time delay in photon propagation, also known as Shapiro delay,
(b) the bending of light trajectories, also known as gravitational lensing effect.

We start from the Schwarzschild solution Eq. 5.32 of the Einstein equations for the empty space-time outside a static spherical body of mass M.

We immediately note that if we let $M = 0$ the Eq. 5.32 spells the line element of the Minkowski space-time in spherical coordinates, while for $M \neq 0$, both time and radial coordinate are distorted. The logic consequence is that, for a clock at a fixed position in space $(r,\ \theta,\ \varphi = constant)$, just as we got for the red-shift of clocks in Chap. 4, we have

$$d\tau^2 = \left(1 - \frac{\mathcal{R}}{r}\right)dt^2 \tag{5.34}$$

while the length of a segment at rest, radially oriented in the gravitational field of the mass M $(t,\ \theta,\ \varphi = constant)$ is computed integrating the following differential:

$$dR^2 = \left(1 - \frac{\mathcal{R}}{r}\right)^{-1}dr^2 \tag{5.35}$$

5.4.1 Shapiro Time Delay

Consider an observer in a fixed point of space coordinates $(r_1,\ \theta_o,\ \phi_o)$ and in the field of a massive object M, sending a photon in the direction of a small body (so that it does not to perturb the metric), radially positioned in (r_2, θ_o, ϕ_o). The photon is reflected back by the body towards the observer, who receives it at a later time. From an experimenter's point of view, this is realized by emitting a radio beam from the Earth, reflecting it off Mercury, or Venus, when it is aligned with the Earth and the Sun (planets in conjunction). We calculate the elapsed time between transmission and subsequent reception of a photon by the observer in the simpler, although less interesting case of inferior conjunction, when the alignment is Sun-Mercury-Earth. The photon travels in radial direction at the speed of light. In this case, being $d\tau = 0$ and $d\theta = d\phi = 0$ in the line element expressed by the Eq. 5.32, we have

$$(1 - \frac{\mathcal{R}}{r})(c\, dt)^2 = \left(1 - \frac{\mathcal{R}}{r}\right)^{-1} dr^2 \tag{5.36}$$

which implies the radial *coordinate velocity* of the light is

$$\frac{dr}{dt} = \pm c\left(1 - \frac{\mathcal{R}}{r}\right) \tag{5.37}$$

This equation can be integrated to get the round trip travel time of the photon

$$\Delta t = \frac{1}{c}\left[-\int_{r_1}^{r_2} \frac{dr}{1 - \frac{\mathcal{R}}{r}} + \int_{r_2}^{r_1} \frac{dr}{1 - \frac{\mathcal{R}}{r}}\right] =$$

$$= \frac{2}{c}\int_{r_2}^{r_1} \frac{dr}{1 - \frac{\mathcal{R}}{r}} \tag{5.38}$$

The experimentally observable quantity is the proper time τ recorded in r_1; so, we need to express this elapsed time Δt in terms of the observer's proper time. Thus, using Eq. 5.34, we have

$$\tau = \left(1 - \frac{\mathcal{R}}{r_1}\right)^{\frac{1}{2}} \Delta t =$$

$$= \frac{2}{c}\left(1 - \frac{\mathcal{R}}{r_1}\right)^{\frac{1}{2}}\left[r_1 - r_2 + \mathcal{R}\, ln\left(1 - \frac{r_1 - \mathcal{R}}{r_2 - \mathcal{R}}\right)\right] \tag{5.39}$$

Let us compare this result with the time interval we expect, simply due to the light travelling from r_1 to r_2 and back:

$$\tilde{\tau} = \Delta R/c = \frac{1}{c}\int_{r_2}^{r_1}\left(1 - \frac{\mathcal{R}}{r}\right)^{-1/2} dr =$$

$$= \frac{2}{c}\left\{\left(\sqrt{r_2(r_2 - 2\mathcal{R})} - \sqrt{r_1(r_1 - \mathcal{R})} + \right.\right.$$

$$\left.\left. + 2\mathcal{R}\, ln\left(\frac{\sqrt{r_2} + \sqrt{r_2 - \mathcal{R}}}{\sqrt{r_1} + \sqrt{r_1 - \mathcal{R}}}\right)\right)\right\}$$

If r_1, $r_2 \gg \mathcal{R}$, the above formula can be simplified:

$$\tilde{\tau} \simeq \frac{2}{c}\Big[r_2 - r_1 + \mathcal{R}\,ln\Big(\frac{r_2}{r_1}\Big)\Big] \tag{5.40}$$

Applying the same approximation to Eq. 5.39, we have the following prediction for the measurement of the photon time delay:

$$\Delta\tau = \tau - \tilde{\tau} \simeq \frac{2}{c}\mathcal{R}\Big[ln\Big(\frac{r_1}{r_2}\Big) - \frac{r_1 - r_2}{r_1}\Big] \tag{5.41}$$

This, evaluated for Earth and Mercury in inferior conjunction, yields too small a value, about 10 ps. More appealing is the case of superior conjunction, when the two planets are on opposite sides of the Sun. The calculation is slightly more complex, because we need to take into account an impact parameter for the light, so that it is not obscured by the Sun; we just quote the result when the impact parameter takes its smallest value, i.e. the radius of the Sun:

$$\Delta\tau \approx \frac{4GM_{\odot}}{c^3}(1 + log(4r_1 r_2/R_{\odot}^2))$$

This value can be as large as 250 μs, easily measurable even in the 1960s. In 1964, Shapiro proposed to perform this measurement using a radio signal sent towards Mercury and Venus, used as reflectors, when these planets were passing behind the Sun, i.e. in superior conjunction. Experimental results will be discussed in the next chapter.

5.4.2 Light Deflection—Gravitational Lensing

One famous prediction of General Relativity is that the null geodesic line in the vicinity of a mass is curved. This implies that the photon trajectory deviates from the straight line when passing near a star. To predict the effect amount we start from the Lagrangian of the generic particle propagating in a gravitational field

$$\mathcal{L} \equiv \frac{1}{2}g_{\mu\nu}\dot{x}^{\mu}\dot{x}^{\nu} \tag{5.42}$$

where the symbol *dot* denote the derivate with respect to the proper time τ for a mass particle and the affine parameter s for the massless photon. We write it, using spherical coordinates, in the vicinity of the static spherical object of mass M:

$$\mathcal{L} = \frac{1}{2}\Big(-c^2\dot{t}^2(1 - \frac{\mathcal{R}}{r}) + \dot{r}^2(1 - \frac{\mathcal{R}}{r})^{-1} + r^2(\dot{\theta}^2 + \sin^2\theta\dot{\varphi}^2)\Big) \tag{5.43}$$

Thus, Euler-Lagrange equations give us four geodesic equations:

$$\frac{d}{ds}\Big(\frac{\partial\mathcal{L}}{\partial\dot{x}^{\mu}}\Big) - \frac{\partial\mathcal{L}}{\partial x^{\mu}} = 0$$

Without loss of generality, we assume that the motion is confined in the equatorial pane , i.e. $\theta = \frac{\pi}{2}$ and we end up with the second equation, corresponding to the index $\mu = 1$, of the form

$$\ddot{r}(1 - \frac{\mathcal{R}}{r})^{-1} + \frac{GM}{r^2}\dot{t}^2 - \dot{r}^2\frac{\mathcal{R}}{r^2}(1 - \frac{\mathcal{R}}{r})^{-2} - r\dot{\varphi}^2 = 0 \qquad (5.44)$$

Integrating two other equations, $\mu = 0$ and $\mu = 3$, in the case $\theta = \pi/2$, we have

$$\frac{\partial \mathcal{L}}{\partial \dot{t}} = (1 - \frac{\mathcal{R}}{r})\dot{t} = k \equiv \text{costant} \qquad (5.45)$$

$$r^2\dot{\varphi} = h \equiv \text{costant} \qquad (5.46)$$

where k and h are integration constants.

Equation 5.44 is a second order, non-linear equation: it is simpler to use instead the relation can be obtained from the invariant interval ds^2 of Eq. 5.32:

$$-c^2\dot{t}^2(1 - \frac{\mathcal{R}}{r}) + \dot{r}^2(1 - \frac{\mathcal{R}}{r})^{-1} + r^2\dot{\varphi}^2 = c^2 \qquad (5.47)$$

and for the photon case ($ds^2 = 0$)

$$-c^2\dot{t}^2(1 - \frac{\mathcal{R}}{r}) + \dot{r}^2(1 - \frac{\mathcal{R}}{r})^{-1} + r^2\dot{\varphi}^2 = 0 \qquad (5.48)$$

The constraint Eq. 5.46 is the analogous of the conservation of angular momentum, while the Eqs. 5.47 and 5.48 lead to relations echoing the conservation of energy.

Substituting, in Eq. 5.47 (for massive particle) and in Eq. 5.48 (for photons), the values of $\dot{\varphi}$ and $\dot{t}$ as given by Eqs. 5.45 and 5.46, we end up with two simpler equations:

$$\left(\frac{du}{d\varphi}\right)^2 + u^2 = E + \frac{2GM}{h^2}u + \frac{2GM}{c^2}u^3 \quad \text{for particles} \qquad (5.49)$$

$$\left(\frac{du}{d\varphi}\right)^2 + u^2 \quad = F + \frac{2GM}{c^2}u^3 \qquad \text{for photons} \qquad (5.50)$$

where $u = 1/r$, $E = c^2(k^2 - 1)/h^2$ and $F \equiv (\frac{ck}{h})^2$.

We now focus on the photon trajectory.

In the limit case $M = 0$ (no field source, no curvature), the Eq. 5.50 has a simple solution $u = u_o \sin\varphi$ with $u_o = 1/b = \sqrt{F}$. This solution represents the straight line path taken by a photon originating from $-\infty$ ($\varphi = 0$) to $+\infty$ ($\varphi = \pi$). The point of the photon trajectory nearest the origin, corresponding to $\varphi = \pi/2$ is at a distance b: in scattering theory it is called *impact parameter*.

In the case $M \neq 0$, we rewrite Eq. 5.50 introducing the reduced variable $\tilde{u} = \frac{u}{u_o}$ and the constant $\epsilon = \frac{GM}{c^2 b} = \frac{\mathcal{R}u_0}{2}$. We are now dealing with a small dimensionless

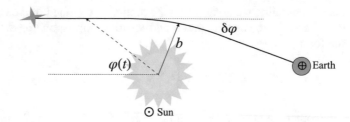

Fig. 5.1 Deflection of light rays by the Sun. The effect is vastly exaggerated

quantity $\epsilon \ll 1$, in the case of photons emitted by a far star and detected by a telescope on the Earth when grazing the Sun image. In this condition, $b = R_\odot$, the typical value for ϵ is $\sim 4 \cdot 10^{-6}$.

The Eq. 5.50 takes the form

$$\left(\frac{d\tilde{u}}{d\varphi}\right)^2 + \tilde{u}^2 = \frac{F}{u_o{}^2} + \epsilon\tilde{u}^3 \tag{5.51}$$

at the point of the photon trajectory closest to the Sun, we have, by symmetry

$$\left(\frac{d\tilde{u}}{d\varphi}\right)_{\varphi=\frac{\pi}{2}} = 0; \quad \tilde{u}(\varphi = \frac{\pi}{2}) = 1; \quad \frac{F}{u_o{}^2} = 1 - \epsilon$$

so that the Eq. 5.51 becomes

$$\left(\frac{d\tilde{u}}{d\varphi}\right)^2 + \tilde{u}^2 = 1 - \epsilon + \epsilon\tilde{u}^3 \tag{5.52}$$

We note that, for $M = 0$, we had $\epsilon = 0$ and $\tilde{u} = \sin\varphi$. We then look for a solution of Eq. 5.52 in the form

$$\tilde{u} = \sin\varphi + \epsilon f(\varphi)$$

where $f(\varphi)$ is a function to be determined. Using the initial condition $\tilde{u}(\varphi = 0) = 0$, that requires $f(\varphi = 0) = 0$, the solution for $f(\varphi)$ to first order in ϵ is

$$f(\varphi) = \frac{1}{2}(1 + \cos^2\varphi - \sin\varphi) - \cos\varphi$$

so that, for $\tilde{u}$, we have (Fig. 5.1)

$$\tilde{u} = (1 - \frac{1}{2}\epsilon)\sin\varphi + \frac{1}{2}\epsilon(1 - \cos\varphi)^2 \tag{5.53}$$

The consequence of this solution is that the end point of the photon trajectory is no longer $\varphi(t \to \infty) = \pi$, but different by a small quantity $\delta\varphi$, i.e. $\pi + \delta\varphi$.

Thus, let us compute the Eq. 5.53 at the end point of the trajectory, i.e. for $\tilde{u} = 0$ and $\varphi = \pi + \delta\varphi$. To first order in the small angular quantity $\delta\varphi$, we have $\sin(\pi + \delta\varphi) \simeq -\delta\varphi$ and $\cos(\pi + \delta\varphi) \simeq -1$, we find

$$\delta\varphi = 2\epsilon = \frac{4\,G\,M}{bc^2} = \frac{2\mathcal{R}}{b} \tag{5.54}$$

The deflection is inversely proportional to the impact parameter b. In the case of light passing through the gravitational field of the Sun, the smallest value that b can assume is the Sun radius. In this case, we have for the total deflection $\delta\varphi = 1.75"$. The method involves photographing the stars around the Sun during a total eclipse (when the stars near the Sun edge can be seen), and comparing the photograph with one of the same star field taken several months later. This methods suffers of several limitations. Among them, we cite

- the relevant changes in conditions which occur when bright sunlight changes to the semidarkness of an eclipse
- the time lapse of several months, which makes it difficult to reproduce similar conditions when taking the comparison photograph
- the smallness of the effect, which pushes photography to its limits

However, this deviation was measured for the first time the 29 May 1919 during a total solar eclipse. The British astronomers Frank Watson Dyson and Arthur Stanley Eddington carried out two observations expeditions, one to the West African island of Principe and the other to the Brazilian town of Sobral. Following the return of the expeditions, the data were presented by Eddington to the Royal Society of London and the news of this experimental results led to worldwide fame for Einstein and his General Theory of Relativity.

5.5   Effects of Gravity on Particle Dynamics: Precession of Periastron

There are two other effects, altering the dynamics of mass particles, that can be explained within the weak field approximation, and that have been indeed observed:

(a) the perihelion advance of the planet orbit
(b) the spin precession

The first important application of GR was to explain the non-Newtonian behaviour of the planets. The issue of the orbit of Mercury had been withstanding among the astronomers of the early twentieth century: the planet deviates from Keplerian orbit for a number of reasons that can be explained by Newtonian gravity: mass quadrupole of the Sun, perturbation by Venus, Jupiter and other planets, etc., all contribute to perturbing Mercury's motion in such a way that the orbit is not a closed ellipse:

its axes gradually rotate, in what is called a *precession*.[5] However, once all these classical contributions are accounted for and subtracted, it remains a discrepancy of about 43 s of arc per century, or $6.6 \cdot 10^{-7}$ rad. per orbit. The direction of the effect is in the forward direction: if we could observe the Mercury orbit from above the ecliptic, watching it move in counterclockwise direction, then the gradual rotation of the major axis is also counterclockwise. An intuitive way to look at this phenomenon is: the planet spends more time near perihelion than it is classically predicted and when it moves away from the sun, the Keplerian orbit is rotated counterclockwise. This effect, tiny in the case of the gravitation interaction between Sun and Mercury, is enhanced enormously in the case of a mass in the vicinity of a black hole, that spends an almost infinite time (due to the slowing of proper time) at its perihelion.

To compute this effect in the GR framework, we consider the equatorial ($\theta = \pi/2$) motion of a test particle in the field of a spherical object of mass M, set at the origin of our coordinates. We restart from the Eqs. 5.50, 5.45 and 5.46, that we repeat here for convenience recalling that the derivatives are computed with respect to the proper time τ of the mass particle.

$$\left(1 - \frac{R}{r}\right)\dot{t} = k \qquad r^2\dot{\varphi} = h$$

$$\left(\frac{du}{d\varphi}\right)^2 + u^2 = E + \frac{2GM}{h^2}u + \frac{2GM}{c^2}u^3 \tag{5.55}$$

where $u = 1/r$ and $E \equiv c^2(k^2 - 1)/h^2$.

Compare this with the corresponding equation of Newtonian dynamics:

$$\left(\frac{du}{d\varphi}\right)^2 + u^2 = E + \frac{2GM}{h^2}u \tag{5.56}$$

obtained assuming the conservation of the classical quantities: E is related to the orbital energy and h is the angular momentum per unit mass. The solution of this equation is

$$u = \frac{GM}{h^2}\left[1 + e\cos(\varphi - \varphi_0)\right] \tag{5.57}$$

where φ_0 is an integration constant and e is the orbital eccentricity $e = 1 + \frac{Eh^4}{G^2M^2}$. Comparing the two Eqs. 5.55 and 5.56, it is evident that they differ only for the non-linear term $\frac{2GM}{c^2}u^3$. The coefficient G/c^2 is depressing this term, confirming that the expected effect is small. In addition the dependence on $u^3 = 1/r^3$ suggests that it is more significant for smaller values of r, i.e. in the perihelion proximity.

---

[5] Astronomers had hypothesized the existence of an additional planet to perturb Mercury's orbit: they had computed its mass, orbital radius (intermediate between Mercury and Venus) and even given it a name: Volcanus. Such planet, obviously, was never found.

In principle, to find the particle trajectory in the equatorial plane, we integrate Eq. 5.55 to obtain $u = 1/r$ verses φ. Except for special cases, this integration is impossible in practice, and therefore we will follow, to compute the perihelion advance, the method originally presented in (Möller 1972).

Aphelion and perihelion occur where $d\varphi/dt = 0$. Imposing this condition in Eq. 5.55, we have

$$\frac{2\,G\,M}{c^2}u^3 - u^2 + \frac{2GM}{h^2}u + E = 0 \qquad (5.58)$$

This is a cubic equation, which admits three roots u_1, u_2, u_3 and two of them u_1, u_2 correspond to the aphelion and perihelion, respectively. This implies that $u_1 \le u \le u_2$ As we did previously, when we computed the lensing effect of the Sun on the photon, we introduce the function $\tilde{u} = u/u_o$ with $u_o = (u_1 + u_2)/2$ and the small dimensionless coefficient $\epsilon = \frac{2\,G\,M}{c^2}u_o$.

We rewrite Eq. 5.58 using $\tilde{u}$ and ϵ and we have

$$\epsilon\tilde{u}^3 - \tilde{u}^2 + \frac{2GM}{u_oh^2}\tilde{u} + \frac{E}{u_o^2} = \epsilon(\tilde{u} - \tilde{u}_1)(\tilde{u}_2 - \tilde{u})(\tilde{u}_3 - \tilde{u}) = 0 \qquad (5.59)$$

where $\tilde{u}_1, \tilde{u}_2, \tilde{u}_3$ are the roots of the new cubic equation in $\tilde{u}$. Applying the same normalization to Eq. 5.55 and using Eq. 5.59, we have

$$\left(\frac{d\tilde{u}}{d\varphi}\right)^2 = \epsilon(\tilde{u} - \tilde{u}_1)(\tilde{u} - \tilde{u}_2)(\tilde{u} - \tilde{u}_3) \qquad (5.60)$$

We need to recall here a property of the generic cubic equation $ax^3 + bx^2 + cx + d = 0$. Its roots x_1, x_2, x_3, satisfy the relation: $x_1 + x_2 + x_3 = -b/a$. Thus, in our case, we have $\tilde{u}_1 + \tilde{u}_2 + \tilde{u}_3 = 1/\epsilon$. Also, we have defined $2u_o = u_1 + u_2$ and as consequence $\tilde{u}_1 + \tilde{u}_2 = 2$. We conclude that

$$\left(\frac{d\tilde{u}}{d\varphi}\right)^2 = (\tilde{u} - \tilde{u}_1)(\tilde{u}_2 - \tilde{u})[1 - \epsilon(2 + \tilde{u})] \qquad (5.61)$$

From the previous equation, we can isolate $\frac{d\varphi}{d\tilde{u}}$ and compute an expansion to first order in ϵ:

$$\frac{d\varphi}{d\tilde{u}} = \frac{1 + \frac{1}{2}\epsilon(2 + \tilde{u})}{\sqrt{(\tilde{u} - \tilde{u}_1)(\tilde{u}_2 - \tilde{u})}} = \frac{\frac{1}{2}\epsilon(\tilde{u} - 1) + 1 + \frac{3}{2}\epsilon}{\sqrt{\beta^2 - (\tilde{u} - 1)^2}} \qquad (5.62)$$

where in last step we have introduced $\beta = \frac{1}{2}(\tilde{u}_2 - \tilde{u}_1)$ and we made repeated use of the relation $\tilde{u}_1 + \tilde{u}_2 = 2$.

By integrating Eq. 5.62 in the interval $\tilde{u}_1$, $\tilde{u}_2$, we can compute the angle $\Delta\phi$ between the aphelion and the following perihelion:

$$\Delta\phi = \int_{\tilde{u}_1}^{\tilde{u}_2} \frac{\frac{1}{2}\epsilon(\tilde{u} - 1) + 1 + \frac{3}{2}\epsilon}{\sqrt{\beta^2 - (\tilde{u} - 1)^2}}d\tilde{u} = \left(1 + \frac{3}{2}\epsilon\right)\pi$$

$2\Delta\phi$ gives the angle between successive perihelions and we conclude that for each orbit the perihelion advancement is

$$< \delta\psi >_{(orbit)} \quad = 3\epsilon\pi = \frac{3\,G\,M}{c^2}\Big(\frac{1}{r_1} + \frac{1}{r_2}\Big)\pi =$$

$$= \frac{6\pi\,GM}{c^2 a(1 - e^2)} \tag{5.63}$$

In the last expression, we used the relations $r_{1,2} = a(1 \pm e)$ relating apoastron (r_1) and periastron (r_2) to the ellipse semi-major axis and eccentricity. For Mercury, which is the planet of our solar system closest to the Sun, Eq. 5.63 predicts an advance of $42.98''$ per century, in excellent agreement with the observed precession.

We conclude by noting that the formula derived above is useful when dealing with weak solar system effects. If we consider binary systems of two neutron stars, as in the case of the Pulsar 1913 + 16, the perihelion advance is much larger, ~ 4.2 degrees per year: that means that, in a single day, we observe the same precession as Mercury's perihelion advances in a century. This is due to a much stronger gravitational field, and this approximate method is no longer applicable: the periastron advance must be computed using a fully relativistic approach.

5.6   Gravitoelectromagnetism

In this section, we shall make use of the similarity between the equations of electromagnetism and gravity outlined in Eq. 5.21 and following. Before venturing into this analogy, we remark that, from a fundamental point of view, e.m. and gravity, even in its linearized form, are two very different theories, the main difference being the equivalence principle: gravitational effects can be locally removed by a choice of reference frame, the e.m. effects can't. Therefore, the analogy must just be taken for a formal likeness of the equations and a useful tool for computation.

We return to the Eq. 5.27, we derived for the metric perturbation in the weak field limit:

$$\tilde{h}_{00} = -\frac{4G}{c^2}\frac{M}{r}; \qquad\qquad \tilde{h}_{0j} = -\frac{2G}{c^3}\frac{(\vec{r} \times \vec{J})_j}{r^3};$$

We can push further the analogy between GR and the electromagnetic theory by defining a four-component gravitation potential

$$\Phi^{(g)} = \frac{c^2}{4}h_{00} = -G\frac{M}{r} \tag{5.64}$$

$$A_j{}^{(g)} = -\frac{c^2}{2}h_{0j} = \frac{G}{c}\frac{\vec{r} \times \vec{J}}{r^3}$$

echoing the 4-vector $A_\mu^{(em)}$ of the classical electromagnetism. From these potentials, we derive a gravitoelectric and a gravitomagnetic field

$$\vec{E}^{(g)} = -\vec{\nabla}\Phi^{(g)} - \frac{1}{2c}\frac{\partial \vec{A}^{(g)}}{\partial t} \quad , \quad \vec{B}^{(g)} = \vec{\nabla} \times \vec{A}^{(g)} \tag{5.65}$$

and the gravitational Lorentz gauge, Eq. 5.19 becomes

$$\frac{1}{2}\vec{\nabla} \cdot \vec{A}^{(g)} + \frac{1}{c}\frac{\partial \vec{\Phi}^{(g)}}{\partial t} = 0 \tag{5.66}$$

We remark that the analogy of the e.m. and weak field gravitation is not perfect. The most relevant difference is a factor $1/2$ present in the gravitational Eqs. 5.66. This factor is a consequence of the GR linear approximation procedure applied to a tensor theory, i.e. a spin-2 field compared to the classical electrodynamics, a spin-1 field.[6]

We must also note that we must assume a negative expression for $\Phi^{(g)}$ in order to preserve the attractive nature of the gravitational interaction.

With these caveats in mind, we can end up with a set of equations similar to the Maxwell's ones (in Gaussian units)

$$\vec{\nabla} \cdot \vec{E}^{(g)} = -4\pi G\rho$$
$$\frac{1}{2}\vec{\nabla} \times \vec{B}^{(g)} = \frac{1}{c}\frac{\partial \vec{E}^{(g)}}{\partial t} - \frac{4\pi G}{c}\vec{j}^{(g)} \tag{5.67}$$
$$\vec{\nabla} \times \vec{E}^{(g)} = -\frac{1}{2c}\frac{\partial \vec{B}^{(g)}}{\partial t}$$
$$\vec{\nabla} \cdot \vec{B}^{(g)} = 0$$

Again, despite the apparent similarity with Maxwell's equations, there are differences, that is worth underlining:

- the factors $1/2$ whenever the gravitomagnetic field is involved: as a mnemonic rule, to keep them in the right places, one could write these equations using the variable $\frac{\vec{B}^{(g)}}{2}$
- the negative sign in the source terms, reminding us that gravity is always attractive.
- the r.h.s. of the third equation should be set to zero, because $\frac{1}{c}\frac{\partial \vec{B}^{(g)}}{\partial t}$ is a term of higher order in v/c, and therefore not compatible with the WFSM assumptions. Preserving this term would imply introducing terms of order $(v/c)^2$ also in the other equations, spoiling their Maxwell-like character.

---

[6] In other words, this means that the effective gravitomagnetic charge is twice the gravitoelectric one.

We push further the analogy, recalling the magnetic potential and field generated by the magnetic moment of a charge current (as stated by Ampére's equivalence), and deriving the corresponding equations for the gravitomagnetic field generated by the angular momentum $\vec{J}$ of a rotating mass current

$$\vec{A}^{(em)} = \frac{\vec{\mu} \times \vec{r}}{r^3}; \qquad\qquad \vec{A}^{(g)} = \frac{G}{c} \frac{\vec{r} \times \vec{J}}{r^3} \qquad (5.68)$$

$$\vec{B}^{(em)} = \frac{3\hat{r}(\hat{r} \cdot \vec{\mu}) - \vec{\mu}}{r^3}; \qquad \vec{B}^{(g)} = -\frac{G}{c} \frac{3\hat{r}(\hat{r} \cdot \vec{J}) - \vec{J}}{r^3} \qquad (5.69)$$

To complete the analogy, we need the equation of motion for a particle in these fields, paralleling the role of Lorentz equation $\vec{F} = q(\vec{E} + \frac{\vec{v}}{c} \times \vec{B})$ in the e.m. case. To this purpose, we shall evaluate the geodesic Eq. 5.8 in the particular limit of the space-time around a rotating body, in the WFSM and stationary case, as described by the metric (Eq. 5.30)

$$ds^2 = \left(1 - \frac{2GM}{c^2 r}\right)c^2 dt^2 - \left(1 + \frac{2GM}{c^2 r}\right)dr^2$$
$$- r^2 d\theta^2 - r^2 \sin^2\theta d\phi^2 - \frac{4GJ}{c^2 r}\sin^2\theta d\phi dt$$

In agreement with our WFSM assumptions ($\mathcal{R} \ll r$ and $v \ll c$), we can assume the interval ds^2 to be dominated by the time-like term: $ds \simeq \sqrt{g_{00}}\, cdt$. As a consequence, we can take $dx^i/ds \simeq v^i/c$ and neglect terms in $v^i v^h/c^2$. Stationarity of the metric also implies $g_{\alpha\beta,0} = 0$. With these approximations, after some algebra, the geodesic equation takes the form:

$$\frac{d^2\vec{r}}{dt^2} = -\vec{\nabla}\Phi^{(g)} + 2\frac{\vec{v}}{c} \times (\vec{\nabla} \times \vec{A}^{(g)}) \qquad (5.70)$$

echoing, again, the equation of motion of a charged particle in the electromagnetic field. The first term on the right-hand side is the Newtonian interaction, responsible for "Keplerian orbits", while the second, a small perturbation $O(v/c)$, represents a precession of the angular momentum, as discussed in the next section.

As an interesting application of this analogy, we consider the precession of the angular momentum of a charged particle orbiting about a magnetic dipole and that of the angular momentum $\vec{S}$ of the massive particle orbiting in the gravitomagnetic field $B^{(g)}$.

In an external magnetic field $\vec{B}^{(em)}$, a magnetic dipole $\vec{\mu}$ feels a torque $\vec{\mu} \times \vec{B}$ and precesses around the $\hat{B}$ direction (Larmor precession). In a similar way, a torque $\mathcal{\vec{M}}$ is applied to the orbiting massive particle

$$\mathcal{\vec{M}} = \frac{d\vec{S}}{dt} = \frac{\vec{S}}{c} \times \vec{B}^{(g)} \equiv \vec{\Omega}_P \times \vec{S} \qquad (5.71)$$

Table 5.2 Precession mechanisms in General Relativity. The effects are referred, for definiteness, to the system where they have been first observed but, obviously, hold for any gravitationally bound system

Effect Name	Source	Interacting with	Measured on
Schwarzchild	Sun mass $M_\odot$	Planet orbital $\vec{L}$	Mercury orbit
De Sitter—geodetic	Earth orbital $\vec{L}_\oplus$	Gyroscope spin $\vec{S}$	Earth + satellite
Lense-Thirring	Earth spin $\vec{J}_\oplus$	Satellite orbital $\vec{L}$	LAGEOS (e.g.) orbits
Pugh-Schiff	Earth spin $\vec{J}_\oplus$	Gyroscope spin $\vec{S}$	GP-B space experiment

so that it precesses with the frequency

$$\boxed{\vec{\Omega}_P = \frac{G}{c^2 r^3}[3(\vec{J} \cdot \hat{r})\hat{r} - \vec{J}]}$$

(5.72)

This formalism is the starting point of Sect. 5.7 where we discuss and review the experimental measurements of the gravitomagnetic phenomena.

Gravitomagnetism deals a lot with spin-spin interactions: to keep notation straight, here and in the following section we use $\vec{J}$ for the spin of the mass generating the gravitational field, and $\vec{S}$ or $\vec{L}$ for the angular momentum (spin or orbital) of the orbiting test object. This distinction will obviously fail when dealing with binary systems of comparable mass and angular momentum, as in Chap. 12.

5.7   Spin Precession in Gravitation

As we have seen in previous sections, the motion of celestial (and man-made) bodies is subject, in GR, to a number of effects that have no correspondence in Newtonian gravity. We shall discuss here the *precessions*, that affect planets, satellites and stars, orbiting a massive central source.

Schiff pointed out (Schiff 1960) that precessions are the only measurements that can really test General Relativity in weak field; both the red shift and the deflection of light (Shapiro time delay was yet to be discovered) can be deduced on the basis of the equivalence principle + special relativity (Table 5.2).

5.7.1   Schwarzschild or Einstein Precession

We have already encountered in Sect. 5.5 the precession of periastron in the field of a central source mass

$$< \delta\psi >_{(Schwarz)} = \frac{6\pi\, GM}{c^2 a(1 - e^2)}$$

The case of Mercury is the most famous, but any orbiting body is subject to this effect. Recent observations have shown Schwarzschild precession in the star S2 orbiting

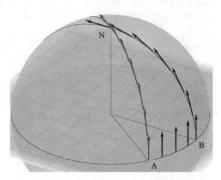

Fig. 5.2 2D analogy of the parallel transport of a vector along a geodesic: we start in position A and parallel transport the vector to N along the meridian $A - N$ (green vectors). If, instead, we move the vector to position B along the parallel $A - B$, and then to N along another meridian $B - N$, the end vector (red) is rotated with respect to the previous transport. Credits: https://i.stack.imgur. com/H2xCM.png

the galactic black hole Sgr A^* (The Gravity Collaboration 2020). The precession has also been accurately measured for geodetic satellites orbiting the Earth (Lucchesi and Peron 2010). We note that this precession changes the angle ω defining the periastron: it is a rotation of the ellipsis axes, and does not change the orbital plane: it is an in-plane effect. Moreover, we remark that it is a gravitoelectric effect, as it depends on the mass M of the central source (in general, on the gravitoelectric potential $\Phi^{(g)}$) and not on its rotation.

5.7.2 Geodetic or de Sitter Precession

The Geodetic Effect, first predicted by W. De Sitter in 1916, is another gravitoelectric precession that affects rotating bodies (either orbiting or spinning) and it consists in the change in the direction of their angular momentum. A spinning body that rotates around a central mass, changes the axis of its spin $\vec{S}$ by a small quantity at each revolution (Fig. 5.2). This is a consequence of the *parallel transport* of a vector along a geodesic in a curved space-time. Another way to look at the problem involves the different flow of time in a frame away from the source (coordinate time t), and on the orbiting body itself (proper time τ).

We put forward the result for the de Sitter precession:

$$\frac{d\vec{S}}{dt} = \vec{\Omega}_{DS} \times \vec{S}; \qquad \text{with} \quad \vec{\Omega}_{DS} \equiv \frac{3}{2c^2} \vec{v} \times \vec{\nabla}\Phi^{(g)} \tag{5.73}$$

and we leave a somehow technical derivation to the next section.

The angular momentum of a spinning gyroscope orbiting a central source mass experiences a precession around a vector normal to the orbital plane. When completed one orbit, the in-plane component S^ϕ does not return to the initial value $+ 2\pi$, but

Fig. 5.3 The de Sitter, or geodetic precession: the Earth and any orbiting body (the Moon, or an artificial satellite) behaves like a giant gyroscope with spin $\vec{S}$ in the static field of the Sun. The spin slowly precesses around the direction $\vec{\Omega}_{DS}$ normal to the ecliptic

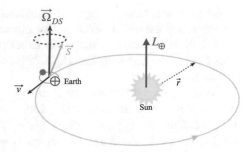

accumulates a small (negative) extra angle, *a precession* (Fig. 5.3):

$$\Delta\phi = 2\pi\left[\left(1 - \frac{3\mathcal{R}}{2r}\right)^{1/2} - 1\right] \simeq -\frac{3\pi GM}{c^2 r} \tag{5.74}$$

This is the *de Sitter or Geodetic Precession* of the angular momentum. It is usually more interesting the *rate* of precession, i.e. the amount of phase delay accumulated in a given time unit (e.g. one year). The phase delay of Eq. 5.74 must then be divided by the period of revolution, that can be computed, to our level of approximation, by Kepler's law: $T = 2\pi\sqrt{r^3/GM}$. We then find

$$\frac{d}{dt}\phi_{DS} = -\frac{3}{2c^2}\frac{(GM)^{3/2}}{r^{5/2}} \tag{5.75}$$

Finally, recalling that the orbital velocity is $v = \sqrt{GM/r}$, we can re-express the above rate as

$$\frac{d}{dt}\phi_{DS} = \frac{3}{2c^2}v\frac{GM}{r^2} \tag{5.76}$$

The precession takes place in the orbital $(\hat{r}, \hat{\phi})$ plane, i.e. around the vector

$$\vec{\Omega}_{DS} = \frac{3}{2c^2}\frac{GM}{r^2}\hat{r} \times \vec{v} = \frac{3}{2c^2}\vec{v} \times \vec{\nabla}\Phi^{(g)} \tag{5.77}$$

normal to that plane. The time evolution of the spin $\vec{S}$ is then described by

$$\frac{d\vec{S}_{DS}}{dt} = \frac{3}{2c^2}\vec{\nabla}\Phi^{(g)} \times \vec{v} \times \vec{S} \tag{5.78}$$

Just as a curiosity, we can identify $\vec{r} \times \vec{v}$ in Eq. 5.77 as the orbital angular momentum $\vec{L}/m$ of the gyroscope and write the precession in yet another way:

$$\frac{d\vec{S}_{DS}}{dt} = -\frac{3GM}{2mc^2 r^3}\vec{L} \times \vec{S} \tag{5.79}$$

that is suggestive of a spin-orbit interaction, like in atomic systems.

We have seen that the angular momentum of a spinning and orbiting body precesses around the radial direction. We discuss in the next section the experimental confirmations obtained by the GP-B experiment, with a spinning quartz sphere in orbit close to the Earth surface: Eq. 5.75 predicts a tiny precession of 30 μrad/year, or 6 arcsec/year.

However, we can consider the Earth + any of its satellites as a gigantic spinning gyro, that moves in the field of the Sun. $\vec{S}$ is, in this case, the angular momentum of the satellite orbiting the Earth, while $\vec{L}$ refers to the Earth (+satellite, negligble) moving around the Sun: as it precesses, $\vec{S}$ will change its orbital plane around the Earth, with a change in the Right Ascension of the Ascending Node (RAAN), the Keplerian parameter defined in Sect. 1.1.

A word of caution is in order about notation: the Right Ascension (or Longitude, in terrestrial coordinates) of Ascending Node is an angle that astronomers traditionally label with the symbol Ω. To avoid confusion between the RAAN Ω and the precession frequency $\vec{\Omega}$ of Eqs. 5.72 and 5.77 we shall use in this chapter the acronym $RAAN$ instead of Ω.

Clearly, the effect is the same for any satellite orbiting the Earth: Eq. 5.75 predicts, using the Sun as the source ($M = M_\odot$, $r = 1\ AU$) a precessional rate

$$\Omega_{DS} = \frac{3}{4}\mathcal{R}_\odot \frac{2\pi}{rT} = 93\ \text{nrad/yr} = 19.2\ \text{marcsec/yr} \qquad (5.80)$$

We should consider, however, that this precession takes place, as mentioned above, around a direction normal to the ecliptic; observing it on an Earth based reference frame, the RAAN we measure the component in the equatorial plane:

$$\frac{d}{dt}RAAN = \Omega_{DS} \cdot \cos(\epsilon) = 17.6\ \text{mas/yr} \qquad (5.81)$$

where $\epsilon = 23^o27$ is the inclination of the equator on the ecliptic. This effect was first observed on the precession of the Moon perigee (Bertotti et al. 1978), using Lunar Laser Ranging, with an accuracy of 10%, later improved by other experiments, to the current level of agreement, between theory and measurement, of 0.2%.

5.7.2.1 Derivation of the Geodetic Precession

We recall that, in Newtonian mechanics and in Special Relativity, a spinning gyro, in absence of external torques, will maintain the direction of its angular momentum $\hat{S}$:

$$\frac{dS^\alpha}{d\tau} = 0$$

Indeed, three orthogonal gyros make a "physical" reference frame: on this principle systems of inertial navigation have been developed, in the last century, for ships and airplanes.

In GR, the above equation of motion, for a free gyro, must be replaced by an equation that take into account the curvature of space-time; by similarity with the geodesic equation, we can convince ourselves that an appropriate relation, satisfying the requirements of Newtonian limit, covariance and linearity, is the so-called *gyroscope equation*:

$$\frac{dS^\alpha}{d\tau} + \Gamma^\alpha_{\beta\gamma} S^\beta u^\gamma = 0 \tag{5.82}$$

where u^γ is the 4-velocity of the gyro along its geodesic. In a Local Inertial Frame (instantaneous, as an orbit is by no means an inertial trajectory), we expect the proper velocity to be $u^\gamma = [c, \vec{0}]$, and the spin to be a space-like 4-vector: $S^\beta = [0, \vec{S}]$, so that the two vectors are orthogonal, and they will be in any other frame.

Let us assume a spherical source mass, so that we can use, in a coordinate frame at rest with respect to distant stars, the Schwarzschild geometry Eq. 5.32

$$ds^2 = \left(1 - \frac{\mathcal{R}}{r}\right) c^2 dt^2 - \left(1 - \frac{\mathcal{R}}{r}\right)^{-1} dr^2 - r^2 (d\theta^2 + sin^2\theta d\varphi^2)$$

The spherical symmetry of the source allows us to assume an equatorial ($\theta = \pi/2$) circular orbit at a Schwarzschild radial coordinate r. The position vector is: $x^\mu = [ct, r, \pi/2, \phi]$ and the velocity is

$$u^\beta = c\left[\frac{dt}{d\tau}, 0, 0, \frac{d\phi}{dt}\frac{dt}{d\tau}\right] = c\frac{dt}{d\tau}[1, 0, 0, \omega]$$

where $\omega = \frac{d\phi}{dt}$ is the orbital coordinate (i.e. measured from far away) angular frequency. Using the normalization condition $g_{\mu\nu}u^\mu u^\nu = c^2$, we find that there is a centrifugal contribution to coordinate time:

$$\left(\frac{dt}{d\tau}\right)^2 = \left(1 + \frac{r^2\omega^2}{c^2}\right)\left(1 - \frac{\mathcal{R}}{r}\right)^{-1} \simeq \left(1 - \frac{\mathcal{R}}{r} - \frac{r^2\omega^2}{c^2}\right)^{-1}$$

$$\implies \frac{dt}{d\tau} = \left(1 - \frac{3\mathcal{R}}{2r}\right)^{-1/2} \equiv A \tag{5.83}$$

to first order in $\mathcal{R}/r$, and where Kepler's law $\omega^2 = GM/r^3$ was used in the last expression. We further simplify the analysis by taking the *initial* spin angular momentum in the $\hat{r}$ direction ($\vec{S}_{in} = [0, S, 0, 0]$), and solve Eq. 5.82 for the time evolution of all its components $\vec{S}(t) = [S^0, S^r, S^\theta, S^\phi]$. We find the time component of S^μ with the help of the above-mentioned orthogonality condition:

$$g_{\mu\nu}S^\mu u^\nu = c^2\left(1 - \frac{\mathcal{R}}{r}\right) u^0 S^0 - r^2 \cdot u^\phi S^\phi = 0 \tag{5.84}$$

that yields

$$S^0 = \frac{r^2\omega}{c^2}\left(1 - \frac{\mathcal{R}}{r}\right)^{-1} S^\phi$$

The other three gyroscope equations are:

$$\frac{dS^r}{d\tau} + \Gamma^r_{00}S^0 u^0 + \Gamma^r_{\phi\phi}S^\phi u^\phi = 0$$

$$\frac{dS^\theta}{d\tau} = 0; \qquad\qquad \frac{dS^\phi}{d\tau} + \Gamma^\phi_{r\phi}S^r u^\phi = 0 \qquad (5.85)$$

we see that, as expected, S^θ is null at all times; we end up with two coupled, first-order differential equations for the time derivatives of the S^r, S^ϕ components of the spin. Computing the Christoffel symbols for the Schwarzschild metric is a tedious but straightforward calculations: they can also be found in many textbooks of GR (see, e.g. Hartle 2003) and in *Mathematica*® notebooks; those relevant for our equatorial problem are

$$\Gamma^0_{r0} = \frac{GM}{c^2 r^2}\left(1 - \frac{\mathcal{R}}{r}\right)^{-1} \qquad\qquad \Gamma^r_{00} = \frac{GM}{c^2 r^2}\left(1 - \frac{\mathcal{R}}{r}\right) \qquad (5.86)$$

$$\Gamma^r_{\phi\phi} = -r\left(1 - \frac{\mathcal{R}}{r}\right) \qquad\qquad \Gamma^\phi_{r\phi} = \frac{1}{r}$$

Substituting these into Eq. 5.85 and recalling the relation Eq. 5.83, we find

$$\frac{dS^r}{d\tau} - \frac{r\omega}{A}S^\phi = 0 \qquad (5.87)$$

$$\frac{dS^\phi}{d\tau} + \frac{A\omega}{r}S^r = 0 \qquad (5.88)$$

It is immediate, by eliminating either variable from the system, to obtain, for both components, the equation of a simple harmonic oscillator with angular frequency ω. The solution of these equations, with initial conditions $\vec{S} = S\hat{r}$ at t = 0, is

$$S^r = \frac{C}{A}\cos(\omega\tau) \qquad\qquad S^\phi = \frac{C}{r}\sin(-\omega\tau) \qquad (5.89)$$

where C is a constant of integration. We also see that the phase of S^r and of S^ϕ evolve in opposite directions with time. These equations, expressed in terms of proper time and LFI, are not very helpful, because we are interested in the spin evolution as seen from large distance, in a *coordinate* reference frame. We then recall the relation 5.83 between t and τ to re-express

$$\omega\tau = \frac{\omega t}{A} = \sqrt{1 - \frac{3\mathcal{R}}{2r}} \cdot \omega t \equiv \omega' t \qquad (5.90)$$

that defines the angular frequency ω' as measured in coordinate time:

$$S^r(t) = \frac{C}{A}\cos(\omega' t) \qquad\qquad S^\phi(t) = \frac{C}{Ar}\left(\frac{\omega'}{\omega}\right)\sin(-\omega' t) \qquad (5.91)$$

We have found that, even if we start with the spin aligned in the $\hat{r}$ direction, a S^ϕ component will arise during the revolution, although depressed by a factor $\frac{\omega'}{\omega}$. At completion of a revolution, after a time $T = 2\pi/\omega$, the phase of component S^ϕ has grown to $\omega'T = 2\pi(1 - \frac{3\mathcal{R}}{2r})^{1/2}$. Therefore, when completed one orbit, S^ϕ does not return to 2π, but there is a small (negative) extra angle, *a precession*:

$$\Delta\phi = \omega'T - 2\pi = 2\pi\left[\left(1 - \frac{3\mathcal{R}}{2r}\right)^{1/2} - 1\right] \simeq -\frac{3\pi GM}{c^2 r} \qquad (5.92)$$

that is Eq. 5.74 from where we started.

5.7.3 The Lense-Thirring Precession

The Lense-Thirring effect, or *dragging of inertial frames* is the perfect application of the GEM formalism developed in the previous section. It consists of a non-Newtonian torque, applied to bodies endowed with spin, generated by the motion (we shall focus on rotation) of the source mass, as shown by the last term of Eq. 5.70. The similarity with the magnetic field generated by a charge in motion is obvious and leads to the name gravitomagnetism.

Gravitomagnetic precession was first studied and formalized in 1918 by the physicists J. Lense and H. Thirring (Lense 1918). The story goes that Lense, describing his results to Einstein, commented that this phenomenon would confirm Einsten's idea about the *dragging of inertial frames* by rotating masses but, he concluded, it would not be experimentally observable on the Moon, because of its extreme weakness. And Einstein allegedly objected that the effect, scaling with distance as $1/r^3$, might one day be observable on an artificial satellite, in an orbit much closer to the Earth.

This effect is observed on a test mass close to the source mass and endowed with angular momentum. We shall thus consider an artificial satellite orbiting the Earth. According to our convention, the angular momentum of the Test Mass is labelled by $\vec{L}$, if orbital (e.g. an orbiting satellite around the Earth), or $\vec{S}$ if spin (rotating around its own axis): these two cases are analysed in the same way, but are at the basis of two very different verification experiments. The Earth angular momentum $\vec{J}$ generates the gravitomagnetic field $\vec{B}^{(g)}$.

As shown in the previous section, we associate to the spinning mass a gravitomagnetic moment given by $\vec{\mu}^{(g)} = \vec{S}/c$ (or $\vec{L}/c$ if orbiting); the gyroscope, moving in the gravitomagnetic field of the Earth $\vec{B}^{(g)}$, feels a mechanical torque as shown in Eq. 5.71 (Fig. 5.4):

$$\frac{d\vec{S}}{d\tau} = \vec{\Omega}_{LT} \times \vec{S} = \frac{G}{c^2 r^3}[3(\vec{J} \cdot \hat{r})\hat{r} - \vec{J}] \times \vec{S} \qquad (5.93)$$

This is the **Lense-Thirring (LT) precession**: a spin-spin interaction between the angular momentum $\vec{S}$ (or $\vec{L}$) of the satellite and the Earth angular momentum $\vec{J}$. We just mention a suggestive interpretation of the LT effect as *dragging of inertial*

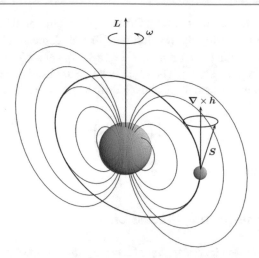

Fig. 5.4 The Earth, spinning with angular velocity ω and angular momentum J, creates a gravitomagnetic field with field lines analogous to those of a magnetic dipole. A gyroscope with spin angular moment S orbits the Earth along a polar geodesic orbit (thick continuous line). The Lense-Thirring effect produces a precession of S around the direction given by the field lines of $\vec{B}^{(g)}$. Credits: courtesy of C. Lämmerzahl

Fig. 5.5 Pictorial representation of the dragging of inertial frame by a rotating massive object. Credits: the GP-B website www.einstein.stanford.edu

frames: the central body drags, in this picture, the space-time along its rotation, in a sort of frictional coupling (without dissipation !). Distortion of the metric thus gives rise to the LT precession. The orbits of satellites are affected by this relativistic phenomenon, hard to detect because it is masked by numerous other effects (Fig. 5.5).

5.8   Lense-Thirring Measurements on the Orbits of the LAGEOS and LARES Satellites

We first consider the LT effect in the case where the angular momentum of the Test Mass is orbital: for a satellite of negligible dimensions and mass m, in a circular orbit of radius r around the Earth, with the help of Kepler's third law

$$\vec{L} = m\vec{r} \times \vec{v} = m\sqrt{GM_{\oplus}r}\,\hat{L} \tag{5.94}$$

$\hat{L}$ is orthogonal to the orbital plane, and therefore a precession implies rotation of this plane. We shall focus again on the Keplerian element defining such plane: the Longitude of Ascending Node.

The instantaneous rate of precession, Eq. 5.72, depends on the satellite position $\vec{r}$ along the orbit, so it makes sense to average the contribution over one complete orbit: it is an exercise in vector algebra, that can be solved in a reference frame centred on the Earth,[7] with $\vec{J}$ along $\hat{z}$, the equator on the $x - y$ plane and the line of nodes along $\hat{y}$:

$$
\begin{aligned}
< \vec{\Omega}_{LT} > &= \frac{G}{r^3 c^2} < (3(\hat{r} \cdot \vec{J}\hat{r} - \vec{J}) > \\
&= \frac{3J}{2r^3 c^2}[(\cos i \sin i), 0, (\sin^2 i - \frac{2}{3})]
\end{aligned} \tag{5.95}
$$

The averaged precession depends on the orbit inclination i:

- For a polar orbit, $i = \pi/2$, we find $< \vec{\Omega}_{LT} > = \frac{GJ}{2c^2 r^3}\hat{z}$
- For an equatorial orbit, $i = 0$, we have $< \vec{\Omega}_{LT} > = -\frac{GJ}{c^2 r^3}\hat{z}$ but, as the satellite angular momentum is also directed along $\hat{z}$, no precession takes place.

Consider now a satellite in a polar orbit around the Earth: in one period of revolution $T = 2\pi\sqrt{r^3/GM}$ the Ascending Node moves by

$$\delta\vec{L} = T < \vec{\Omega}_{LT} > \times \vec{L} = \frac{\pi J_{\oplus}}{c^2}\sqrt{\frac{G}{r^3 M_{\oplus}}} = 4.3 \cdot 10^{-11}\left(\frac{R_{\oplus}}{r}\right)^3 \, rad/orbit$$

a really tiny amount. Fortunately, this effect accumulates with time so that, for a satellite in LEO (Low Earth Orbit $r \gtrsim R_{\oplus}$), in 1 year the orbital plane rotates by:

$$\frac{\Delta RAAN}{\Delta t} = 0.25 \, \mu rad/year$$

i.e. 1.6 m. It will take 23 million years to complete a rotation of the orbital plane.

We have derived these results in the GEM framework, by analogy with the field of a magnetic moment. The same results can be obtained, in a more formal way, by computing the change of direction of the $\vec{L}$ vector under parallel transport in the field of a rotating mass. The LT effect is responsible for precession of both the longitude Ω of the ascending node and of the argument ω of the perigee.

---

[7] We can also evaluate it in a frame with the orbit on the $x - y$ plane and S along $\hat{z}$; we would find $< \vec{\Omega}_{LT} > = \frac{J}{r^3 c^2}[\frac{1}{2}\sin i, 0, -\cos i]$.

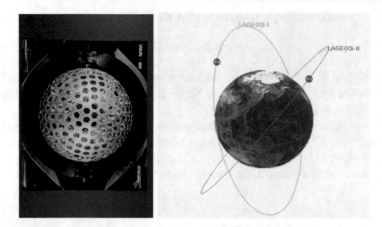

Fig. 5.6 Left: Picture of the LAGEOS II: the satellite is composed of two aluminium hemispheres, 60 cm diameter and 117 kg total mass, hosting an internal Brass core of 190 kg. 43% of the surface is covered by 426 corner cube retroreflectors (CCR): 422 made of Silicon and 4 made of Germanium, to reflect infrared beams; each CCR is 1.9 cm in diameter and 33.2 g in mass. Image credits: NASA/GSFC. Right: schematic diagram of the orbits of the two LAGEOS satellites

As Einstein's intuition had predicted, this tiny precession is observable on the motion of some of the geodetic satellites described in Sect. 1.7. Such measurement was first proposed by Everitt and Van Patten but then pursued by Ciufolini and co-workers (2004) on the orbit of the laser-tracked satellites LAGEOS and LAGEOS II and, more recently on LARES,[8] specially built and launched for this purpose.

Such secular effects on the orbits are really small, less than 2 m per year on the node of satellites like LAGEOS and LAGEOS2: the orbit of these objects is accurately known thanks to the Satellite Laser Ranging (SLR) technique (see Sect. 1.7). The SLR allows the determination of the position of the satellites with a precision that has improved over the decades: it is now of the order of few mm for the *normal points* (resulting from averaging and processing); the rms error on the distance is 2–3 cm on observation times 2 weeks long. Just like we saw for the Schwarzschild precession, the multipoles of the Earth Newtonian field produce a classical precession on $RAAN$, many orders of magnitude larger than the LT effect. The lowest moment, the quadrupole, produces the largest contribution to this classical precession (Fig. 5.6):

$$\frac{d}{dt}RAAN^{classical} = \frac{3}{2}\sqrt{\frac{GM_\oplus}{a^3}}\left(\frac{R_\oplus}{a}\right)^2 \frac{\cos i}{1-e^2}\sqrt{5}\bar{C}_{2,0}$$

where i is the orbit inclination on equatorial plane. This accounts for $126^o/yr$ for LAGEOS, and up to $-624^o/yr$ for LARES.[9] Although this value is known and can be subtracted from the measured precession, the incertitude on $\bar{C}_{2,0}$ yields an error comparable with the searched for LT effect, and therefore blurs the result.

---

[8] LAser RElativity Satellite, ASI 2012.
[9] See Sect. 1.6 for the definition of the $\bar{C}_{l,m}$ coefficients.

The ideal condition to measure the precession of the line of nodes would be to have a pair of satellites in orbits with supplementary angles inclinations i: in this way, one would obtain perfect cancellation of the dominant, classical precession. Unfortunately, these satellites were conceived and launched for other (geodetic) applications (the GR measurements are a sort of free bonus), and a different inclination was eventually chosen for LAGEOS-2. Nevertheless, I. Ciufolini and E. Pavlis in 2004 combined the data to extract the classical precession effect. Using the measurements on the Right Ascension of the Ascending Node of the two satellites, and a sophisticated orbit-tracing program, they solved for a system of equations with two unknown: the LT precession and $\bar{C}_{2,0}$.

This was possible thanks to a detailed model of the Newtonian field, based on geodetic surveys of satellites like CHAMP and GRACE, as mentioned in Sect. 1.6. With this technique, Ciufolini and Pavlis verified the relativistic precession to coincide with the prediction of GR within an error bar of 6% (Ciufolini and Pavlis 2004).

On February 13, 2012, a third satellite, named LARES, was launched by ASI, the Italian Space Agency, on the inaugural launch of the vector VEGA, on an orbit actually supplementary to that of LAGEOS, but at a much lower elevation: about 1450 km above the Earth surface, verses 5800 km for the two LAGEOS. Unlike the predecessors, LARES is composed of one solid sphere of Tungsten.

Observation of the orbits of three satellites allows to remove from the problem the errors contributed by both the quadrupole and the octupole moment, thus refining the measurement. Recent analyses (Ciufolini et al. 2016; Lucchesi et al. 2020) have thus reduce the incertitude on the LT measurement to 5% and 1% respectively.

5.8.1 The Pugh-Schiff Effect: Gravity Probe B

The precession of a gyroscope that is, at the same time, *orbiting and spinning* in the field of a *rotating* central mass is known as the Pugh-Schiff effect. This problem was examined by L. Schiff in a famous two-page paper (Schiff 1960. He considered at the same time three precessions of the spin $\vec{S}$:

$$\frac{d\vec{S}}{dt} = (\vec{\Omega}_{DS} + \vec{\Omega}_{LT} + \vec{\Omega}_{Th}) \times \vec{S} \qquad (5.96)$$

where $\vec{\Omega}_{DS}$ and $\vec{\Omega}_{LT}$ are the de Sitter and Lense-Thirring precession rates, as given by Eqs. 5.77 and 5.95:

$$\vec{\Omega}_{DS} = \frac{3GM}{2c^2r^3}[\vec{r} \times \vec{v}] \qquad \vec{\Omega}_{LT} = \frac{GI_\oplus}{c^2r^3}[3\hat{r}(\hat{r} \cdot \vec{\omega}_\oplus) - \vec{\omega}_\oplus]$$

where $I_\oplus \vec{\omega}_\oplus = \vec{J}_\oplus$ and $\vec{v}$ is the satellite velocity. $\vec{\Omega}_{Th}$ is a special relativity effect, similar to Thomas precession in atomic physics, due to interaction with any non-gravitational force $\vec{F}$:

$$\vec{\Omega}_{Th} = \frac{\vec{F} \times \vec{v}}{2mc^2}$$

Fig. 5.7 A sketch of the
GP-B experiment: the
gyroscope, rotating around
the Earth on a polar orbit,
changes its direction by
6.6 $arcsec/yr$ in the orbit
plane due to geodetic
precession and by
0.039 $arcsec/yr$, normal to
that plane, due to LT
precession. Credits: (Everitt
et al. 2011)

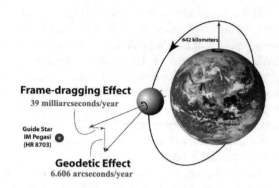

Such force is, by necessity, present in any laboratory experiment: this is the driving motivation for measuring the first two precessions in an orbiting, free-falling experiment.

The rotation frequency of a gyro is constant, and the spinning particle behaves like a clock. This frequency exhibits Doppler and gravitational shifts when observed from outside, just like that of a more conventional clock.

Computation of the resulting precession in the general case is quite involved, but a manageable result can be obtained if we choose a polar orbit ($\vec{v}, \vec{r}, \vec{\omega}_\oplus$ in the same plane) and the radial initial direction for $\vec{S}$.

An intuitive explanation of the precession can be given in Schiff's words (Schiff 1960): ... *think of the moving earth as "dragging" the metric with it to some extent. At the poles, this tends to drag the spin around in the same direction as the rotation of the earth. But at the equator, since the gravitational field falls off with increasing r, the side of the spinning particle nearest the earth is dragged more than the side away from the earth, so that the spin precesses in the opposite direction.*

W.M. Fairbank accepted the challenge and started the experiment Gravity Probe B (GP-B): the development, carried out at Stanford University under the lead of F. Everitt, lasted over 45 years.

The goals of GP-B, for an altitude of 642 km above Earth surface, were

- the measurement of $\Omega_{LT} = 0.041''/$ year due to rotation and with *nominal* precision of 1%
- the measurement of $\Omega_{DS} = 6.6''$/year with precision $\sim 10^{-5}$

Clearly the LT precession, being 160 times smaller than the DS one, would stand no hope of being detectable: nevertheless, with the chosen orbit and spin direction, the two effect are orthogonal: the resulting precessions of the gyroscope spin in two orthogonal directions can thus be separated and detected (Fig. 5.7).

The satellite was launched on April 20, 2004, (the project had begun in 1961) and placed on a polar orbit at 642 km altitude. It contained a superfluid He ($T = 1.8$ K) cryostat with 4 spherical quartz rotors (3.8 cm in diameter), coated with a superconducting Niobium layer (Niobium becomes superconducting at temperatures $T_c < 9$

K). Each of these small spheres was spun by He gas jets, and became a superconducting gyroscope, with a magnetic moment parallel to their angular momentum. The tiny magnetic fields produced by the rotors were picked up by superconducting coils coupled with SQUID magnetometers (see Sect. 8.2.5), based on Josephson effect. When the spheres preceded, a current change was induced in the coils and measured through the SQUIDs. In order to detect both, orthogonal precession signals, the whole spacecraft, and the coils with it, was continuously rotated around its axis.

The satellite was also equipped with a telescope aimed at the distant star IM-Pegasi which constituted the basic element of a control system that commanded the satellite trim correctors. The aiming system was based on VLBI measurements of the IM-Pegasi motion relative to distant quasars. In this way, the co-movable reference system was realized, having axes with fixed orientation with respect to the distant stars. Precessions of the gyroscopes were measured with respect to this star-fixed reference.

Data were collected during the period August 28, 2004–August 14, 2005, when all the liquid Helium had finally evaporated and the rotors were no longer superconducting. The data were analysed for a long time, trying to achieve the announced goal of verifying the predicted LT precession effect to a 1% precision. Among the many problems faced, we mention the effort to subtract the residual effect due to the electric dipole of the spheres: apparently, in the process of evaporating niobium on quartz, the level of uniformity of the metal layer was insufficient to prevent the formation of "charge patches" that produced a residual electric dipole moment. The detailed modelling of the electric dipole presented considerable difficulty in influencing the error in the measurements of the precession effect.

After long and intense efforts, the experimental results were officially announced in May 2011 (Everitt et al. 2011): the measured precessions were in accordance with General Relativity, although with larger incertitudes than expected:

- 0.28% for the geodetic precession effect (de Sitter term)
- 19% for the Lense-Thirring precession.

Despite the partial accomplishment of the initial goals, GP-B was a very successful experiment: it did verify the Schiff precession, and tested, for the first time in space, cryogenics, superconductivity, attitude control by microthrusters and other sophisticated techniques.

5.8.2   The Moon as a Gyroscope

The Earth-Moon system can also be seen as a *gyroscope* with its axis perpendicular to the Moon orbital plane. The de Sitter precession on this system is about 2 arcsec per century. The Earth-Moon system also exhibits a gravitomagnetic precession. The Earth, moving around the Sun, produces gravitomagnetic field, and the Moon moves through this field, experiencing a Lorentz-like force perpendicular to both its velocity and the gravitomagnetic field. This effect produces $\sim$6 m/yr amplitude terms in the Moon's orbit as evaluated in the Solar System Barycenter (SSB) frame.

The quality of Lunar Laser Ranging (see Sect. 1.9.2) data has settled to the level of $\sim$2 cm range uncertainty, so that the de Sitter effect has been measured by LLR (Turyshev and Williams 2007) within 0.6% accuracy.

Analysis of the relative motion of the Earth-Moon system has provided a successful probe to explore many other gravitational effects (Merkowitz 2010):

- the best test, to date, of the Strong Equivalence Principle: $\eta_s < 4.5 \cdot 10^{-4}$
- the best test, to date, of the time constancy of Newton's gravitational constant: $\dot{G}/G < 10^{-12} \, s^{-1}$
- a test of the Inverse Square Law: $\alpha < 10^{-10}$ at the $\lambda \sim 10^8$ m length scale
- WEP: $\Delta a/a < 1.3 \cdot 10^{-13}$
- gravitomagnetism to $\sim$0.1%.

References

Bertotti, B., Bender, P., Ciufolini, I.: New test of general relativity: measurement of de sitter geodetic precession rate for lunar perigee. Phys. Rev. Lett. **58**, 1062 (1987)

Brans, C., Dicke, R.H.: Mach's principle and a relativistic theory of gravitation. Phys. Rev. **124**, 925 (1961)

Ciufolini, I., et al.: A test of general relativity using the LARES and LAGEOS satellites and a GRACE Earth gravity model. Eur. Phys. J. C **76**, 119 (2016)

Ciufolini, I., Pavlis, E.C.: A confirmation of the general relativistic prediction of the Lense-Thirring effect. Nature **431**(7011), 958 (2004)

Einstein, A.: Die Grundlage der allgemeinen Relativitätstheorie (The foundation of the General Theory of Relativity). Annalen der Physik **49**, 769–822 (1916)

Everitt, C.W.F., et al.: Gravity probe B: final results of a space experiment to test general relativity. Phys. Rev. Lett. **106**, 221101 (2011)

Hartle J.B.: Gravity: An Introduction to Einstein's General Relativity. Addison Wesley (2003)

Jackson J.D.: Classical Electrodynamics. Wiley (1975)

Lense, J., Thirring, H.: Uber den Einflub der Eigenrotation der Zentralkorper auf die Bewegung der Planeten und Monde nach der Einsteinschen Gravitationstheorie (On the Influence of the Proper Rotation of Central Bodies on the Motions of Planets and Moons According to Einstein's Theory of Gravitation). Physikalische Zeitschrift 19, 156 (1918)

Lucchesi, D.M., et al.: A 1% measurement of the gravitomagnetic field of the Earth with laser-tracked satellites. Universe **6**, 139 (2020)

Lucchesi, D.M., Peron, R.: Accurate measurement in the field of the earth of the general-relativistic precession of the LAGEOS II pericenter and new constraints on non-Newtonian gravity. Phys. Rev. Lett. **105**, 231103 (2010)

Merkowitz, S.M.: Tests of gravity using lunar laser ranging. Living Rev. Relat. **13**, 7 (2010)

Möller, C.: The Theory of Relativity, 2nd edn. Oxford University Press, Oxford (1972)

Ni, W.T.: A new theory of gravity. Phys. Rev. D **7**, 2880 (1973)

Schiff, L.: Possible new experimental test of general relativity theory. Phys. Rev. Lett. **4**, 215 (1960)

Schiff, L.: On experimental tests of the general theory of relativity. Am. J. Phys. **28**, 349 (1960)

The Gravity Collaboration: Detection of the Schwarzschild precession in the orbit of the star S2 near the Galactic centre massive black hole. A&A **636**, L5 (2020)

Turyshev, S.G., Williams, J.G.: Space-based tests of gravity with laser ranging. Int. J. Mod. Phys. D **16**, 2165 (2007)

Gravity at the Second Post Newtonian Order

6

6.1 Gravitation as a Metric Theory

General Relativity is an extremely successfull theory, as proven by the experimental tests described in the previous chapter. However, there are reasons to look for and to develop other theories of gravitation: for example, the issue of dark matter and dark energy. No dark matter has been detected so far: obviously, a modified gravitational theory that could explain, beside all the classical effects, also the data on galaxy rotation curves would be most welcome. A large number of theories have been formulated, beginning with Eddington in 1922; many of those are mathematical extensions of GR.

In this chapter we shall overview the Parametrized Post Newtonian (PPN) formalism, that attempts to describe these theories in a unified fashion. This parametrization allows a direct comparison of these theories with the experimental evidence that can support or disprove them. We will also address the issue of how to effectively compare the predictions of different metric theories with the experimental observations. Although recent observations on pulsars binaries, as well as gravitational waves emitted by black holes, provide some insight on gravity in the strong field regime, most of the available measurements are relative to weak field effects observed in the Solar System. Hence they are also relative to phenomena that can be described by approximated solutions of the general equations of the theory; these should reduce, to the lowest order in the WFSM expansion, to the predictions of the Newtonian theory and, to the successive order, to the predictions of GR.

We have seen, in Chap. 3 that, according to Schiff's conjecture, any viable theory of gravity that admits the Weak Equivalence Principle must also sustain Einstein Equivalence Principle. This allows us to restrict our attention only to *metric* theories of gravity. All these theories, regardless of the number of fields involved, are characterized by these features:

© The Author(s), under exclusive license to Springer Nature Switzerland AG 2022   113
F. Ricci and M. Bassan, *Experimental Gravitation*, Lecture Notes in Physics 998,
https://doi.org/10.1007/978-3-030-95596-0_6

- A symmetric metric tensor $g_{\mu\nu}$ describes the geometry of space-time, such that $ds^2 = g_{\mu\nu}dx^\mu dx^\nu$.
  This determines lengths and proper times in the ways we already saw from Special and General Relativity.
- Such geometry is locally approximated by a Lorentz space-time.
- Matter in free fall only responds to the metric, according to the same equation of motion, i.e. the geodesic equation:

$$\frac{d^2x^\mu}{ds^2} + \Gamma^\mu_{\ \nu\lambda}\frac{dx^\nu}{ds}\frac{dx^\lambda}{ds} = 0 \qquad (6.1)$$

- A stress-energy tensor exists, with null covariant derivative: $T^{\mu\nu}_{\ ;\mu} = 0$

We adopt these four properties as the definition of metric theory.

In metric theories of gravity, both matter and non-gravitational (NG) fields couple only to the space-time metric $g_{\mu\nu}$.

We must now define our playground.

Consider a local reference frame, moving in a space-time described by a given metric theory of gravity: this *local* frame must be sufficiently large to contain a system of gravitating matter and its associated gravitational fields. It could be the solar system, a black hole or a torsion pendulum. On the other hand this frame should be sufficiently small that inhomogeneities in the gravitational fields, generated on the *outside* it, be small throughout its volume. We shall call this a *quasi-local Lorentz frame*. This implies that the metric, which is coupled directly or indirectly to the other fields of the theory, has structure and evolution depending also on the boundary values taken by these fields. The *outside* gravitational environment in which the local system resides can therefore influence the metric generated by the local system via the boundary values of the auxiliary fields.

In general, gravitational and NG fields *modulate* the process by which matter generates $g_{\mu\nu}$, by coupling among them or to $g_{\mu\nu}$. Many alternative theories have been formulated, and they differ only in how the metric is generated: indeed, if EEP is valid, matter only couples to the metric. We just mention here the main classes, with some famous examples:

* $\{g_{\mu\nu}\}$ In purely tensorial theories, like Einstein's General Relativity, there exists a unique gravitational field $g_{\mu\nu}$. It is generated by the stress-energy tensor which contains contributions of matter and other fields. What differs from theory to theory are the field equations, i.e. the particular way in which matter, and possibly other fields, generate the metric.

In these theories, local gravitational physics is independent of the position and velocity of the local reference frame: being $g_{\mu\nu}$ the only field coupled to the environment, it is always possible to find a coordinate system in which $g_{\mu\nu}$ takes the form $\eta_{\mu\nu}$ at the boundary between the local system and the external environment, neglecting the tidal potentials. One class of such theories postulates a gravitational Lagrangian density that is a general function of the Ricci scalar, rather than the

Ricci scalar itself; these are called $f(R)$ theories, devised to alter the behaviour of gravity on cosmological scales. Another class of theories adds quadratic and higher-order curvature terms to the general relativistic Lagrangian density; this alters the behaviour of the metric on short scales, and the higher-order terms are sometimes interpreted as representing quantum corrections to classical general relativity.

* $\{g_{\mu\nu}, \Phi\}$ In scalar-tensor theories, matter and fields also generate one or more scalar fields. In such theories, the local gravitational physics can depend on position of the frame but is independent of its velocity. The scalar field $\Phi(x_\mu)$ may depend on the space-time position, but on the velocity, because a scalar field is Lorentz-invariant. The scalar field can vary in time because of cosmological evolution, or it can vary in space because of the proximity of matter outside the quasi-local Lorentz frame.

* $\{g_{\mu\nu}, K^\mu, B^{\mu\nu}, \ldots\}$ In the vector-tensor theories, other fundamental fields contribute to generate the metric: vector, tensor or both. The local gravitational physics may have both position and velocity-dependent effects. For example we consider a time-like vector field K^μ, whose value depends on the distribution of matter in the universe: in a given quasi-local Lorentz reference frame it has only a time component K^0. Consider now a Lorentz boost to a frame moving with velocity v relative to the first: the asymptotic form of K^μ has now spatial components $K^j \propto K^0 v^j$, and these velocity-dependent components can then contribute to the form of the local metric.

* *Bimetric and stratified* theories: In Ni's theory (Ni 1973) for example, it is assumed the existence of a flat metric in all the Universe where a proper time exists. This flat metric contributes, with matter and non gravitational fields, to generate a scalar field Φ. All these fields then combine to generate the *physical metric* $g_{\mu\nu}$, the metric that is relevant for the equivalence principle.

For a comprehensive list and description of the many theories proposed since 1922, we refer to the classical and thorough source (Will 2018).[1]

6.2   Notation

In this chapter we introduce the PPN formalism, that extends the WFSM limit to a further degree of approximation, largely following the reasoning outlined in Will (1974). Such formalism, mostly developed by Nordtvedt and Will in the 1970s, has evolved in the years to a standardized notation, that is adopted in most textbooks and papers. These papers use the *natural, or theorist's units*, where $G = c = 1$. Here, we shall try and maintain our notation with SI units and all universal constants at their place, with one exception: the symbol U is used for a sort of dimensionless,

---

[1] There is also a useful page on Wikipedia: https://en.wikipedia.org/wiki/Alternatives_to_general_relativity.

negative Newtonian potential: it is related to Φ, that we used in previous chapters, by the definition

$$U(\vec{x}) \equiv \frac{G}{c^2} \int \frac{\rho \, d^3 x'}{|\vec{x} - \vec{x}'|} = -\frac{\Phi(\vec{x})}{c^2} \tag{6.2}$$

For a simple, central source mass, $U = GM/rc^2$ or, in natural units, $U = M/r$. We have kept this symbol to adhere to the standard PPN notation. Among the "evolutions" that the notation has undergone, there is the change in the metric signature: in Will (1974) the Minkowski metric was $[+, -, -, -]$, while in subsequent literature, such as Will (2014), the opposite signature has been adopted. We shall here mantain, as in previous chapters, the signature $[+, -, -, -]$, the so called *West Coast* or *Landau Lifshitz* convention. This is also the signature adopted in High Energy Physics, because it yields a positive invariant mass.[2] The reader should keep this caveat in mind when comparing these pages with the current literature.

We introduce here the expansion parameter

$$\epsilon \equiv v/c \ll 1$$

where v is the velocity of the body under scrutiny, *the test mass* with respect to the local reference frame.

The 1PN approximation of GR in the Weak Field Slow Motion hypothesis, that we discussed in Chap. 5, assumed $v^2/c^2 \ll 1$ and $U \ll 1$: the virial theorem assures us that kinetic and potential energy are of the same order, i.e. $O(\epsilon^2)$. Being the lowest order of correction, expansion of the metric elements up to $O(\epsilon^2)$ is called *linear*.[3]

For consistency, also the stress-energy tensor elements were expanded to order ϵ^2: we considered a perfect fluid, described by:

$$T^{\mu\nu} = \left(\rho + \rho\Pi + \frac{p}{c^2}\right) u^\mu u^\nu + pg^{\mu\nu} \tag{6.3}$$

where p is the fluid pressure, ρ the mass density and $\rho\Pi c^2$ the internal energy (thermal, radiative, nuclear, compressional...) density: Π is the ratio of internal energy density to rest mass density. This had effect on both the pressure and the internal energy density:

$$\frac{p}{\rho c^2} \sim O(\epsilon^2); \quad \Pi \sim O(\epsilon^2)$$

---

[2] See also https://wikipedia.org/wiki/Sign_convention.
[3] For this reason, some authors refer to the linear approximation as $v^2/c^2 = O(\epsilon)$: this choice leads to non-integer order for odd powers of v/c. The one way to avoid notation ambiguity: disregard the parameter ϵ and state "expanded to the order $(v/c)^n$".

Stated differently, we require both pressure and internal energy to be small with respect to the gravitational energy of the system:

$$\frac{p}{\rho c^2} \leq U; \quad \Pi \leq U$$

These assumptions are reasonable: $p/\rho c^2 \sim 10^{-5}$ in the Sun, $\sim 10^{-10}$ on the Earth. Similar values hold for Π.

On the other hand, in compact objects the expansion in powers of ϵ may fail: in a neutron star $\Pi \sim 0.2$ while $p/\rho c^2$ can be as large as 0.5 and $U \sim 0.2$.

Finally, as a consequence of $v \ll c$, the time derivative are small with respect to the space derivatives:

$$dx \ll cdt \quad \text{that is} \quad \partial/\partial ct \sim O(\epsilon) \cdot \partial/\partial x_i \tag{6.4}$$

6.3 Newtonian Limit—A Recap

In Newtonian physics, the motion of a particle in the gravitational potential U, defined in Eq. 6.2, is described by:

$$\vec{a} = \frac{d^2\vec{x}}{dt^2} = c^2\vec{\nabla}U, \quad \text{or} \quad a^j = c^2\frac{dU}{dx_j} = c^2 U^{\cdot j}$$

this must be recovered as a weak field limit for any gravitational theory.

The geodesic equation can be rewritten in a local (x^j, t) frame, applying the requirement of Eq. 6.4:

$$a^j = -c^2\Gamma_{00}^j = c^2\frac{1}{2}g^{jk}g_{00,k}$$

The Newtonian limit is recovered if $g^{jk} = \delta^{jk}$ and $g^{00} = 1 - 2U$. This simple substitution yields to the Newtonian formulation, that is adequate to describe many phenomena in the Solar System, with accuracy up to 10^{-5}. However, further corrections must be introduced to explain the precession of perihelion of Mercury ($\sim 5 \ 10^{-7}$ rad/orbit) or the other *classical* tests of GR. We learned in the previous chapter that the Schwarzschild metric of GR

$$g_{00} = 1 - 2U, \quad g_{rr} = (-1 - 2U),$$
$$g_{\theta\theta} = -r^2, \quad g_{\phi\phi} = -r^2\sin^2\theta, \quad g_{0i} = 0$$

is adequate for this purpose: an additional correction factor had to be introduced for g_{rr}. Here, both corrections are in terms of U, i.e. of order ϵ^2. We conclude that both Newtonian and first order post-Newtonian (1PN) physics are described by corrections $O(\epsilon^2)$ to the Minkowski metric.

The geodesic equation (6.1) can be derived from a variational principle:

$$\delta \int_A^B dt \sqrt{g_{\alpha\beta} \frac{dx^\alpha}{dt} \frac{dx^\beta}{dt}} = \delta \int_A^B dt \sqrt{g_{00}c^2 + 2g_{0j}cv^j + g_{jk}v^j v^k} = 0 \quad (6.5)$$

We can therefore interpret the expression under square root as a Lagrangian for a single particle in a gravitational field described by the metric $g_{\alpha\beta}$:

$$L = c^2 \left[g_{00} + 2g_{0j} \frac{v^j}{c} + g_{jk} \frac{v^j}{c} \frac{v^k}{c} \right] \quad (6.6)$$

knowing that, in the Newtonian limit, we must find: $L = m(c^2 - 2U - \frac{1}{2}v^2)$ this shows again that corrections to the metric are terms U, $(v/c)^2$, i.e. $O(\epsilon^2)$.

6.4   A General Theory at Second Post-Newtonian Order

We cannot rule out the hypothesis that gravitation is more complicated than GR and is governed by a more complex theory: what we saw so far could then be simply a first approximation to the real metric, and we can hypothesize further, higher order correction terms, that would characterize the other, competing theories.

Indeed, the first attempts at such generalization were made by Eddington in 1922, Robertson and Schiff; they simply hypothesized, for a central, spherical source, a power expansion to the next order in U:

$$g_{00} = 1 - 2\alpha U + 2\beta U^2 + \cdots$$
$$g_{0j} = 0$$
$$g_{jk} = -(1 + 2\gamma U)\delta_{jk} + \cdots \quad (6.7)$$

In this formalism, β accounts for the amount of non-linearity of the theory, while γ measures the amount of space curvature produced by the source. We shall see that this physical interpretation is retained for the corresponding coefficients of the general PPN formulation. The parameter α is normally neglected because it can always be absorbed in the coupling constant G; e.g. for an isolated central mass: $2\alpha U = 2[\alpha G']M/rc^2 \equiv 2GM/rc^2$, where G corresponds to our present knowledge of the gravitational constant.

Looking for a more general formulation of the theory, what form should these additional terms have? The first-order correction to Minkowski metric is the gravitational potential U, we can expect that further terms would be similarly fabricated "potentials", fanciful combinations of mass density and distance. We shall then require the following rules:

1. In order to build a metric $g_{\mu\nu}$ generated by matter fields, without knowledge of the particular field equation, we will include terms up to order ϵ^4. Recalling that

(Eq. 6.4) time derivatives and velocities v are $O(\epsilon)$ with respect to spatial ones and to c respectively, this means that an extension of Eq. 6.6 to $O(\epsilon^4)$ demands to expand g_{00} up to order ϵ^4, g_{0j} to order ϵ^3 and g_{ij} to order ϵ^2:

$$g_{00} \simeq 1 + O(\epsilon^2) + O(\epsilon^4) + \cdots$$
$$g_{0j} \simeq -1 + O(\epsilon^3) + \cdots$$
$$g_{ij} \simeq -\delta_{ij} + O(\epsilon^2) + \cdots$$

For consistency, we will also expand the stress-energy tensor of a perfect fluid, Eq. 6.3 up to order ϵ^4.

2. The metric elements can be dimensionless functions of matter properties p, ρ, $\rho\Pi$ and of its velocity v, but not of their gradients.
3. The metric generated by an isolated source must be locally Lorentzian. That means $|g_{\mu\nu} - \eta_{\mu\nu}| \to 0$ when $r \to \infty$.
4. The general expression of a PPN metric should be valid in any quasi-Lorentzian frame. This implies that the functional dependence of the metric on the potentials and velocities (some theories postulate a preferred frame of the universe) of the theory should be the same in any quasi-Lorentzian frame. In other words, the metric must be invariant for spatial translations, for spatial rotations, for Lorentz boosts. A way to satisfy this requirement is to impose that all terms in g_{00} be scalars under translations, rotations and boosts. Similarly, terms in g_{0j} and g_{ij} should be vectors and tensors, respectively, under the above transformations.

According to these guidelines, we shall add the following terms to the metric[4] :

- Expand g_{ij} to order ϵ^2. According to the above rule 4, g_{ij} must behave as a three-dimensional tensor under rotations. The only terms of $O(\epsilon^2)$ are

$$U\delta_{ij} \quad \text{and} \quad U_{ij} = \frac{G}{c^2} \int \frac{\rho(\vec{x}', t)(x - x')_i (x - x')_j}{|\vec{x} - \vec{x}'|^3} d^3 x' \tag{6.8}$$

- Expand g_{0j} to order ϵ^3. The following terms satisfy the requirement of behaving as 3-vectors.

$$V_j = \frac{G}{c^3} \int \frac{\rho(\vec{x}', t)v'_j}{|\vec{x} - \vec{x}'|} d^3 x' \tag{6.9}$$

$$W_j = \frac{G}{c^3} \int \frac{\rho(\vec{x}', t)\vec{v}' \cdot (\vec{x} - \vec{x}')(x - x')_j}{|\vec{x} - \vec{x}'|^3} d^3 x' \tag{6.10}$$

$$w_j U \qquad w^i U_{ij} \tag{6.11}$$

---

[4] All these functions are found in the literature with $G = c = 1$.

W_j and V_j are two possible *potentials* whose existence is allowed by our rules. The vector w_j, assumed to be $O(\epsilon^1)$, represents the velocity of the adopted reference frame with respect to the rest frame of the Universe. So, $w_j U$ and $w_i U_{ij}$ are also $O(\epsilon^3)$.

- Expand g_{00} to order ϵ^4. This metric element should be a scalar under rotation; therefore it can be composed of any of the following scalars, $O(\epsilon^4)$ terms

$$\Phi_1 = \frac{G}{c^4} \int \frac{\rho' v'^2}{|\vec{x} - \vec{x}'|} d^3 x' \qquad \Phi_2 = \frac{G}{c^2} \int \frac{\rho' U'}{|\vec{x} - \vec{x}'|} d^3 x'$$

$$\Phi_3 = \frac{G}{c^2} \int \frac{\rho' \Pi'}{|\vec{x} - \vec{x}'|} d^3 x' \qquad \Phi_4 = \frac{G}{c^4} \int \frac{p'}{|\vec{x} - \vec{x}'|} d^3 x'$$

$$\mathcal{A} = \frac{G}{c^4} \int \frac{\rho' [\vec{v}' \cdot (\vec{x} - \vec{x}')]^2}{|\vec{x} - \vec{x}'|^3} d^3 x'$$

$$\mathcal{B} = \frac{G}{c^4} \int \frac{\rho'}{|\vec{x} - \vec{x}'|} (\vec{x} - \vec{x}') \cdot \frac{d\vec{v}'}{dt} d^3 x'$$

$$\Phi_W = \frac{G^2}{c^4} \int \rho' \rho'' \frac{\vec{x} - \vec{x}'}{|\vec{x} - \vec{x}'|^3} \cdot \left(\frac{\vec{x} - \vec{x}'}{|\vec{x} - \vec{x}'|} - \frac{\vec{x} - \vec{x}'}{|\vec{x} - \vec{x}'|} \right) d^3 x' d^3 x'' \qquad (6.12)$$

plus the combinations of two terms $O(\epsilon^2) \cdot O(\epsilon^2)$, or $O(\epsilon^3) \cdot O(\epsilon^1)$:

There are a number of functional relations linking these potentials; we write here a few that will be useful later on:

$$\nabla^2 V_j = -\frac{4\pi G}{c^3} \rho\, v_i \qquad\qquad V_{j,k} = -U_{,0}$$

$$\nabla^2 \Phi_1 = -\frac{4\pi G}{c^4} \rho\, v^2 \qquad\qquad \nabla^2 \Phi_2 = -\frac{4\pi G}{c^2} \rho U$$

$$\nabla^2 \Phi_3 = -\frac{4\pi G}{c^2} \rho\, \Pi \qquad\qquad \nabla^2 \Phi_4 = -\frac{4\pi G}{c^4} p \qquad (6.13)$$

With these *building blocks* we can define the most general metric. It can be shown (see Will 2018 for details) that the freedom of choice of coordinates allows us to impose some simplification: in the so called *Standard PPN gauge*, the spatial part of the metric g_{ij} is diagonal and isotropic, and the number of potentials can be reduced: so we set $\mathcal{B} = 0$. We then deal with ten independent potentials:

$$U, \quad U_{ij}, \quad V_j, \quad W_j, \quad \Phi_W, \quad \Phi_1, \quad \Phi_2, \quad \Phi_3, \quad \Phi_4, \mathcal{A}$$

Some theories admit a preferred reference frame, the *rest frame of the Universe*, so also w_j must be retained. The most general metric will be composed of a linear combination of these potentials, with ten coefficients to be determined by experiments:

$$g_{00} = 1 - 2U + 2\beta U^2 - (2\gamma + 2 + \alpha_3 + \zeta_1 - 2\xi)\Phi_1$$
$$- 2(3\gamma - 2\beta + 1 + \zeta_2 + \xi)\Phi_2 - 2(1 + \zeta_3)\Phi_3$$
$$- 2(3\gamma + 3\zeta_4 - 2\xi)\Phi_4 + (\zeta_1 - 2\xi)\mathcal{A}$$
$$+ (\alpha_1 - \alpha_2 - \alpha_3)w^2 U + \alpha_2 w^i w^j U_{ij}$$
$$- (2\alpha_3 - \alpha_1)w^j V_j - 2\xi\Phi_W + O(\epsilon^6)$$

$$g_{0j} = \frac{1}{2}(4\gamma + 3 + \alpha_1 - \alpha_2 + \zeta_1 - 2\xi)V_j + \frac{1}{2}(1 + \alpha_2 - \zeta_1 + 2\xi)W_j$$
$$+ \frac{1}{2}(\alpha_1 - 2\alpha_2)w^j U + +\alpha_2 w^k U_{jk} + O(\epsilon^5)$$

$$g_{ij} = -(1 + 2\gamma U)\delta_{ij} + O(\epsilon^4) \tag{6.14}$$

We could choose to pair each potential with a parameter, but we can just as well pair them with linear combinations of these ten coefficients. This second choice, although less intuitive, allows us to give physical meaning to the ten PPN parameters:

$$\gamma, \quad \beta, \quad \xi, \quad \alpha_1, \quad \alpha_2, \quad \alpha_3, \quad \zeta_1, \quad \zeta_2, \quad \zeta_3, \quad \zeta_4$$

The metric in every particular theory can be obtained by setting these parameters to particular values, often zero.

The stress-energy tensor will also be expanded beyond what stated in Eq. 6.3, to include the gravitational energy U:

$$T^{00} = \rho\left(1 + \Pi + \frac{v^2}{c^2} - 2U\right)c^2$$

$$T^{0j} = \rho\left(1 + \Pi + \frac{v^2}{c^2} - 2U + \frac{p}{\rho c^2}\right)cv^j$$

$$T^{jk} = \rho\left(1 + \Pi + \frac{v^2}{c^2} - 2U + \frac{p}{\rho c^2}\right)v^j v^k + p\delta^{jk}(1 + 2\gamma U) \tag{6.15}$$

In the following list, we summarize the physical meaning of the 10 PPN parameters, and their role in the metric theories of gravity:

- γ measures how much curvature in space is generated by an unitary mass (g_{ij})
- β measures the degree of non-linearity in the superposition law of the theory (g_{00})
- ξ measures the degree of LPI violation: it is a measure of the dependence on the position with respect to a preferred reference frame; in the theories where $\xi \neq 0$, it yields an anisotropy in the local gravitational constant induced by the mass distribution of the universe.
- The three α_j are called the *preferred frame parameters*. They indicate if the metric depends on the velocity with respect to a preferred reference system, typically a *universe rest frame*: setting $\alpha_1 = \alpha_2 = \alpha_3 = 0$, the direct dependence on $\vec{w}$ disappear from the metric. Note that this is not in contradiction with Special

Relativity because this anisotropy is created by gravitational terms while Special
Relativity holds only when gravity is absent.

- Theories that are *semi-conservative*, i.e. where energy and momentum are conserved, require $\alpha_3 = \zeta_1 = \zeta_2 = \zeta_3 = \zeta_4 = 0$. We should add that the momentum P^μ is conserved in any Lagrangian theory.
  Conservation of the angular momentum $J^{\mu\nu}$ is assured only if also $\alpha_1 = \alpha_2 = 0$.
  A *fully conservative* theory, where energy, momentum and angular momentum are conserved, requires thus all the α_j and all the ζ_μ to vanish.

The process of applying the PPN formalism to a theory is lengthy and algebraic-intensive:

1. Identify the variables, which may include:
   (a) dynamical gravitational variables such as the metric $g_{\mu\nu}$, scalar field ϕ, vector field K_μ, tensor field $B_{\mu\nu}$ and so on; (b) geometrical variables such as a flat background metric $\eta_{\mu\nu}$, a cosmic time t, and so on; (c) matter and non-gravitational field variables.
2. Pick a coordinate system whose metric becomes Minkowskian far from the local distribution of matter. It doesn't matter where the coordinate system is moving or at rest relative to the Universe, although for many theories calculations are simplified using the rest coordinate frame of the universe.
3. Set the boundary conditions: typically, a homogeneous isotropic cosmology, with isotropic coordinates in the rest frame of the universe. Compute to zeroth order the metric and fields:
   $g_{\mu\nu}^{(0)} = diag(c_0, -c_1, -c_1, -c_1)$, $\phi^{(0)}$, $K_\mu^{(0)}$, $B_{\mu\nu}^{(0)}$ $g_{\mu\nu}^{(0)}$ must not necessarily correspond to Minkowski's flat space, but we shall restrict our study to the simple case $g_{\mu\nu}^{(0)} \equiv \eta_{\mu\nu}$.
4. Compute new variables $h_{\mu\nu} = g_{\mu\nu} - g_{\mu\nu}^{(0)}$ using, where needed, $\phi - \phi^{(0)}$, $K_\mu - K_\mu^{(0)}$ and $B_{\mu\nu} - B_{\mu\nu}^{(0)}$
5. Substitute these forms into the field equations, operating to a consistent order. Find solutions for $h_{\mu\nu}$. Substitute the perfect fluid stress tensor for the matter sources.
6. Solve for h_{00} to $O(\epsilon^2)$. This shall tends to zero far from the system: compute the form $h_{00} = -2U$. Solve for h_{ij} to $O(\epsilon^2)$ and h_{0j} to $O(\epsilon^3)$.
7. Solve for h_{00} to $O(\epsilon^4)$. This is the most complex step, involving all the nonlinearities in the field equations. The stress–energy tensor must also be expanded to sufficient order.
8. Convert to local quasi-Cartesian coordinates.
9. Read off the PPN parameter values by comparing the result for $g_{\mu\nu}$ with the values of Eq. 6.14.

6.5 PPN Applied to General Relativity

We now apply the above recipe to the particular, but very important case of GR, to recover the equations derived in Chap. 5. The starting point is the Einstein field equations

$$R_{\mu\nu} - \frac{1}{2}g_{\mu\nu}R = \frac{8\pi G}{c^4}T_{\mu\nu} \tag{6.16}$$

If we multiply this equation by $g^{\mu\nu}$, we easily obtain $R = -\frac{8\pi G}{c^4}T$ with $T = g^{\mu\nu}T_{\mu\nu}$. This allows us to rewrite Eq. 6.16 in the form

$$R_{\mu\nu} = \frac{8\pi G}{c^4}\left(T_{\mu\nu} - \frac{1}{2}g_{\mu\nu}T\right) \tag{6.17}$$

In the WFSM approximation, with

$$g_{\mu\nu} \simeq \eta_{\mu\nu} + h_{\mu\nu}; \qquad \text{with } h_{\mu\nu} << 1$$

we have (see Eq. 5.15), to the required $O(\epsilon^4)$:

$$R_{00} = \frac{1}{2}\nabla^2 h_{00} - \frac{1}{2}(h_{jj,00} - 2h_{j0,j0}) + \frac{1}{2}h_{00,j}\left(h_{jk,k} - \frac{1}{2}h_{kk,j}\right) - \frac{1}{4}|\nabla h_{00}|^2 + \frac{1}{2}h_{jk}h_{00,jk}$$

$$R_{0j} = -\frac{1}{2}(\nabla^2 h_{0j} - h_{k0,jk} + h_{kk,0j} - h_{kj,0k}) \tag{6.18}$$

$$R_{ij} = -\frac{1}{2}(\nabla^2 h_{ij} - h_{00,ij} + h_{kk,ij} - h_{ki,kj} - h_{kj,k,i})$$

while the stress-energy tensor is given by Eq. 6.3.

First, we compute h_{00} and T_{00} to $O(\epsilon^2)$:

$$R_{00} = \frac{1}{2}\nabla^2 h_{00} \qquad T_{00} = T = \rho c^2 \tag{6.19}$$

Plugging these into Einstein's equation (6.17) gives

$$\nabla^2 h_{00} = \frac{8\pi G}{c^2}\rho \quad \text{that has solution: } h_{00} = -2U \tag{6.20}$$

To compute h_{ij} to $O(2)$, recall that indexes are raised and lowered using the flat-space metric: $h_\alpha^\mu \equiv \eta^{\mu\beta}h_{\beta\alpha}$. Impose the three gauge conditions:

$$h_{i,\mu}^\mu - \frac{1}{2}h_{\mu,i}^\mu = 0 \tag{6.21}$$

Again, Eq. 6.17 yields

$$\nabla^2 h_{ij} = \frac{8\pi G}{c^4} \rho \delta_{ij} \quad \text{that has solution:} \quad h_{ij} = -2U\delta_{ij} \qquad (6.22)$$

To expand h_{0j} to $O(\epsilon^3)$, we impose one additional gauge condition:

$$h^{\mu}_{0,\mu} - \frac{1}{2}h^{\mu}_{\mu,0} = -\frac{1}{2}h_{00,0} \qquad (6.23)$$

and Eq. 6.17 takes the form

$$\nabla^2 h_{0j} + U_{,0j} = -\frac{16\pi G}{c^2}\rho v_j \qquad (6.24)$$

The solution of Eq. 6.24 is not straightforward and we refer to textbooks as Weinberg (1972) or Poisson and Will (2014). In essence, one has to relate the derivatives of U to the potentials V_j, W_j defined in Eqs. 6.9 and 6.10 to finally obtain the solution

$$h_{0j} = \frac{7}{2}V_j + \frac{1}{2}W_j \qquad (6.25)$$

To iterate the expansion to $O(\epsilon^4)$, on the same gauge condition, we make use of the solutions of $h_{\mu\nu}$ to order $O(\epsilon^2)$. We eventually find

$$R_{00} = \frac{1}{2}\nabla^2(h_{00} - 2U^2) + 4U\nabla^2 U$$

and

$$T_{00} - \frac{1}{2}g_{00}T = \frac{1}{2}\rho c^2\left(1 + \frac{v^2}{c^2} - U + \frac{1}{2}\Pi + \frac{3}{2}\frac{p}{\rho c^2}\right)$$

Plugging again the last two expressions into the Einstein equations (6.17) and making use of the relationships Eq. 6.13 for the potentials Φ_j $(j = 1 \ldots 4)$ one eventually reaches the solution

$$h_{00} = -2U + 2U^2 - 4\Phi_1 - 4\Phi_2 - 2\Phi_3 - 6\Phi_4 \qquad (6.26)$$

In conclusion, the GR metric in the PPN formulation is, from Eqs. 6.22, 6.25, 6.26:

$$g_{00} = 1 - 2U + 2U^2 - 4\Phi_1 - 4\Phi_2 - 2\Phi_3 - 6\Phi_4$$

$$g_{0j} = \frac{7}{2}V_j + \frac{1}{2}W_j \qquad (6.27)$$

$$g_{ij} = -(1 + 2U)\delta_{ij}$$

Comparison with the general form of the PPN metric, Eq. 6.14, shows, by inspection, that General Relativity is returned with the choice: $\beta = 1$; $\gamma = 1$ and all other PPN parameters equal to zero.

6.5.1 A Few Words About Brans–Dicke Theory

We shall just mention here the most famous among the many gravitational theories that have been proposed as competitors to GR: the Jordan–Brans–Dicke theory.[5]

R. Dicke and his student C. Brans, building on previous work of P. Jordan, proposed (Brans 1961) a scalar-tensor theory of gravity that incorporated Mach's principle into gravitation. This principle states that inertial forces are gravitational effects of distant matter: who decides whether a reference frame is inertial or accelerated ? According to Mach, inertial frames are those that are not accelerated with respect to the distant stars. In this way, one could envision a preferred reference frame, in contrast with Lorentz invariance. The issue is almost philosophical and not resolved yet. The consequence of this assumption is that matter must play an additional role, beside determining the geometry. Hence, the simplest and most famous of the many wordings of Mach's principle:

"matter there influences inertia here".

Mach's principle had a profound influence on the development of Einstein's thought; however, as Brans and Dicke noted, the principle is imperfectly satisfied in General Relativity, so they moved to remedy this situation. To this purpose, another formulation of Mach's principle comes useful:

Newton's Gravitational Constant G is a Dynamical Field.

This formulation is influenced by Dirac's Large Number Hypothesis (see also Sect. 6.6.4): it states that, calling M_u the estimated mass of the universe and $2R_u$ its diameter, the fact that

$$\frac{GM_u}{R_u c^2} \sim 1$$

cannot be a coincidence, resulting from cancellation of extremely large ($O(10^{40})$) numbers. So, the theory allows $1/G \sim M_u/R_u c^2$ to be a scalar field $\phi(x_\mu)$, rather than a constant.

The Brans–Dicke theory is thus a scalar-tensor theory containing a parameter ω, that measures the coupling of the two fields.

For GR, the field equations can be derived from the variational principle (Landau and Liftschitz 1951)

$$\delta \int d^4 x \sqrt{|g|} \left[R + \frac{16\pi G}{c^4} \mathcal{L}^m \right] = 0 \qquad (6.28)$$

with R the curvature scalar, g the determinant of the metric and $\mathcal{L}^m$ the Lagrangian density of matter, including all non gravitational fields. Brans and Dicke extended this action to include the additional scalar field ϕ, allowing Newton's constant to vary.

---

[5] An excellent, readable overview of this theory and related experimental tests can be found online: http://www.scholarpedia.org/article/Jordan-Brans-Dicke_Theory, by Carl H. Brans himself.

They added the standard Lagrangian density of a scalar field, $\mathcal{L}^\phi \propto \phi_{,\mu}\phi^{,\mu}$. Moreover, they chose a constant, dimensionless coupling parameter[6] ω. By multiplying Eq. 6.28 by $G^{-1} = \phi$, we obtain the resulting action:

$$\delta \int d^4x \sqrt{|g|} \left[\phi R + \frac{16\pi}{c^4} \mathcal{L}^m + \omega \frac{\phi_{,\mu}\phi^{,\mu}}{\phi} \right] = 0 \tag{6.29}$$

The factor $1/\phi$ in the denominator of the third term is introduced so that the coupling constant ω be dimensionless. The field equations are derived by varying $\phi, \phi_{,\mu}$ in Eq. 6.29:

$$R_{\mu\nu} - \frac{1}{2} g_{\mu\nu} R = \frac{8\pi}{\phi c^2} T_{\mu\nu} + \frac{\omega}{\phi^2} \left[\phi_{,\mu}\phi_{,\nu} - \frac{1}{2} g_{\mu\nu}\phi_{,\alpha}\phi^{,\alpha} \right] +$$
$$+ \frac{1}{\phi} [\phi_{,\mu;\nu} - g_{\mu\nu}\Box\phi]$$

where ϕ can play either the role of additional "matter" source, if positioned, as here, on the right hand side of the equation, or the role of additional gravitational field, if we move it to the left. By expanding this equations in the weak field limit, we would find:

$$g_{00} = 1 - \frac{2M}{\phi_0 c^2 r} \left(1 + \frac{1}{3 + 2\omega} \right)$$

$$g_{ij} = -\left[1 + \frac{2M}{\phi_0 c^2 r} \left(1 - \frac{1}{3 + 2\omega} \right) \right] \delta_{ij}$$

$$\phi(x_\mu) = \phi_0 \left(1 + \frac{1}{3 + 2\omega} \right)$$

where ϕ_0 is the asymptotic value of the scalar field away from our source M.

By comparison with the usual PPN value $g_{00} = 1 + 2GM/rc^2$ we derive the present, i.e. measured here and now, value of Newton's constant:

$$G_{now} = \frac{2\omega + 4}{2\omega + 3} \frac{1}{\phi_0}$$

Moreover, comparison with g_{ij}^{PPN} of Eq. 6.14 yields:

$$\gamma = \frac{\omega + 1}{\omega + 2} \tag{6.30}$$

In the limit $\omega \to \infty$ we have $\gamma \to 1$. Therefore, for large values of ω, the predictions of Brans–Dicke become more and more indistinguishable from those of GR. Dicke hypnotized a value $\omega \sim 5$, to fit the data of Mercury's perihelion precession (see following section). Recent tests (Bertotti et al. 2003) push the limit toward $\omega > 40000$. This determined the decline of interest in Brans–Dicke theory.

---

[6] More complex, later scalar-tensor theories will have $\omega(\phi)$.

6.6   Experimental Limits of the PPN Parameters

Using the PPN expansion of each theory in the equations of motion we find clues and traces to perform experimental verifications. Particularly, we report here the fundamental relations that are the bases to deduce the bounds of the PPN parameters in some of the *classical* observational tests of general relativity. We will then mention the tests of violation of the strong equivalence principle (SEP). The analytical description of these effects within the PPN formalism is carried on following the logic paths illustrated in the previous chapter. The significant change is the use of the PPN metric $g_{\mu\nu}$ given by (6.14) instead of the Schwarzchild metric. For a complete derivation of the following formulas we refer to Will's textbook (Will 2018).

6.6.1   Limits on the Parameter γ

Light Deflection
Historically, the first confirmation of General Relativity came from observations of the deflection of photon trajectories in the proximity of the solar gravitational field, as discussed in Sect. 5.4.2. We consider the case of an observer at rest on the Earth, who receives two light rays: the first from a targeted source, the second from another star, used as reference and located in a different position of the sky. as sketched in Fig. 6.1. φ is the angle between the directions of the two incoming rays, being φ_0 the value when there is no gravitational perturbation: that is, when the massive object interposed between the target source and the observer is far from the line of sight. The difference $\Delta\varphi = \varphi - \varphi_0$ is the measurement of the light deflection due to lensing: the calculation in the PPN framework is carried out similarly to Scct. 5.4.2, inserting the PPN metric elements into the starting Eq. 5.43.

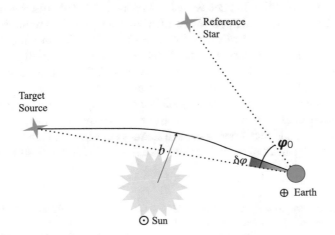

Fig. 6.1 Sketch of the geometry for the measurements of light deflection

The light deflection computation yields the rather simple relation:

$$\Delta\varphi = \frac{1+\gamma}{2}\frac{4GM}{bc^2}\frac{1+\cos\varphi_0}{2} \tag{6.31}$$

For light grazing the Sun, that means:

$$\Delta\varphi_\odot^{max} = \frac{1+\gamma}{2}\,1.75'' \tag{6.32}$$

In analogy to Eddington's pioneering experiment, the measurements are carried out during solar eclipses, when the solar disc is obscured and it is possible to observe stars with angular position very close to the Sun. Values of the deviation $\Delta\varphi_\odot^{max}$ are averaged by observing the position of several star in the sky. The main limitation to this type of measurement is due to fluctuations in the refractive index of the atmosphere. This effect can be avoided using astrometric data taken by a telescope in space. In fact, the most accurate measurements to date of the light bending effect, is a byproduct of the satellite mission Hipparcos,[7] a project aimed at accurately measuring the positions of celestial objects on the sky. The Hipparcos limit on γ was obtained by accumulating accurate measurements of star coordinates at various elongations from the Sun. The optical observations, taken over 37 months, did not rely on Solar eclipses. The data were taken at very large angular distance from the Sun between 47 and 133 degrees. At the first glance this could be a drawback, but apparently this method allows a better randomisation of the systematic errors. The Hipparcos limit on γ is (Froeschlé et al. 1997):

$$\gamma = 0.997 \pm 0.003 \qquad \text{- Hipparcos, 1993}$$

The follow-up astrometric mission is GAIA: this ESA satellite was launched on December 19th, 2013 and it is orbiting around the Sun-Earth L2 Lagrange point.[8] There is great expectation to achieve, using its data, a dramatic improvement in the γ limit with respect to the Hipparcos result of $\sigma_\gamma = 3 \times 10^{-3}$. Simulations predict a gain of at least two order of magnitude (Vecchiato 2014), under the assumption that the error in a single measurement of light deflection effect will be on average 100 µas for at least 1 million stars, observed about 400 times during 5 years. GAIA is indeed expected to observe and catalog more than a billion stars.

Actually, the measurement is best carried on observing electromagnetic radiation of longer wavelength, such as radio waves or microwaves. In fact, the Sun radio emission is weak, and many well localized sources emitting in the radio band can be observed applying an interferometric technique. The technique consists of combining

---

[7] Hipparcos: *HIgh Precision PARallax COllecting Satellite* ESA 1989–1993. Named after the Greek astronomer Hipparchus of Nicaea, known for applications of trigonometry to astronomy.
[8] The Lagrange points of the Sun-Earth system are 5 points in the orbital plane where the sum of both gravitational forces and of the centrifugal force cancels out. L2 is about 1.5 Gm beyond the Earth along the Sun-Earth axis.

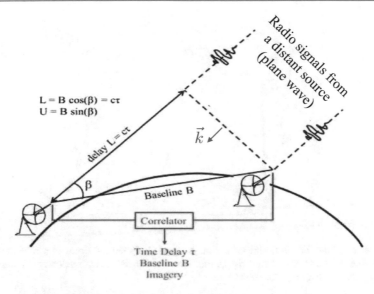

Fig. 6.2 Schematic of the operation for a VLBI antenna array: cross-correlation of the two recorded time series yields the delay τ in reception. Hence, via the relation $\cos \beta = c\tau / B$ one can measure either the baseline B or the beam inclination β

the signals collected by two radio-telescopes located tens of kilometres apart: the *Long Baseline Interferometry* (LBI), As sketched in Fig. 6.2, by cross-correlating the outputs of two antennas, one can determine the time delay τ. However, an exact and synchronized timekeeping at the two (or more) stations is required: this is achieved by linking the clock that time-stamps the observational data to a superstable oscillator, like a H maser and, more recently, to the GPS network, for long-term stability.

Knowledge of τ and of the length of the ideal line joining the two antennas (the *baseline B*), gives, through the relation $c\tau = B \cos \beta$ the inclination angle β of the incoming plane wave. Conversely, observation of radio signals from a known source can determine the baseline length B. This method is easily extended to a network of several antennas (in analogy with multi-beam interferometry), where the *phase closure* technique can be applied. In oversimplified terms, the sum of the signal phase differences from one antenna to the next must be equal to the total phase difference between the first and the last:

$$\Delta\phi_{1,2} + \Delta\phi_{2,3} + \cdots + \Delta\phi_{n-1,n} - \Delta\phi_{1,n} = 0$$

Applying this method the overall phase noise contribution of each observing station contributions cancel out (Jennison 1958).

In 1975 Fomalont and Sramek (1976) used a baseline of 35 km, obtaining as limiting value

$$\gamma = 1.007 \pm 0.009 \qquad \text{- LBI, 1975}$$

With an angular resolution of $0.01''$ they observed distant sources such as the quasars $0111 + 02$, $0119 + 11$, and $0116 + 08$. The main limitation of this method was

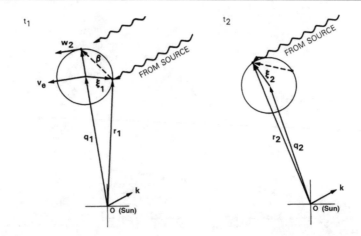

Fig. 6.3 Corrections for the timing of the radio signal. Between the times of arrival at the first VLBI station (left) and the second (right), the Earth has translated (with velocity v_e) and rotated (with tangential velocity w_2 at station 2)

related to the modelling of the solar corona, whose effects are subtracted by observing electromagnetic signals at different frequencies (in their case 2.695 and 8.085 GHz).

Clearly, the accuracy of these measurements increases with the baseline B. This consideration led to establish the *Very Long Base Interferometry* (VLBI) where the antennas are distant thousands of km. With the increase of the baseline length, additional care must be taken in time-keeping. The measurement is carried out in the Solar System Barycentric (SBB) reference frame, to "make simple" the path of the radio beam: therefore the timing must take into account both rotation and revolution of the Earth, as shown in Fig. 6.3.

At present, the US-operated Very Long Base Array (see Fig. 6.4) can rely on ten antennas with a maximum baseline (Mauna Kea, Hawaii to St. Croix, US Virgin Islands) of 8611 km, not counting the possible extension to a site in Germany.[9] The recent best result obtained with this technique is (Lambert and Le Poncin-Lafitte 2011):

$$\gamma = 1 - (8 \pm 12)\ 10^{-5} \qquad \text{- VLBI, 2011}$$

---

[9] The US-National Radio Astronomy Observatory has an interesting website: https://public.nrao.edu/telescopes/vlba/.

Fig. 6.4 The VLBA network of radio antennas. Credits: NRAO/AUI/NSF

Fig. 6.5 Sketch of the geometry for the measurements of propagation delay

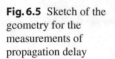

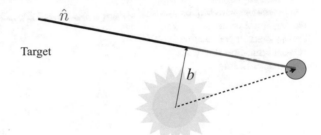

Propagation Delay of Radio Echo

Still in the domain of classical tests of relativity, we discussed the *Shapiro time delay*, i.e. the delay in the propagation time of electromagnetic signals traveling through a gravitational field. Here too, we just report the final expression of the delay time in the PPN framework. This relation allows us to establish new bounds on the parameters γ. The guiding principle for the experiments is to measure the round-trip travel time of an electromagnetic signal sent towards a target that plays the role of a mirror. Such experiment is carried on within the Solar System, and the reflecting target is a planet or a satellite: therefore, the retarding gravitational field of interest is that generated by the Sun. Consider a reference frame with the origin at the Sun center, as shown in Fig. 6.5. Let $\vec{x}_\oplus$ and $\vec{x}_t$ be the position vectors of the Earth and the target body. $\hat{n}$ the unit vector oriented from the target to the Earth and $r_\oplus, r_t$ are the distances along $\hat{n}$ from the point P of closest approach to the Sun to the Earth and to the target star, respectively. Performing the calculation with the PPN metric, the Shapiro delay for a round trip is

$$\delta t = \frac{1+\gamma}{2}\frac{4GM_\odot}{c^3}\ln\frac{(r_\oplus + \vec{x}_\oplus \cdot \hat{n})(r_t - \vec{x}_t \cdot \hat{n})}{b^2} \tag{6.33}$$

The measurement is best performed when the target is in opposition (superior conjunction) with the Earth and the Sun. Then $\vec{x}_\oplus \cdot \hat{n} \simeq r_\oplus$, $\vec{x}_t \cdot \hat{n} \simeq -r_t$, $b \simeq R_\odot$ and, considering Mars as the target:

Fig. 6.6 Top: The measurements of the Shapiro time delay during superior conjunction (Mercury-Sun-Earth) of 1967: on the vertical axis, the excess *range* time. Predictions are based on orbits determined from other data. Credits: reprinted figure with permission from Shapiro et al. (1968). Copyright 1968 by the American Physical Society. Bottom: The Cassini *range-rate* measurement of the Shapiro time delay during superior conjunction (Cassini-Sun-Earth) of 2002. Courtesy of L. Iess

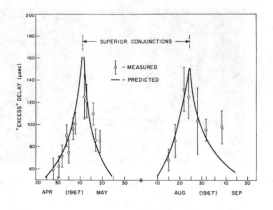

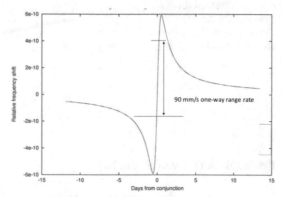

$$\delta t = \frac{1+\gamma}{2}\frac{4GM_\odot}{c^3}\ln\frac{4r_\oplus\, r_t}{b^2} = 247\,\mu s \qquad \text{if } R_\odot \simeq b$$

We should also consider the time derivative of this equation, that is related to the Doppler frequency shift of the received signal:

$$\frac{\delta f}{f} = \frac{d\,\delta t}{dt} = -4\frac{1+\gamma}{2}\frac{GM_\odot}{c^3 b}\frac{db}{dt} \qquad (6.34)$$

The first measurements were performed reflecting radio beams off the surface of Mercury, Venus and Mars. Figure 6.6, top, shows results obtained in 1967 by Shapiro and coworkers observing two superior conjunctions of Mercury. Planets however have rough surfaces: problems related to the orography can produce errors in the propagation time of the order of 5 μs. These pioneering measurements thus confirmed GR ($\gamma = 1$) at the 5% level.

Artificial satellites offer a good alternative: the uncertainty in their position is ~50 m (150 ns) due to spurious accelerations caused by the position control system or by the solar wind. To overcome this problem, it is more common to reflect off artificial satellites orbiting around a planet, where their passive orbit can be precisely predicted and monitored.

Prior limits on γ were derived by analysing the echo radar delays obtained from ranging to the Mariner 9 spacecraft,[10] the first artificial satellite sent in orbit around Mars. The smallest uncertainty associated to γ values was about 2%.

A serious problem with the experiments carried out in solar conjunction is related to density fluctuations in the solar corona: the radio beams grazing the Sun pass through regions where rapid changes in the electron density cause the delay to vary in a non-controllable fashion, adding a hard-to-model delay of up to 100 µs. A two-fold substantial improvement in the Shapiro delay measurement was achieved with the Viking (Shapiro 1977) mission, devoted again to the exploration of Mars. In this case NASA sent to the Red Planet two spacecrafts, Viking[11] 1 and 2, each made of two main parts: an orbiter designed to photograph the surface of Mars while orbiting it, and a lander designed to study the planet from the surface. The two Vikings began orbiting Mars in the summer of 1976 and spent about a month surveying landing sites. They then released their landers, that touched down on flat lowland sites in the northern hemisphere of Mars about 6,500 km apart. Scientists used the radar echo to the landers to test GR: radio signals were sent to the landers on Mars, reflected back to the orbiter and from these to the Earth. The improvement in accuracy, over the Mariner experiment, of the echo delays measurement, transponded by the spacecraft, was essentially due to the fact that the two Viking landers were implanted on the surface of Mars so that their trajectories were precisely determined. In addition, while the two Viking landers transmitted to the orbiters only in S-band ($\sim$2.3 GHz), the spacecrafts transmitted signals to Earth in both S- and X-band ($\sim$8.4 GHz), both coherent with the single S-band signal received from the lander. This dual-band, one-way ranging allowed estimation of the contribution to the echo delays from the solar-corona plasma.

Mars passed in superior conjunction with the Sun and the Earth on November 25, 1976, few months after the arrival of the Vikings. From those data the following limit was obtained (Reasenberg et al. 1979):

$$\gamma = 1.000 \pm 0.002 \qquad \text{Viking ranging, 1976}$$

Note that the distance of Mars from the Sun is $d_{\odot-Mars} = 1.56$ AU, therefore the round trip time from the Earth to Mars, when in opposition, is $2 \cdot d_{\oplus-Mars} = 2.56\ AU \cdot 2/c = 2552$ s. The detection of the Shapiro delay, 247 µs, requires with a time accuracy of one part in 10^7. Measuring γ at 0.2 % level means measuring the transit time to better than $2 \cdot 10^{-10}$. This difficult requirement poses a threefold challenge to experimenter:

- clock stability, (amply satisfied by H masers and, more recently by atomic clocks, now in the 10^{-15} range),
- ability to describe and predict the orbits to better than $c\delta t \simeq 150$ m,

---

[10] Mariner 9, NASA and JPL; May 1971–Oct. 1972.

[11] Viking 1 and 2, NASA and JPL; launched in 1975, they reached Mars in the summer 1976. The end of activity differs for the two orbiters (1980, 1978) and the two landers (1982, 1980).

– precise knowledge of the positioning for the ranging stations on Earth, as discussed in Sect. 1.7.

The best limit, to date, on the γ parameter was set by the analysis of the radio signal tracking the Cassini[12] spacecraft, during its trip to Saturn.

The mission contributed to studies of Jupiter for six months in 2000 before reaching its destination, Saturn, in 2004 and starting a series of flybys of Saturn's moons. That same year it released the Huygens lander on Saturn's moon Titan to conduct a study of the moon's atmosphere and surface composition. On its way toward Saturn, Cassini passed superior conjunction with Sun and Earth on July 7, 2002: the radio signals passed near the Sun at a distance of $d = 1.6R_\odot$. Bertotti, Iess and Tortora exploited a sophisticated and innovative radio link, with two frequencies, 7.175 GHz (X-band) and 34.316 GHz (Ka-band), uplink (transmitted from ground to the spacecraft) and three frequencies downlink, to suppress the corona noise down to 10^{-4} of the relativistic signal. While the experiment with the Vikings measured the excess transit time (range measurement), the analysis carried out on the data of Cassini focused on the time derivative of the range data (range-rate), as shown in Eq. 6.34. Indeed, in superior conjunction, the distance b from the Sun, and hence the delaying effect, reaches a minimum, so that its time derivative is most sensitive. This procedure was possible only because the corona noise was efficiently suppressed. Indeed, the signal in the lower pane of Fig. 6.6 (Cassini) is proportional to the time derivative of the upper one. They have produced (Bertotti 2003) what is still today (2021) the best measurement of propagation delay, constraining γ :

$$\gamma = 1 + (2.1 \pm 2.3) \cdot 10^{-5} \qquad \text{- Cassini, 2002}$$

Finally, we mention the use of the Shapiro time delay also in the measurement of radio waves emitted by quasars, or other extragalactic sources. It is obvious that the classical Shapiro formula, Eq. 6.33, requiring knowledge of the target position, is of little use when looking at a source at cosmological distances. However, if one simultaneously observes two sources with a similar impact parameter b, and computes the *difference* of the two Shapiro delays, many common terms drop out and the relations simplifies to (Hellings 1986):

$$\Delta\tau_{grav} = (1 + \gamma)\frac{GM_\odot}{c^3} log \frac{|\vec{r}_1| + (\vec{r}_1 \cdot \hat{k})}{|\vec{r}_2| + (\vec{r}_2 \cdot \hat{k})}$$

where $\vec{r}_j$ is the position of the jth VLBI station, in Solar System Barycentric coordinates, and $\hat{k}$ is the unit vector in the direction of the source: if the source is extragalactic, the direction is the same for both antennas. With this method, a dedicated 1-day (as opposed to years) observation session was devoted to two sources, 0229+131

---

[12] Cassini–Huygens, ESA/ NASA / ASI. 1997–2017. Named after the XVII century astronomers Giovanni Domenico Cassini, discoverer of four Saturn moons and Christiaan Huygens, discoverer of Titan.

and 0234+164, whose position is known to better than 200 μas. By observing them at angular distances from the Sun of 1.15° to 2.6°, the team (Titov et al. 2018) achieved a limit:

$$\gamma - 1 = (2.72 \pm 0.92) \cdot 10^{-4} \qquad \text{- delay + VLBI, 2018}$$

This accuracy approaches, but does not reach, that of Cassini reported above.

6.6.2   Limits on the PPN Parameter β

Estimates for the parameter β can be deduced from the measurements of the precession in the perihelium of the planets, but we will need the prior knowledge of the value of γ, that we have derived in the last subsection.

Let us now return to the precession of Mercury's perihelion $\omega_{Mercury}$, that we analyzed, in the framework of GR, in Sect. 5.5.

From the observational point of view, this precession measured from an Earth-based observer is actually very large: 5599.74″ But we need to subtract a number of classical corrections:

- the precession of Earth equinoxes, 5025.64 (a period of 25 700 years): this has nothing to do with Mercury precession, is just a change of coordinates to refer the measurement to an inertial frame (ECI frame, see Chap. 14),
- perturbations to the orbit due to Venus (277.8″), Jupiter (153.6″), Earth (90.0″) and other planets (10.0), inferred from Newtonian mechanics and observation of the orbits,

  When all these terms are considered, we are left with a residual precession of 0.1035″/revolution, or 42.98″/century, almost exactly the result that Einstein had computed.[13]

$$\dot{\omega}_{Mercury}^{GR} = \frac{6\pi G M_{\odot}}{a(1 - e^2)c^2 T} = 42.98''/\text{century}$$

T, e and a are the period, the eccentricity and the semi-major axis of the planet's orbit. This was the first, amazing success for his new theory[14] : the excess motion of Mercury's perihelion of 43″ per century had been observed by Le Verrier in 1859 and had remained enigmatic up to the formulation of GR.

So, GR prediction exactly matches observation: what else is there to learn from Mercury precession? As mentioned in Sect. 1.6, this stunning result could be spoiled by one additional correction:

---

[13] Einstein had computed 43″/century, compatible with the less precise knowledge of c and a of those times.

[14] Einstein wrote:"for a few days, I was beside myself with joyous excitement".

– perturbation from the Sun non-sphericity, i.e. by its quadrupole moment:

$$\dot{\omega}^{J2\odot}_{Mercury} = -\frac{3\pi R_\odot{}^2 \, J_{2\odot}}{a^2(1-e^2)^2 \, T} = 1.275 \; 10^5 \cdot J_{2\odot} \; \text{sec/century} \qquad (6.35)$$

The dimensionless parameter $J_{2\odot}$ is defined in Eq. 1.18, but can also be expressed, for a spheroid like the Sun, as $J_{2\odot} = \frac{I_c - I_A}{M_\odot R_\odot^2}$ with I_C and I_A the two principal inertia moments of the Sun. Therefore, a $J_{2\odot}$ value of the order of 10^{-5} or more could give a substantial contribution to the excess precession, and thus invalidate the "perfect match" between theory and measurement. This objection was moved by R. Dicke, who had just proposed his theory alternative to GR. He actually carried out a measurement (Dicke and Goldenberg 1967), with a result, $J_{2\odot} = (2.47 \pm 0.23) \, 10^{-5}$, that seemed to support this interpretation.

This was sufficient motivation to work out the derivation with the PPN metric:

$$\dot{\omega} = \frac{6\pi G M_\odot}{a(1-e^2)c^2 T} \cdot \left(\frac{2+2\gamma-\beta}{3} + \frac{\mu}{M} \cdot \frac{2\alpha_1 - \alpha_2 + \alpha_3 + 2\zeta_2}{6} \right)$$
$$+ \frac{3\pi R_\odot{}^2}{T a^2 (1-e^2)^2} J_{2\odot} \qquad (6.36)$$

μ and M are the reduced mass and the total mass of the Sun-Mercury system.

The first factor, multiplying the parenthesis, is the GR result. Inside the parenthesis we have two terms: the first is the one that allows to evaluate β. Its value is unity for GR. The second term in Eq. 6.36 depends on the parameters α_i and ζ_2. However, this term is weighted by the ratio μ/M between reduced and total mass of the system. For Mercury this is

$$\frac{\mu}{M} \sim \frac{M_{Mercury}}{M_\odot} \sim 2 \cdot 10^{-7}$$

Therefore, its contribution to the precession is well below the experimental uncertitude. The last term is due to the Sun quadrupole moment, that had raised warm expectations among the supporters of the Brans–Dicke theory. However, many measurements of the solar $J_{2\odot}$ were made after Dicke's (see Sect. 1.6), and improved data have dispelled these allegations, pushing the corresponding β value toward the GR limit. Recent determinations based on helioseismology, mostly from the space missions SOHO[15] and RHESSI[16] give

$$J_{2\odot} = (2.2 \pm 0.2) \cdot 10^{-7}$$

---

[15] Solar and Heliospheric Observatory, NASA. Orbiting the L1 Lagrange point; 1995–ongoing.
[16] Reuven Ramaty High Energy Solar Spectroscopic Imager, NASA. Geocentric 2002–2018.

way too low to produce any appreciable contribution to $\dot{\omega}$. An interesting historical review of the quest for $J_{2\odot}$ and its implications in relativity can be found in Rozelot and Damiani (2011).

Once all the classical effects have been accounted for, a bound for the parameter β is obtained. A first bound was (Shapiro 1976):

$$\frac{1}{3}(2 + 2\gamma - \beta) = 1.003 \pm 0.005 \quad \text{- Mercury perihelion, 1966--1976}$$

A significant boost in the accuracy of the measurements of the perihelion advance was obtained thanks to the data of the mission MESSENGER.[17] In 2011, the first artificial satellite was set in orbit around Mercury. It orbited for four years, ending its life by a controlled crash of the spacecraft on the Mercury surface. The range and Doppler (i.e. range rate) data of MESSENGER improved our knowledge of Mercury's orbit around the Sun. Although the most striking relativistic feature, the perihelion precession, only depends on the combination $(2 + 2\gamma - \beta)$, the detailed dynamics of the planet has more complex dependence (see Will 2018) that allow to separately pinpoint the two parameters. Analysis of the MESSENGER data yielded these bounds on γ and β (Verma et al. 2014):

$$\gamma - 1 = (-0.3 \pm 2.5) \times 10^{-5}$$
$$\beta - 1 = (\ 0.2 \pm 2.5) \times 10^{-5} \qquad \text{- MESSENGER, 2014}$$

The parameter β is also sensitive to the Nordtvedt effect, that we will discuss in the next section. Using MESSENGER data and assuming the existence of the Nordtvedt effect,[18] the limit on $|\beta - 1|$ derived, was set to

$$\beta - 1 = (-1.6 \pm 1.8) \times 10^{-5}$$

The MESSENGER data also yielded an improved estimate for the solar quadrupole moment (Genova et al. 2018)

$$J_{2\odot} = (2.246 \pm 0.022) \times 10^{-7} \qquad \text{- MESSENGER, 2008--2015}$$

consistent with the results from helioseismology.

---

[17] MErcury Surface, Space ENvironment, GEochemistry, and Ranging, NASA 2004–2015.

[18] They assumed spatial isotropy so that the Nordtvdet parameter $\eta \simeq 4\beta - \gamma - 3$ (see next section).

6.6.3   Bounds for Other PPN Parameters

To obtain experimental bounds for the other PPN parameters we need to consider other aspects of metric theories. To reveal a possible violation of the strong equivalence principle, (SEP), we apply the PPN formalism to self-gravitating systems, i.e. systems in which the gravitational energy of the body itself is not negligible.

We recall that EEP already incorporates, besides the concept of universality of free fall (UFF), the principles of Lorentz local invariance (LLI—non-existence of preferred reference frames) and of Lorentz position invariance (LPI—invariance of position in the space-time of non-gravitational experiments). The SEP expands the LPI and LLI concepts to experiments with a strong self-gravitating contribution. General Relativity incorporates SEP, while other theories can violate it.

Consider a SEP violation, where gravitational energy falls with different acceleration than other energies: a body of mass m and gravitational self-energy E_g, in a gravitational potential Φ, falls with acceleration

$$a = -\left(1 - \eta \frac{E_g}{m}\right)\nabla\Phi$$

The parameter of violation η quantifies the difference between the inertial and gravitational mass of the test body

$$\frac{m_g}{m_i} = 1 + \eta \frac{E_g}{m_i c^2}$$

and can be computed in the PPN framework:

$$\eta = \left(4\beta - \gamma - 3 - \frac{10}{3}\xi - \alpha_1 - \frac{2}{3}\alpha_2 - \frac{2}{3}\zeta_1 - \frac{1}{3}\zeta_2\right) \tag{6.37}$$

To observe such violations, we need systems with a large ratio E_g/m

$$\frac{E_g}{m} = \frac{1}{\int \rho(\vec{r})d^3r}\left(\int\int G\frac{\rho(\vec{r})\rho(\vec{r}')}{|\vec{r} - \vec{r}'|}d^3r'd^3r\right) \tag{6.38}$$

where ρ is the body mass density. For laboratory systems, the values of the gravitational self-energy are of the order $E_g/mc^2 \sim 10^{-27}$. To have larger values of this ratio, we should consider systems such as the Moon ($\sim 2 \cdot 10^{-11}$), the Earth ($\sim 4.6 \cdot 10^{-10}$), or even better, the Sun ($\sim 10^{-5}$). If SEP is violated, the difference in self-gravitational energy content of the Earth and the Moon will force the two bodies to fall differently towards the Sun. As a consequence, we would observe a polarization of the Earth-Moon orbit. This is known as the *Nordtvedt effect* (Nordtvedt 1968). A proper calculation of the deformation in the Earth-Moon orbit yields the following magnitude for the effect:

$$\delta r = 13.1\,\eta\,cos(\omega_{\text{Å}} - \omega_{\odot})t$$

where δr is measured in metres $\omega_{\text{Å}}$ e $\omega_{\odot}$ are the angular frequencies for the Moon and the Sun revolutions, assumed as circular and observed from a reference frame in Earth. A violation of SEP would displace the lunar orbit along the Earth–Sun line by an amount δr, producing a range signature having a 29.53 day synodic period (different from the lunar orbit period of 27.32 days).

Measurements of Lunar Laser Ranging have been performed regularly since 1968, and discussed in Sect. 1.9. Their accuracy is better than 1 cm, that is, 50 ps for the round-trip travel time of an e.m. signal from the Earth to the Moon.

A comprehensive model took into account the classical orbit, the perturbation effects due to the classical field from the Sun and from the other planets, the tidal interactions, lunar librations, atmospheric effects, etc. The analysis of the residuals yielded the following limit (Hofmann et al. 2010):

$$|\eta| = (-0.6 \pm 5.2) \times 10^{-4} \qquad \text{- LLR, 1970–2009}$$

The study of the orbit of Mercury is the other method well-suited to test SEP via the Nordtvdet effect. A SEP violation would cause an indirect perturbation on Mercury's orbit. It has been noted that the current planetary ephemerides studies are based on the hypothesis that the gravitational and inertial masses are equal to compute the Solar Barycentric System (SSB) and a violation of the SEP may lead to an intrinsic mismodeling of the SSB position.

The data confirm the validity of the strong equivalence principle with a significantly refined uncertainty of the Nordtvedt parameter (Genova et al. 2018)

$$\eta = (-6.6 \pm 7.2) \times 10^{-5} \qquad \text{- MESSENGER, 2008–2015.}$$

6.6.4 The Inconstancy of the Universal Constants

Violations of the invariance for LLI and LPI are related, in the PPN framework, to the values of the parameters

$$\alpha_1, \alpha_2, \alpha_3, \xi$$

We recall that LPI states that identical experiments must have identical results regardless of where and when they are performed: i.e. they must not depend on their location x^{μ} in space-time. Therefore, non-zero values of these parameters would imply a local variation of G in time and/or its dependence on the position in space. The idea that G, as well as other universal constants, might not be really constants is rather old. Indeed, it goes back to the 1930s, when Paul Dirac (1938) presented his *Large Number Hypothesis—LNH*. Probably he was influenced by some numeric coincidences, such as the ratio between the Universe radius cT_U to the classical radius of the electron $r_e = \frac{e^2}{4\pi\epsilon_o m_e c^2}$:

$$\frac{cT_U}{r_e} \sim 10^{40}$$

being so similar to the ratio of the electrical to the gravitational forces between a proton and an electron:

$$\frac{e^2}{4\pi\epsilon_o Gm_p m_e} \sim 10^{40}$$

In the LNH Dirac argued that five dimensionless combinations of universal constants (pure numbers) should be considered as variable parameters characterising the state of the universe. One of the consequences of this phenomenological approach is that the gravitational coupling constant should change as the universe evolves, as it comes out by supposing that the numeric coincidence cited above hides a fundamental physics principle.[19] It is then reasonable to speculate that the time dependence of G be directly connected to the Universe expansion rate given by the Hubble parameter $H_0 = 70 \pm 5 \, \text{km}/(\text{s Mpc})$,[20] i.e. $\dot{G}/G \propto H_0$. We must also note that, if fundamental constants are time dependent, the present mystery of the so-called dark energy effect should have to do with it, because dark energy dominates the present expansion of the universe.

The Dirac's argument opened a research field with a significant number of experiments of different types, such as the search of anomalies in the Earth tides, anomalous contributions to the planets and lunar orbits, self-accelerations of pulsars or variations of the Earth's spin. These measurements, from which upper bounds of the time derivative of G (and thus on the PPN parameters), can be roughly classified in three categories:

- Experiments measuring the variation of G on time scales comparable with the age of the universe, $t \sim 10^{10}$ years: in this class, we include studies related to the primordial nucleosynthesis of elements and the anisotropies of the background cosmic radiation. They yield the upper bound:

$$\frac{\dot{G}}{G} \lesssim 10^{-13} \quad yr^{-1}$$

- Observations over times of the order of $10^9 - 10^{10}$ years, not necessarily including the early universe. In this category we classify the bounds obtained by palaeontological or stellar astrophysics observations, with the latter including also the observations of helioseismology, pulsars in binary systems and globular clusters. The bounds obtained are

$$\frac{\dot{G}}{G} \sim 10^{-11} - 10^{-12} \quad yr^{-1}$$

---

[19] Dirac focused on Newton's constant G: since then, possible time dependence of light velocity, Planck constant and fine structure constant have also been considered.

[20] Several estimates of H_0 are available today, and there is *tension* among them. We chose an uncertitude large enough to contain all determinations.

- The third class consists of experiments performed on the time scale of human life, i.e., of the order of decades. These include possible variations in the planetary orbits, the revolution motion of a binary system containing a pulsar and/or oscillations of white dwarfs. The bounds obtained are:

$$\frac{\dot{G}}{G} \sim 10^{-10} - 10^{-11} \quad yr^{-1}$$

Although the results of the first category place the most stringent bounds on the long term variations of G, it must be noted that they depend on the cosmology model used and on the relative model for the variation of G.

The best result to date is derived from a global fit performed on orbits of planets and satellites (Verma et al. 2014):

$$\frac{\dot{G}}{G} = (-5 \pm 3) \cdot 10^{-14} \, yr^{-1} \qquad \text{- Solar System Orbits}$$

However, like in all observations in the Solar System, the measured quantity is actually the product $GM_\odot$:

$$\frac{1}{GM_\odot} \frac{d}{dt} GM_\odot = \frac{\dot{G}}{G} + \frac{\dot{M}_\odot}{M_\odot} \leq (-6 \pm 4) \, 10^{-14} yr^{-1}$$

thus, that limit depends then on the estimate of Sun's mass loss $\frac{\dot{M}_\odot}{M_\odot} \sim (-1.12 \pm 0.25) \, 10^{-13} yr^{-1}$ due to solar radiance and wind.

Actually, the change in time of any fundamental constant would be an evidence of LPI violation. There is a wide variety of measurements on this subject (Uzan 2003). We just mention here the bound set on the fine structure constant $\dot{\alpha}_{em}/\alpha_{em} \leq 2 \cdot 10^{-8}$ derived by the geophysical surveys on the abundances of U isotopes measured in a natural nuclear reactor, active some 10^6 years ago, located in Oklo, Gabon.

6.6.5 Bounds of the Parameters Related to Conservation Laws

We finish this rather long section on the experimental bounds of the PPN parameters, reporting some results about the tests of conservation laws. These tests involve the parameters

$$\alpha_3, \zeta_1, \zeta_2, \zeta_3, \zeta_4.$$

A limit on ζ_3 can be derived from the Lunar Laser Ranging tracking of the Moon orbit. We discussed in Chap. 1 the experiment of Bartlett-Van Buren, that probed the difference between active and passive mass: it set a bound of $m_A/m_p - 1 \leq 4 \times 10^{-12}$ (Bartlett and Van Buren 1986). If we assume this violation to be associated with the electrostatic energy contribution of the nucleus E_{el}, we can translate this limit on a limit for ζ_3 :

$$\zeta_3 = 2\frac{m_A - m_P}{E_{el}} \leq 1 \cdot 10^{-8}$$

A significant limit on the $|\zeta_2 + \alpha_3|$ is deduced from the observations of binary systems of neutron stars such as PSR1913+16, that we shall describe in detail in Chap. 12. The violation of the conservation of linear momentum in a binary system s would cause a self-acceleration of the binary system's centre of mass

$$\vec{a}_{cm} = -\frac{1}{2}(\zeta_2 + \alpha_3)\frac{m_1 - m_2}{a^3}\frac{m_1 m_2}{m_1 + m_2}\frac{e}{(1 - e^2)^{3/2}}\vec{n} \tag{6.39}$$

where m_1 and m_2 are the star masses, a the semi-major axis of the orbit, and $\vec{n}$ is a unitary vector oriented from the center of mass to the point of periastron of m_1. From the observed data of the pulsar PSR1913+16 it has been deduced one of the best upper bounds for the quantities

$$|\alpha_3 + \zeta_2| = 4 \cdot 10^{-5}$$

Finally, we recall a consideration by Will (2018): in any reasonable theory there must be a relation between the quantities $\rho v^2, \rho\Pi, p$ (something like Bernoulli's theorem in classical fluid dynamics), that leads to a linear relation among the PPN parameters: $6\zeta_4 = 3\alpha_3 + 2\zeta_1 - 3\zeta_3$. This allows us to reduce the number of parameters, disregarding ζ_4.

6.7   Gravitomagnetism and PPN Parameters

The gravitomagnetic effect can also be studied using the PPN metric. Consider the PPN predictions for the two precessions studied in Sects. 5.6 and 5.8:

- the *geodesic precession* depends on the curvature of space-time and therefore involves just the parameter γ. In PPN, the GR prediction is modified as follows:

$$\vec{\Omega}_{\text{Geo}} = \frac{1}{2}(1 + 2\gamma)\vec{v} \times \vec{\nabla}\Phi \tag{6.40}$$

- the *Lense-Thirring precession*, is related to the dragging of inertial frames caused by the spin-spin coupling with the central body. In the PPN farmework, it depends on the parameters γ and α_1:

$$\vec{\Omega}_{\text{LT}} = -\frac{1}{2}\left(1 + \gamma + \frac{\alpha_1}{4}\right)\frac{G}{r^3 c^2}(\vec{L} - 3(\vec{n} \cdot \vec{L})\vec{n}) \tag{6.41}$$

At present, measurements of these precessions have been performed with accuracy approaching 1%, as described in Sect. 5.8, and therefore cannot compete with other determinations of $\gamma - 1$ and give weak indications for α_1. We note that the precession is a measurable effect in some binary neutron star systems containing a pulsar. Radio telescopes are detecting an increasing number of such systems, and there are prospects for improving limits on PPN parameters also through the observation of gravitomagnetic effects.

Chapter 12 deals with these amazing stellar "laboratories".

6.8   Conclusions

We have briefly reviewed over fifty years of measurements and observations of the deflection and delay of e.m. radiation, precession, geophysical and astronomical effects. They have all been repeatedly performed, with ever-increasing accuracy, and they are all leading to the conclusion that:

The values of the parameters of the PPN formalism are consistent with those expected by General Relativity.

In Table 6.1 we report a summary of the situation as of 2020. Nevertheless, experimental efforts to find a flaw in the predictions of Einstein's theory continue, with the hope of opening the way to a description of the universe that can address the fundamental questions, such as the nature of dark matter and dark energy, that still remain unanswered.

In the foreseeable future, improvements in PPN testing are expected, on the weak field side, from new space projects, with improved sensitivity for tracking and gravity measurements.

Many missions have been proposed, approved or launched, such as

Table 6.1 Upper bounds, measured in the *weak field regime*, i.e. in the Solar System, for the PPN parameters (as of 2020). Since the values of all these parameters are consistent with zero, we report the experimental error, as a bound to each value. See Imperi et al. (2018), Will (2018) and references therein

Parameter	σ	Solar system observation
$\gamma - 1$	$2.3 \cdot 10^{-5}$	Time delay—Cassini radio tracking
$\beta - 1$	$3.9 \cdot 10^{-5}$	Mercury precession—MESSENGER Tracking
ξ	10^{-3}	Terrestrial tides—Gravimeters
α_1	$6 \cdot 10^{-6}$	Solar System precession—LLR
α_2	$2 \cdot 10^{-6}$	Precession of solar spin—Ecliptic alignment
α_3	$2 \cdot 10^{-7}$	Perihelion shift—Lunar Laser Ranging
ζ_3	10^{-8}	Newton's 3rd law—Bartlett and Van Buren (Chap. 1)
η	$3 \cdot 10^{-4}$	Nordtvedt Effect—Moon Orbit—LLR
$\dot{G}/G$	$4 \cdot 10^{-14}$	Mars ephemeris—MRO ranging

* GAIA (*Global Astrometric Interferometer for Astrophysics*) launched by ESA in 2013. It is the successor of Hipparcos, and one of its scientific targets is to measure light deflection with an accuracy of 10^{-6};
* BepiColombo, named after the Italian scientist Giuseppe (Bepi) Colombo, is the successor of MESSENGER: it was launched in 2018 by ESA and JAXA and is presently on its way toward Mercury, and starting from 2026, it will orbit the planet. This mission, that employs improved radio tracking combined and accelerometer measurements will probe the gravitational field near Mercury. It is expected to substantially improve the present limits on the parameters: γ, β, α_1, α_2, $J_{2\odot}$ and $\dot{G}/G$;
* MRO—Mars Reconnaissance Orbiter, launched by NASA in 2005 and orbiting Mars since 2006;
* Juno, launched by NASA in 2011, still orbiting Jupiter;
* JUICE (Jupiter Icy Moons Explorer) planned for launch in 2022 and arrival at Jupiter in 2029, it will explore the Jovian system. JUICE will carry on detailed observations through repeated flybys to the giant gaseous planet Jupiter and of its largest moons Callisto and Europa and finally orbiting Ganymede;
* Europa Clipper—NASA's complementary mission to JUICE, will orbit Jupiter and study the gravity of Europa with repeated flybys;
* VERITAS (Venus Emissivity, Radio Science, InSAR, Topography, and Spectroscopy) is a proposed NASA mission to Venus.

These missions have probed, or will probe, effects that we discussed here, like Mercury's perihelion precession and Shapiro delay, as well as others, like planetary perturbations on the ranging, Lense-Thirring effect, Compton wavelength of the graviton and non-gravitational forces (De Marchi and Cascioli 2020). It is worth noting that all these missions, as well as CASSINI and MESSENGER mentioned above, carry numerous complex instruments for a variety of different measurements. The relativistic measurements, typically based on the analysis of tracking and timing data, are a byproduct of much more expensive and complex projects.

The strongest input for setting bounds (or finding the values, should they turn out to be different from zero) on the PPN parameters, is coming in recent years from pulsar observations and timing and from gravitational wave detection. In both cases, observations are carried out in object where the gravity is much stronger: $U \sim 10^{-3} - 10^{-1}$ near pulsars, as opposed to $U \sim 10^{-5}$ in the Solar System. Therefore, the expansion in powers of U and other potentials can easily break down, and a full general-relativistic analysis might be required. The PPN parameters as coefficients of such expansion might therefore lose meaning. For this reason, the PPN parameters as measured in strong field are labeled with a "hat" ˆ in the literature. In the Chap. 12 we will overview the limits set in the strong field regime.

References

Bartlett, D.F., Van Buren, D.: Equivalence of active and passive gravitational mass using the moon. Phys. Rev. Lett. **57** (1986)

Bertotti, B., Iess, L., Tortora, P.: A test of general relativity using radio links with the cassini spacecraft. Nature **425**, 374–376 (2003)

Brans, C., Dicke, R.H.: Mach's principle and a relativistic theory of gravitation. Phys. Rev. **124**, 925 (1961)

De Marchi, F., Cascioli, G.: Testing general relativity in the solar system: present and future perspectives. Class. Quantum Grav. **37**, 095007 (2020)

Dicke, R.H., Goldenberg, H.: Mark solar oblateness and general relativity. Phys. Rev Lett. **18**, 313 (1967)

Dirac, P.A.M.: A new basis for cosmology. Proc. R. Soc. Lond. Ser. A **165**, 199–208 (1938)

Fomalont, E.B., Sramek, R.A.: Measurements of the solar gravitational deflection of radio waves in agreement with general relativity. Phys. Rev. Lett. **36**, 1475 (1976)

Froeschlé, M., Mignard, F., Arenou, F.: Determination of the PPN parameter gamma with the HIPPARCOS data. In: ESA Symposium Hipparcos - Venice 97, May 1997, Venice, Italy, pp. 49–52

Genova, A., Mazarico, E., Goossens, S., Lemoine, F.G., Neumann, G.A., David, E., Smith, D.E., Zuber, M.T.: Solar system expansion and strong equivalence principle as seen by the NASA MESSENGER mission. Nat. Commun. **9**, 289 (2018)

Hellings, R.W.: Relativistic effects in astronomical timing measurements. Astron. J. **91**, 650 (1986)

Hofmann, F., Müller, J., Biskupek, L.: Lunar laser ranging test of the Nordtvedt parameter and a possible variation in the gravitational constant. Astron. Astrophys. **522**, L5 (2010)

Imperi, L., Iess, L., Mariani, M.J.: An analysis of the geodesy and relativity experiments of Bepi-Colombo. Icarus **301**, 9–25 (2018)

Jennison, R.C.: A phase sensitive interferometer technique for the measurement of the Fourier transforms of spatial brightness distributions of small angular extent. MNRAS **118**, 276–284 (1958)

Lambert, S.B., Le Poncin-Lafitte, C.: Improved determination of γ by VLBI. A&A **529**, A70 (2011)

Landau, L., Liftschitz, E.: Classical Theory of Fields. Addison-Wesley, Reading (1951)

Ni, W.T.: A new theory of gravity. Phys. Rev. D **7**, 2880 (1973)

Nordtvedt, K.: Testing relativity with laser ranging to the Moon. Phys. Rev. **170**, 1186 (1968)

Poisson, E., Will, C.M.: Gravity. Cambridge University Press, Cambridge (2014)

Reasenberg, R.D., et al.: (1979) Viking relativity experiment - verification of signal retardation by solar gravity. App. J. **34**, L219–L221 (1979)

Rozelot, J.P., Damiani, C.: History of solar oblateness measurements and interpretation. Eur. Phys. J. H. **36**, 407–436 (2011)

Shapiro, I.I., et al.: Fourth test of general relativity: preliminary results. Phys. Rev. Lett. **20**, 1265 (1968)

Shapiro, I.I., Counselman, C.C., III., King, R.W.: Verification of the principle of equivalence for massive bodies. Phys. Rev. Lett. **36**, 555 (1976)

Shapiro, I.I., et al.: The viking relativity experiment. J. Geophys. Rev. **82**, 4329–4334 (1977)

Titov, O., et al.: Testing general relativity with geodetic VLBI. A&A **618**, A8 (2018)

Uzan, J.P.: The fundamental constants and their variation: observational and theoretical status. Rev. Modern Phys. **75**, 403 (2003)

Vecchiato, A.: Astrometric tests of general relativity in the solar system: mathematical and computational scenarios. J. Phys.: Conf. Ser. **490**, 012241 (2014)

Verma, A.K., Fienga, A., Laskar, J., Manche, H., Gastineau, M.: Use of MESSENGER radioscience data to improve planetary ephemeris and to test general relativity. A&A **561**, A115 (2014). 1306.5569

Weinberg, S.: Gravitation and Cosmology: Principles and Applications of the General Theory of Relativity. Wiley, New York (1972)

Will, C.M.: The theoretical tools of experimental gravitation, pp. 1–110. In: Bertotti, B. (ed.) Proceedings of the International School Enrico Fermi "Experimental Gravitation". Academic, Cambridge (1974)

Will, C.M.: The confrontation between general relativity and experiment. Living Rev. Relativ. **17**, 4 (2014)

Will, C.M.: Theory and Experiment in Gravitational Physics. Cambridge University Press, Cambridge (2018)

Further Reading

Wambsganss, J.: Gravitational Lensing in Astronomy. Living Reviews in Relativity **1**, 12 (1998)

Will, C.M., Nordtvedt, K. Jr.: Conservation laws and preferred frames in relativistic gravity. I. Preferred-frame theories and an extended PPN formalism. Ap. J. **177**, 774 (1972)

7.1 Introduction: History and Perspectives

The first mention of the remarkable concept of gravitational wave dates back to the end of XIX century, when Heaviside published his book *Electromagnetic Theory* (Heaviside 1893) and assumed that gravitational radiation should exist, based on an analogy with electromagnetic propagation.

In 1905 Henry Poincaré, in his long paper *La Dynamique de l'Electron* for the proceedings of the *Mathematical Circle of Palermo* (Poincaré 1905), wrote that all forces have to be subject to the Lorentz transformations, as in the electromagnetic case. If this condition is also imposed on gravitation, it must follow that the gravitational interaction is also propagated at the speed of propagation of light. It should be emphasized that this work was written before the famous A. Einstein's paper on Special Relativity.

The explicit derivation of a wave propagation of the gravitational interaction is presented by A. Einstein, on June 22nd, 1916 in Berlin, at the meeting of the Königlic Preussichen Akademie der Wissenchaften (Einstein 1916). In that occasion Einstein proved that the equations of the gravitational field can be linearized under the condition of small perturbations $|h_{\mu\nu}| \ll 1$ of the Minkowski metric $\eta_{\mu\nu}$.

We recall once more (see previous chapters) that the elementary space-time interval ds^2 is related to the metric tensor by the relation:

$$ds^2 = g_{\mu\nu}dx^\mu dx^\nu \qquad (7.1)$$

The original version of this chapter was revised: Chapter have been updated with the correction. The correction to this chapter can be found at https://doi.org/10.1007/978-3-030-95596-0_15

F. Ricci and M. Bassan, *Experimental Gravitation*, Lecture Notes in Physics 998, https://doi.org/10.1007/978-3-030-95596-0_7

with

$$g_{\mu\nu} = \eta_{\mu\nu} + h_{\mu\nu} \qquad \text{and} \qquad |h_{\mu\nu}| \ll 1 \tag{7.2}$$

In this approximation, the field equations are linearized and the quantities $h_{\mu\nu}$ are derived by writing the analogous solution of the delayed potentials of electromagnetism (see Chap. 5)

A year later Einstein wrote a second paper (Einstein 1918). This is a complete review of topics such as the absorption of incident waves on mechanical systems. There we can find the expression (with a computation error, edited only later on) of the power radiated in gravitational waves by a moving system and the relative luminosity L_G:

$$L_G = \frac{G}{5c^5} \sum_{kh} \left(\frac{d^3}{dt^3} D_{kh} \right)^2 \tag{7.3}$$

where G and c are the usual universal constants and D_{kh} (with $k, h = 1, 2, 3$) is the quadrupole moment of the emitter:

$$D_{kh} = \int_V \rho \left(x_k x_h - \frac{1}{3} \delta_{kh} x^2 \right) dV \tag{7.4}$$

The issue of whether GW are real and detectable, and not a mere mathematical artifact, had been debated since Einstein's 1916 paper for about 40 years.[1] It was tackled in the 1957 Chapel Hill conference "On the Role of Gravitation in Physics", when F. Pirani proved that, in the presence of a gravitational wave, a set of freely-falling particles would experience actual motions with respect to one another. Thus, gravitational waves must be real and measurable. Pirani (1956) considered 2 test masses linked by a spring and H. Bondi suggested (1959) to also insert a dashpot, and extract energy from the GW. While interferometers eventually substituted the spring with a laser (then, yet to be invented) beam, it was J. Weber of the University of Maryland who began the extraordinary effort of bringing the problem of gravitational waves in the field of experimental physics (Weber 1960) (Fig. 7.1).

Then, the research in the field of experimental gravitation experiments intensified as well as, in parallel, the theoretical activities. During the 1960s R. Penrose publishes his work on the spinorial theory of Relativity and General (Penrose 1960; Penrose and Rindler 1984) and later, the scalar-tensor theory of Brans–Dicke (1961) was formulated as an alternative to Einstein's theory. As mentioned in Chap. 6, many other theories have since been proposed: their benchmark is the observable post-Newtonian effects, translated in terms of *PPN* parameter values.

This theoretical framework is also used to analyze the observations conducted on binary systems such as that in which the post-Newtonian effects are significantly

---

[1] Einstein himself, in a 1937 paper (flawed, and not accepted by Physical Review Letters) had lifted doubts on the existence of GW.

Fig. 7.1 Joseph Weber and one of his resonant gravitational antennas, equipped in the central section with piezoelectric transducers. *Credits* Special Collections and University Archives, University of Maryland Libraries, https://hdl.handle.net/1903.1/32732

larger. The binary systems turn out to be perfect laboratories to test PN gravitational effects. Indeed, the first ever detection of the effects of gravitational wave emission (the change in rotational period due to energy loss) was observed in the first and most famous of this binaries, PSR 1913+16 discovered by Taylor and Hulse (1975).

This observation, although indirect, of the effect of gravitational waves gave new momentum to the activity toward direct detection. The cryogenic resonant antennas were developed in the years between 1975 and 2005: Chap. 8 describes these detectors. In the first decade of the new century, interferometers with arms of km length began to operate, both in the USA and in Italy. In the following decade the construction of the second generation interferometers, with advanced configuration, was finalized and on September 14th, 2015 the two aLIGO interferometers have detected the first signal of a gravitational wave signal, emitted by the coalescence of two black holes. The announcement of the discovery was given in the USA and in Italy by LIGO and Virgo collaborations: it was February 11th, 2016, one hundred years after the publication of the article by A. Einstein on General Relativity. On the same day, a landmark scientific paper was published on Physical Review Letters (Abbott et al. 2016), containing the results and the details of the data analysis jointly carried out by the LIGO and Virgo collaborations over five frentic months. For this long awaited discovery, the Nobel prize for physics was assigned in 2017 to three of the leading figures of the LIGO project: B. C. Barish, K. S. Thorne and R. Weiss. Then, in August 2017 the three instruments sent an "alarm" to the astronomical community: they had detected a signal from the coalescence of two Neutron Stars (NS), located in an area of ∼30 square degrees in the southern sky at a distance from the Earth ranging from 85 to 160 millions of light years. The hunt of the optical counterpart started immediately afterwards, carried on by an impressive number of telescopes covering the entire electromagnetic spectrum. The source is identified as a Kilonova (Metzger 2010) in the galaxy NG4393, marking the birth of multi-messenger astronomy.

At the moment of writing, while the the two LIGOs in the US and Virgo in Europe continue observations, detecting gravitational signals at a rate of about one per week, a fourth interferometer of kilometer scale is about to join the network: KAGRA, in Japan, will have new advanced features like underground location and cryogenic mirrors. A fifth interferometer in India will also be ready in 2024–25. These systems will constitute a planetary network aimed at writing a new chapters of Gravitational Astronomy and Gravitodynamics. Chapter 9 analyzes in detail these amazing instruments, capable of measuring strain as tiny as $h \sim 10^{-21}$.

In the early 2030s, the spaceborne detector LISA (Laser Interferometer Space Antenna) will scan the sky, observing waves in the mHz frequency band that carry a huge potential of astrophysical information, complementary to that in the 20–1000 Hz band observed by Earth-based instruments.

Many key technologies for LISA have been successfully verified by a dedicated space mission of the European Space Agency: LISA-Pathfinder has flown in 2016–2017 exceeding the expected performance in terms of noise immunity and readout accuracy. Details are reported in Chap. 11.

Finally, the next decade will also witness the birth of gravitational wave observation in the nHz range through Pulsar Timing Arrays, thanks to the coordinated effort of IPTA, a world-wide network of radio telescopes. An overview of these observatories and related techniques can be found in Chap. 12.

7.2   Gravitational Wave Properties

We can start our discussion from Eq. 5.20, where we have shown how the modified metric perturbation $\tilde{h}_{\mu\nu}$ satisfies the D'Alembert equation[2]

$$\frac{\partial^2 \tilde{h}_{\mu\nu}}{\partial x^\lambda \partial x_\lambda} = \frac{16\pi G}{c^4} T_{\mu\nu} \tag{7.5}$$

with

$$g_{\mu\nu} = \tilde{h}_{\mu\nu} - \frac{1}{2}\eta_{\mu\nu}\tilde{h} + \eta_{\mu\nu} \tag{7.6}$$

Lorentz's gauge condition, $\frac{\partial}{\partial x^\mu}\tilde{h}^\mu_\nu = 0$ that we used to obtain Eq. 7.5, does not yet univocally bind the choice of the reference system. Indeed, we must consider that any change of coordinates

$x'^\alpha = x^\alpha + \epsilon^\alpha(x)$ subject to the condition:

$$\frac{\partial^2 \epsilon_\mu}{\partial x^\lambda \partial x_\lambda} = 0 \tag{7.7}$$

---

[2] Note that, with our choice of the signature $(+, -, -, -)$, $\partial x^\lambda \partial x_\lambda$ has opposite signs with respect to the usual definition of D'Alembertian operator $\Box = \partial x^j \partial x_j - 1/c^2 \partial_t \partial_t$.

will transform the $\tilde{h}_{\mu\nu}$ field according to:

$$\tilde{h}_{\mu\nu} \to \tilde{h}_{\mu\nu} - \frac{\partial \epsilon_\mu}{\partial x^\nu} - \frac{\partial \epsilon_\nu}{\partial x^\mu} + \eta_{\mu\nu} \frac{\partial \epsilon^\lambda}{\partial x^\lambda} \tag{7.8}$$

In conclusion, the 16 components of the tensor $h_{\mu\nu}$ are reduced to 10 because the tensor is symmetric. Moreover, the choice of Lorentz's gauge eliminates 4 degrees of freedom, and the choice of reference system (Eq. 7.8) 4 more, so that we conclude that there are only two independent components of $h_{\mu\nu}$.

To study the properties of waves, we consider the problem of propagation in empty space, i.e. where the energy-momentum tensor is null. The field equations (7.5) take the form:

$$\frac{\partial^2 \tilde{h}_{\mu\nu}}{\partial x^\lambda \partial x_\lambda} = 0 \tag{7.9}$$

The simplest solution of this equation is a plane wave, of the form:

$$\tilde{h}_{\mu\nu} = Re\left[A_{\mu\nu} \exp(ik_\alpha x^\alpha)\right] \tag{7.10}$$

Here the analogy with the case of electromagnetism is almost perfect. We easily get:

$$k_\alpha k^\alpha = 0$$

that shows that k^α is a null vector: this is typical of fields with zero mass exchange particles (gravitons) and proves that the wave propagates at the speed of light. Imposing the Lorentz condition $\frac{\partial \tilde{h}_\nu^\mu}{\partial x^\mu} = 0$ to the plane wave solution we get

$$k^\alpha A_{\nu\alpha} = 0$$

We now use the freedom offered by the choice of reference frame to set the trace of the perturbation to zero:

$$\tilde{h}_\mu^\mu = \tilde{h} = 0$$

Note that if $\tilde{h}_{\mu\nu}$ is zero-trace, or *traceless*, it coincides with $h_{\mu\nu}$, so that we can, from now on, drop the tilde. We now use the remaining three conditions to set:

$$h_{0k} = 0$$

One of the Lorentz conditions now reads

$$\frac{\partial h_{00}}{\partial x_0} + \frac{\partial h_{0k}}{\partial x_k} = 0$$

and, being the three h_{0k} set to zero, this implies that h_{00} is constant in time: it corresponds to the static (Newtonian) background gravitational field, and can be set to zero for the wave solution: now all $h_{0\mu} = 0$, and the Lorentz conditions reduce to $\frac{\partial h_{ij}}{\partial x_i} = 0$. This last condition, applied to Eq. 7.10, yields $k^i A_{ij} = 0$, showing that gravitational waves are *transverse*, i.e. have no component in the direction of propagation.[3]

Summarizing, the above constraints allow us to have:

$$h = 0; \quad h_{\mu\nu} = \tilde{h}_{\mu\nu}; \quad h_{\mu0} = 0; \quad h^k_k = 0 \quad \frac{\partial h_{ij}}{\partial x_i} = 0$$

In this gauge $h_{\mu\nu}$ is a traceless tensor in which only two spatial components survive. This is called the *TT gauge* (Transverse and Traceless); in the case of a plane wave propagating along the x direction, the most general solution can be cast in the form:

$$h^{TT}{}_{\mu\nu} = \begin{pmatrix} 0 & 0 & 0 & 0 \\ 0 & 0 & 0 & 0 \\ 0 & 0 & h_{yy} & h_{yz} \\ 0 & 0 & h_{yz} & -h_{yy} \end{pmatrix} \tag{7.11}$$

The generic plane wave can be expressed as a linear superposition of just two components:

$$h_{yy} = h_+ = Re\left[A_+ \exp(i(\omega t - x/c))\right] \tag{7.12}$$

$$h_{yz} = h_\times = Re\left[A_\times \exp(i(\omega t - x/c))\right] \tag{7.13}$$

corresponding to two independent polarization states. The reason of these symbols will be clear shortly. The two states are usually represented through the tensors $e_+, e_\times$:

$$e_+ = \begin{pmatrix} 0 & 0 & 0 & 0 \\ 0 & 0 & 0 & 0 \\ 0 & 0 & 1 & 0 \\ 0 & 0 & 0 & -1 \end{pmatrix} \qquad e_\times = \begin{pmatrix} 0 & 0 & 0 & 0 \\ 0 & 0 & 0 & 0 \\ 0 & 0 & 0 & 1 \\ 0 & 0 & 1 & 0 \end{pmatrix} \tag{7.14}$$

and the two scalar quantities $a_+, a_\times$ so that[4]

---

[3] This is analogous to $\vec{k} \cdot \vec{A} = 0$ stating the transversality of e.m. waves.

[4] The two polarization tensors are sometimes expressed through the matrix product of two unitary vectors $\vec{e}_y, \vec{e}_z$

$$\vec{e}_y = \begin{pmatrix} 0 \\ 0 \\ 1 \\ 0 \end{pmatrix} \quad \vec{e}_z = \begin{pmatrix} 0 \\ 0 \\ 0 \\ 1 \end{pmatrix}$$

thus obtaining an equivalent representation of the wave polarization state

$$e_+ = \vec{e}_y \otimes \vec{e}_y - \vec{e}_z \otimes \vec{e}_z$$

$$e_\times = \vec{e}_y \otimes \vec{e}_z + \vec{e}_z \otimes \vec{e}_y$$

where $\otimes$ indicates the product of the transpose of the first vector times the second vector.

$$A_+ = a_+ e_+ \qquad A_\times = a_\times e_\times \tag{7.15}$$

By taking into account the metric perturbation Eq. 7.11, a space-time interval becomes, in presence of gravitational waves traveling along the x axis:

$$ds^2 = c^2 dt^2 - dx^2 - dy^2[1 - h_+(z,t)] - dz^2[1 + h_+(z,t)]$$
$$+ 2dzdy(1 + h_\times(z,t)) \tag{7.16}$$

To give a pictorial representation of these polarization states we consider the effect of the wave on test masses placed on a plane perpendicular to the direction of propagation and we draw the field lines of the pseudo-force that is exerted on these masses. Consider two freely gravitating test masses A and B: in order to describe how their trajectories in space-time will evolve under the influence of curvature, we need to recall the Eq. 5.9 of geodesic deviation:

$$\frac{\partial^2 \xi^\mu}{\partial s^2} + u^\alpha u^\beta \xi^\gamma R^\mu{}_{\alpha\beta\gamma} = 0 \tag{7.17}$$

To further simplify the problem we set the origin of the reference frame on particle A, so that $\xi^i = x^i_B$, and express the variation of the geodesic line in terms of changes in proper time $d\tau$. Within the usual WFSM approximation $g_{\mu\nu} = \eta_{\mu\nu} + h_{\mu\nu}$ we get from Eq. 7.17

$$\frac{d^2 x^i_B}{d\tau^2} \simeq \frac{1}{2} \frac{\partial^2 h^{i(TT)}_k}{\partial \tau^2} x^k_B \tag{7.18}$$

Spelled out for a wave propagating along the x axis, Eq. 7.18 reads (we now drop the (TT) superscript):

$$\frac{d^2 y_B}{d\tau^2} = \frac{1}{2}\left(\frac{\partial^2 h_+}{\partial \tau^2} y_B + \frac{\partial^2 h_\times}{\partial \tau^2} z_B\right)$$
$$\frac{d^2 z_B}{d\tau^2} = \frac{1}{2}\left(\frac{\partial^2 h_\times}{\partial \tau^2} y_B - \frac{\partial^2 h_+}{\partial \tau^2} z_B\right) \tag{7.19}$$

Thus, the motion of the particle B seen from A appears as due to an acceleration field proportional to the second time derivative of h. Integration of Eq. 7.18 yields:

$$x^i_B(\tau) = x^k_B(0)\left(\delta^i_k + \frac{1}{2}h_k\right) \tag{7.20}$$

If, for example, our test-particles A and B are located along the y axis, with unperturbed distance l_{AB}, and the incoming (along x) wave has polarization h_+, Eq. 7.20 simply means:

$$\frac{\Delta l_{AB}}{l_{AB}} = \frac{h_+}{2}$$

Fig. 7.2 The effect of a
plane gravitational wave,
impinging normally on the
plane of the page, on a set of
placed in a ring. The effects
of polarization + and × are
represented

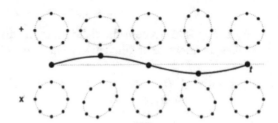

Two test particles along the z axis would move accordingly, but with a π phase difference ($h_{zz} = -h_{yy}$). On the other hand, if the $h_\times$ polarization is involved, it can be seen from Eq. 7.19 that the same effect would be experienced by two test particles positioned at $\pi/4$ with respect to the x, y axes.

This analysis allows us to derive the effect of a gravitational wave on a system of point-like masses distributed along a circle normal to the propagation direction. In Fig. 7.2 we show the field lines for the two polarization states of the wave, with the typical quadrupolar structure of a tidal force.

If we try to extend the wave-particle duality of quantum mechanics to the gravitational interaction, we can hypothesize the existence of the *graviton*, analogous to the photon for the electromagnetic field. Since Einstein's theory predicts that the waves propagate with velocity c (as indicated by the D'Alembert operator of Eq. 7.9), it follows that also the graviton must have zero rest mass. Furthermore, from the polarization states, we can infer another characteristic of the exchange particle, its spin. Although rigorous derivations can be given (Weinberg 1965) to show that the graviton must have spin 2, we'll use two heuristic analogies with the spin 1 of the photon:

(a) the source of gravity is the stress-energy tensor, a rank-2 tensor, just as the source of electromagnetic interaction is a rank-1 tensor, the charge-current density 4-vector;

(b) referring to the simple case of linear polarization: in general, the spin of a particle is related to the symmetry of the polarization of the wave: the number N_s of independent spin states is given by of 2π divided by the rotation angle that restore the original polarization: this angle is π for the photon and $\pi/2$ for the graviton (see Figs. 7.2 and 7.3). This means $N_s = 2$ for the photon and $N_s = 4$ for the graviton: as Lorentz invariance forbids the transversal spin (s $= 0$) for massless particles, this yields $s = 1$ for the photon and $s = 2$ for the graviton.

To show that the gravitational wave carries energy and momentum, we start again from Einstein's field equations (5.1) maintaining for the metric the perturbative approach $g_{\mu\nu} = \eta_{\mu\nu} + h_{\mu\nu}$. In practice, we can calculate the part of the Ricci tensor $R_{\mu\nu}{}^{(1)}$ that is linear in $h_{\mu\nu}$, and write the field equations in the form

$$R_{\mu\nu}{}^{(1)} - \frac{1}{2}\eta_{\mu\nu}R^{(1)} = \frac{8\pi G}{c^4}\left[T_{\mu\nu} + t_{\mu\nu}\right] \tag{7.21}$$

The source of the field is expressed by the term $T_{\mu\nu} + t_{\mu\nu}$ and it depends on $h_{\mu\nu}$, as discussed after Eq. (5.15). At the same time, the metric perturbation is generated

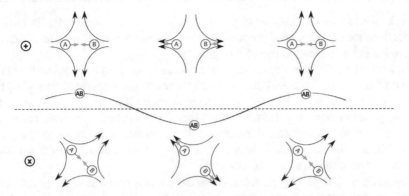

Fig. 7.3 The lines of force of the + and × polarization of gravitational waveforms. It can be easily seen that the two polarization differ by a rotation of $\pi/4$

by the total mass and by the flow of energy and momentum so that we reasonably interpret $t_{\mu\nu}$ as the energy-momentum tensor of the wave itself.

$$t_{\mu\nu} = \frac{c^4}{8\pi G}\left[R_{\mu\nu} - \frac{1}{2}g_{\mu\nu}R - R_{\mu\nu}^{(1)} + \frac{1}{2}\eta_{\mu\nu}R^{(1)}\right] \tag{7.22}$$

$t_{\mu\nu}$ is deduced by a series expansion for small values of h, up to h^2:

$$t_{\mu\nu} \simeq \frac{c^4}{8\pi G}\left[R_{\mu\nu}^{(2)} - \frac{1}{2}\eta_{\mu\nu}\eta^{\alpha\beta}R_{\alpha\beta}^{(2)}\right] \tag{7.23}$$

The explicit expansion is a rather laborious but instructive exercise and we refer the reader to the textbook of Weinberg (1972) for details. Mediating over many wavelengths of radiation results in a relatively simple expression

$$< t_{\mu\nu} >= \frac{c^4}{16\pi G}k_\mu k_\nu\left(h_+^2 + h_\times^2\right) \tag{7.24}$$

from which it is finally possible to derive the expression of the wave intensity (or flux):

$$F = \frac{c^3}{16\pi G}\left[\left(\frac{dh_+}{dt}\right)^2 + \left(\frac{dh_\times}{dt}\right)^2\right] \tag{7.25}$$

7.3 Waves in Other Gravitational Theories

We have previously seen that the verification and comparison of different theories of gravitation are based on the post-Newtonian phenomena, i.e. in the WFSM limit and on comparison of the relative values of the PPN parameters. We now need to remark that alternative theories of General Relativity are associated with different predictions for the wave field properties. We will limit ourselves here to discuss the predictions of a few metric theories of gravity.

In a metric theory the role of non-gravitational fields is to help define the space-time curvature associated with the metric. Matter can generate these fields and they can help generate the metric, but they cannot directly interact with the matter that only responds to the metric itself. It follows that the metric and the equation of motion for matter are the milestones of the theory. Then the difference between alternative metric theories is the particular way that matter (and other non-gravitational fields) have to generate the metric of space-time.

Some theories of gravity postulate the existence of a dynamic scalar gravitational field Φ in addition to the metric tensor g and the Brans–Dicke theory (Brans and Dicke 1961) can be considered a special case in this category. We have here two arbitrary functions of Φ: the cosmological function $\lambda(\Phi)$ and a function of the coupling $\omega(\Phi)$. In the case of the Brans–Dicke theory we assume $\omega = constant$ e $\lambda = 0$ and in the $\omega \to \infty$ limit the theory returns General Relativity.

The cosmological function expresses the characteristic length of interaction l of the scalar field. Indeed, for an isolated system, it can be shown that the gravitational potential contains a Yukawa-like term as well as the Newtonian one:

$$U(\vec{x}, t) = \int \frac{\rho(\vec{x}, t)\big((a + b\exp(-|\vec{x} - \vec{\xi}|/l)\big)}{|\vec{x} - \vec{\xi}|} \vec{d}^3\xi$$

We note that scalar-tensor theories are conservative and, in general, for large values of ω their predictions differ from those of Einstein's theory for corrections of the order of $O(\frac{1}{\omega})$ both for the post-Newtonian limit and for the gravitational wave characteristics.

A different case is represented by vector-tensor theories where a gravitational 4 vector field K_μ is added to the metric tensor. The difference between the theories of this category is at the end, the value of four parameters. In the limit where these parameters tend to zero, the theories collapse towards the Einstein theory.

These semi-conservative theories, in the post-Newtonian limit, produce effects compatible with the existence of privileged reference systems. Moreover, the linearized field equations are more complex than the einsteinian and scalar-tensorial ones, and the tensor propagation solution of $h_{\mu\nu}$ is influenced by the value of the K field of cosmological background. In general there are as many as ten different wave solutions, each with its own characteristic polarization and velocity condition.

In general, there are two ways in which velocity of gravitational waves v_g can differ from the speed of light. The first is its dependence on cosmological parameters and the second on the local distribution of matter. Observations in the solar system limit some of the parameters that define the value of v_g to differ by less than 10^{-3} from the value of the speed of light. A crucial test for these theories was provided by the comparison of the arrival times of the gravitational wave and electromagnetic signals emitted in the same astrophysical process and detected on Earth. The GW event, due to the merging of two neutron stars at 40 Mpc from the Earth, was detected on August 17th,

2017, by the LIGO-Virgo network. The electromagnetic signal was received within 1.7 s by the two satellites Fermi and Integral equipped with Gamma ray detectors. The limits set on the ratio $(c_{em} - c_{GW})/c_{em}$ is ranging between -3×10^{-15} and $+7 \times 10^{-16}$ (Abbott et al. 2017b).

As for the polarization of gravitational radiation, it must be said that the Einstein prediction of having only two independent states is probably unique with respect to other metric theories. The most general prediction is of six polarization states which are expressed by means of six Riemann tensor components R_{0i0j} which determine the forces applied to the detector.

Eardley et al. (1973) developed a classification method based on the study of the symmetry properties of the amplitudes of gravitational waves when a rotation is applied around the wave propagation direction (helicity).

In principle, assuming a wave coming from a single source, with a known direction of propagation, and using the information obtained by two non parallel interferometers, it is possible to discriminate the class of metric theories that fits the polarization state of the detected wave.

7.4 The Emission of Gravitational Waves

Describing the generation of gravitational waves in Einstein's theory is a complex task. To facilitate this study, we divide the space around the source into three regions characterized by different distance scales:

- a generation zone with $r \leqslant r_I$, where r_I is the typical internal radius of the source,
- a local area with $r_I \leqslant r \leqslant r_{ex}$ where r_{ex} defines the extension of the local zone and finally
- the wave zone (propagation zone) for $r \geq r_{ex}$.

The boundary lengths, r_I and r_{ex}, are also related to the wavelength of the emitted radiation. The internal radius must be larger than the *reduced* wavelength of the radiation, that is $r_I \gg \lambda_g/2\pi$; at the same time r_I is the edge of a region where the gravity field of the source is relatively weak. $r_I \gg L$ where L is a measure of the linear dimensions of the source. For the external radius we will have

$$r_{ex} - r_I \gg \lambda/2\pi$$

However, if r_{ex} is too large we should not ignore the phase shift $\delta\Phi$ due to the gravitational field of the source when the wave is passing through the local area. So, we must also impose that

$$\delta\Phi = \frac{2\pi R_g}{\lambda} \ln\left(\frac{r_{ex}}{r_I}\right) \ll 1 \tag{7.26}$$

We need all these conditions to handle propagation in the local area as in the case of quasi-flat metric space-time. We use again, as a starting point, the field equations (7.9) and the Lorentz gauge condition:

$$\frac{\partial^2 h_{\mu\nu}}{\partial x^\alpha \partial x_\alpha} = -\frac{16\pi G}{c^4} T_{\mu\nu}; \qquad\qquad \frac{\partial h^{\mu\alpha}}{\partial x_\alpha} = 0$$

and again we proceed in close analogy with the electromagnetic case:

$$\frac{\partial^2 A_\mu}{\partial x^\alpha \partial x_\alpha} = \frac{4\pi}{c} J_\mu \qquad\qquad \frac{\partial A^\mu}{\partial x_\mu} = 0$$

where A_μ is the 4-potential and J_μ is the 4-current. When we combine the first equation and the second in both systems we get

$$\frac{\partial T^{\mu\alpha}}{\partial x_\alpha} = 0 \qquad\qquad\qquad \frac{\partial J^\mu}{\partial x_\mu} = 0 \qquad\qquad (7.27)$$

Both equations represent conservation laws: the invariance of charge in the electromagnetic case and the laws of conservation in mechanics, in particular the conservation of 4-momentum, in the case of gravity. Both solutions are given in terms of retarded potential. However, when we perform multipolar expansion of the solution, in the case of gravity, due to the conservation of momentum, the dipolar term is zero and the first non-zero contribution to the radiation field is the quadrupolar one. It follows that the simplest calculation techniques related to the problem of gravitational radiation generation are based on the so-called quadrupolar formalism. In quadrupole approximation, it can be shown that the metric tensor expressed in the TT gauge, is a function of the second time derivative of the mass quadrupole moment:

$$h^{(TT)}{}_{jk} = \frac{2}{r} \frac{\partial^2}{\partial t^2} \left[D_{jk}(t - r/c) \right]^{(TT)} \qquad\qquad (7.28)$$

where r is the distance from the central point of the source, t is the proper time measured by an observer at rest with respect to the source, $(t - r/c)$ is the retarded time and D_{jk} is the quadrupole moment already defined in Eq. 7.4. Moreover, if the internal motion of the source is slow, it is possible to neglect, in the local area, the propagation delay (see Eq. 7.26) and to approximate the field considering the Newtonian potential

$$U = \frac{c^2}{2}(1 - g_{00}) \qquad\qquad (7.29)$$

From Eq. 7.28 we conclude that the most powerful sources should be characterized by large internal masses in high-speed non-spherical motion. However, we have the most interesting case when the gravity field of the source is very intense, i.e. when the quadrupole approximation it is not sufficient. To deal with this problem, various formalisms and various levels of approximations were introduced, in order to reduce the high non-linearity of the Einstein equations, and to obtain reasonable predictions on the emission. Clearly, the tool of choice for the study of phenomena associated with high-intensity fields is now numerical relativity, that today has access to more and more powerful supercomputing facilities.

From the quadrupole formula we infer the luminosity of a source of gravitational waves, Eq. 7.3:

$$L_G = \frac{G}{5c^5} \sum_{kh} \left(\frac{d^3}{dt^3} D_{kh} \right)^2 \tag{7.30}$$

Here we see that L_G depends on the third derivative of the quadrupole moment D_{hk} and is proportional to the coefficient $G/5c^5 = 5.5 \cdot 10^{-54}$ m^{-2}kg^{-1}s^3. The coefficient is extremely small, and rules out, as we shall show in the next section, any hope of generating on Earth a measurable amount of g.w. We are then forced to consider as detectable sources of gravitational waves only those of astrophysical origin.

Assuming an experimental point of view, we should correlate the luminosity with the impinging flux on Earth F and the value of the amplitude of the metric perturbation tensor h. If we assume the simplifying hypothesis of isotropic emission we get:

$$F = \frac{L_G}{4\pi r^2} \tag{7.31}$$

where r is here the distance of the detector from the source. It is reasonable to expect most of the sources to rotate around an axis, and thus emit with a non-symmetric radiation pattern, so the factor 4π is an indicative value. When we consider the simple case of a wave with a single Fourier component of angular frequency ω, the relation between F and h is deduced from Eqs. 7.28, 7.30 and 7.31 and we conclude that

$$h = \frac{1}{\omega} \sqrt{\frac{16\pi G}{c^3} F} \tag{7.32}$$

Note that a value of $h \sim 10^{-20}$ corresponds to a large value of continuous energy flow to Earth (about 20 W /m^2 for an frequency of 1 kHz) and a giant amount of power emitted at astronomical distances: of the order of 10^{35} W for a source 1 pc away.

Having introduced the basic relations for the emission, we now review the astrophysical processes that emit gravitational radiation that could be detectable to Earth. We divide the sources into two broad categories:

- emitters of continuous signal
- emitters of transient signals.

Examples of continuous sources are the double star systems that rotate around each other (rather, around their center of mass) or compact stars with some degree of asymmetry, quickly rotating around their axis.

In this same category we also include stochastic signals of cosmological nature, the gravitational analog of the microwave cosmic background of electromagnetic radiation.

In the class of impulsive emitters we classify all the stellar collapse processes, like the explosions of supernovae and the coalescence of binary systems.

7.5   Continuous Sources

In principle, one would hope to build in the laboratory the simplest source of continuous signal, as a simple mechanical device: for example, we might consider a rod or a dumbbell rotating around an axis perpendicular to its length, at angular velocity ω. The calculation of the rod quadrupole moment in Cartesian coordinates is relatively simple and we suggest it as an exercise. Applying the definition of luminosity (7.31), we obtain

$$L_G = \frac{32}{5}\frac{G}{c^5}I^2\omega^6 \tag{7.33}$$

where I is the moment of inertia with respect to the rotation axis. As a practical example, we calculate the emission of a rotor actually built to calibrate the EXPLORER gravitational wave detector, installed at CERN (Astone et al. 1991, 1993). A similar rotor was also used to set a limit on the existence of a Yukawa-like "fifth force" (Astone et al. 1998). This rotor is a 14 kg aluminum bar with a profile that minimizes centrifugal forces. The bar rotates at 462 Hz, close to the tear-apart limit of the material due to centrifugal forces. Its moment of inertia, relative to the rotation axis, is $I = 0.15\,\text{kg m}^2$. This rotor emits radiation at twice the rotation frequency, but its luminosity is a mere $L_{GW} = 2.4 \cdot 10^{-33}$ W. This value is so small that it makes it impossible, in the light of current technology, to attempt and detect the emitted gravitational wave. It is therefore clear that only an astrophysical (i.e. very massive and compact) objects rotating at high angular velocity can provide significantly large signal; and even in this case, the signals are still (to date) not large enough to be measurable.

7.5.1   Rotating Neutron Stars

Rotating neutron stars can be sources of gravitational waves, if their crust has any distortion. We shall consider them here only for what concerns them as GW sources. More details can be found about these peculiar stars in Chap. 12. These compact objects are the result of the evolutionary process of a star of large mass: once the nuclear processes that sustain it are exhausted, the star tends to collapse under the action of gravitational force. In this phase the protons react, through *inverse β decay*,

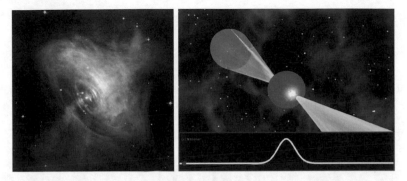

Fig. 7.4 Composite (X-Ray and optical) image of the pulsar in the Crab nebula. *Credits* NASA/HST/CXC/ASU/J. Hester et al. On the left a sketch that highlights the lighthouse effect. Courtesy M. Kramer

with the electrons to form neutrons and anti-neutrinos. The neutrinos carry an enormous amount of energy towards the upper layers of the star, propagating a devastating shock wave and expelling the star mantel in a gigantic explosion. The light pulse that is generated is known as the supernova, which hides the collapsed nucleus inside, that is, a neutron star of radius of the order of 10 km and mass of the order of 1.5 $M_\odot$.

While drastically decreasing its radius, the collapsed object begins to rotate more quickly on itself due to the conservation of angular momentum. Some of these stars have associated a very high magnetic field ($10^7 - 10^{11}$ T). When the axis of rotation does not coincide with the magnetic axis, the star emits electromagnetic waves along the direction of the field. Such a star is called a Pulsar (*pulsating radio source*) and emits radio waves for the combined effect of rapid rotation and intense magnetic field on charged particles in its atmosphere. The waves are emitted within a small cone around its magnetic axis: this, on the other hand, is tilted with respect to the axis of rotation, so that the radio wave sweeps the sky at each turn, as happens for the light emitted from the lighthouse of a port. An observer on Earth intercepts the beam at regular time intervals and then receives radio pulses with a recurrence period equal to that of the star's rotation (Fig. 7.4).

In stars with higher angular velocity the decrease of the rotation period (*spin-down*), when observed, is generally extremely small. This implies a low emission of gravitational waves and consequently a high symmetry of the star. However, given the fact that the intense electromagnetic emission of a pulsar can only be explained by the presence of an extremely strong multipolar magnetic field, its asymmetry can contribute significantly to its structural deformation Moreover, the observation of sudden changes in the rotation period of some pulsars is interpreted as a change in the structure of the crust (a star-quake) and a settlement of its residual distortion.

To evaluate the luminosity of this potential source of gravitational waves, we consider a star of mass M, having a simple ellipsoidal geometry with eccentricity e

$$e = \frac{(a - b)}{\sqrt{a \cdot b}}$$

where a and b the two main axes in the equatorial plane.

$$I_3 = \frac{M}{5}(a^2 + b^2)$$

is the main moment of inertia with respect to the axis perpendicular to the equatorial plane. The axis of rotation of the star is rotated with respect to the axis of I_3 of an angle $\phi = \Omega t$, where Ω is the angular velocity.

Consider the matrix of the main moments of inertia:

$$I_{jk} = \begin{pmatrix} I_1 & 0 & 0 \\ 0 & I_1 & 0 \\ 0 & 0 & I_3 \end{pmatrix} \tag{7.34}$$

Although the non-degenerate case $I_1 \neq I_2$ lets us hope for irregularities ("mounts" or "valleys") that can contribute to the emission of g.w., their difference is invariably very small, allowing us to assume $I_1 \sim I_2$.

To calculate the mass quadrupole moments in the TT gauge, we transform it in the co-rotating reference system of the star, through the rotation matrix

$$\Re_{jk} = \begin{pmatrix} cos\phi & -sin\phi & 0 \\ sin\phi & cos\phi & 0 \\ 0 & 0 & 0 \end{pmatrix} \tag{7.35}$$

We obtain:

$$\Re_{ik}\Re_{jl}I_{kl} = \begin{pmatrix} I_1cos^2\phi + I_2sin^2\phi & -sin\phi cos\phi(I_2 - I_1) & 0 \\ -sin\phi cos\phi(I_2 - I_1) & I_2cos^2\phi + I_1sin^2\phi & 0 \\ 0 & 0 & I_3 \end{pmatrix} \tag{7.36}$$

From this, using the relation $e \simeq \frac{I_1 - I_3}{I_3}$, correct up to terms in e^3, we can compute the quadrupole moments matrix D_{ij}:

$$D_{ik} = \frac{eI_3}{2} \begin{pmatrix} cos2\phi & -sin2\phi & 0 \\ sin2\phi & -cos2\phi & 0 \\ 0 & 0 & 0 \end{pmatrix} \tag{7.37}$$

The wave behaviour can be synthesised using the compact expression

$$h^{TT}{}_{jk} = \frac{2G}{rc^4}\Lambda_{jklm}\left[\frac{d^2}{dt^2}D_{lm}\left(t - \frac{r}{c}\right)\right]$$

where we introduced the *transverse projector* Λ_{jklm}[5]:

----

[5] In practice Λ_{jklm} is obtained by applying twice the 2×2 tensor $\mathcal{P}_{ij} = \delta_{ij} - n_i n_j$, which projects a vector in the plane perpendicular to the direction of the unit vector $\vec{n}$: $\Lambda_{jklm} = \mathcal{P}_{jk}\mathcal{P}_{lm} - \frac{1}{2}\mathcal{P}_{jk}\mathcal{P}_{lm}$.

$$\Lambda_{jklm} = \left[(\delta_{jk} - n_j n_k)(\delta_{kl} - n_k n_l) \right] - \left[\frac{1}{2}(\delta_{jk} - n_j n_k)(\delta_{ml} - n_m n_l) \right]$$

$$h^{TT}{}_{jk} = h_o \begin{pmatrix} -\cos(2\Omega t_r) & -\sin(2\Omega t_r) & 0 \\ -\sin(2\Omega t_r) & \cos(2\Omega t_r) & 0 \\ 0 & 0 & 0 \end{pmatrix} \tag{7.38}$$

where $t_r = t - \frac{r}{c}$ is the retarded time and the amplitude h_o is

$$h_o = \frac{4G\Omega^2}{c^4 r} I_3\, e \tag{7.39}$$

Note that *the gravitational signal is emitted at twice the rotation frequency* $\omega_g = 2\Omega$, its amplitude depends linearly on e and quadratically on the rotation period $T = 2\pi/\Omega$.

A convenient rearrangement of this equation gives the amplitude in terms of typical parameters:

$$h_o = 4.21 \times 10^{-24} \left[\frac{1\,\text{ms}}{T} \right]^2 \left[\frac{1\,\text{kpc}}{r} \right] \left[\frac{I_3}{10^{38}\,\text{kg m}^2} \right] \left[\frac{e}{10^{-6}} \right]$$

Finally the luminosity is given by the equation

$$L_{GW} = \frac{32}{5} \frac{G\, I_3^2\, e^2\, \Omega^6}{c^5} \tag{7.40}$$

An interesting example is the PSR 0532 pulsar in the Crab nebula, rotating at $\Omega \sim 2\pi$ 29.67 Hz at a distance $r = 2$ kpc from the Earth,[6] with e estimated between 10^{-6} and 10^{-8}: we obtain a value of h_o in the range $10^{-26} - 10^{-28}$.

Despite the low value of the metric perturbation, the group of the Tokyo University and the KEK laboratory (Tsubono 1991) tried to measure the gravitational emission using a resonant detector, specially shaped to resonate at 59.35 Hz and, not detecting any signal, set a first upper limit: $h < 2 \cdot 10^{-21}$.

In recent years the LSC and Virgo collaborations, jointly analyzing the data of the interferometers of the LIGO and Virgo project, set a limit of $h \le 2 \cdot 10^{-25}$. It corresponds to an ellipticity of the order of 10^{-4}. This is a first important result of astrophysical nature, since the value is a factor 7 lower than the limit to the total energy flow lost by the pulsar as derived by the radio observation of its rotational frequency slow down.

The neutron stars can rotate at very high angular speeds. In fact, Pulsars with rotational frequency up to 716 Hz have been observed. These are objects produced by the growth mechanism of old neutron stars or by the collapse of a white dwarf

---

[6] 1 pc $\simeq 3.86 \cdot 10^{16}$ m.

in a double star system and we know that about 50% of the stars belong to multiple systems. Current theoretical estimation predict that there should be more than 10^6 double stars in our galaxy.

R. Wagoner suggested (Wagoner 1984) that high-frequency monochromatic waves could be generated when the neutron star reaches a point of instability, at which hydrodynamic waves are generated on the surface, thus determining the emission of gravitational radiation. In this case the emission counterbalances the increase in rotational speed and it should be proportional to the flow of X-rays coming from the star. Some observations in the X band indeed tend to confirm the existence of growth of angular velocity; however, this process would critically depend on the viscosity of the star.

7.5.2   The Binary Systems

We want to study the continuous wave emitted by two massive bodies rotating around their center of mass and describing a circular orbit of radius a, with angular frequency ω.

The system is conveniently described the masses of the two bodies m_1 and m_2 or, by its *total mass* M_{tot} and its *reduced mass* μ:

$$M_{tot} = m_1 + m_2; \qquad\qquad \mu = \frac{m_1 m_2}{m_1 + m_2}$$

We shall use both notations interchangeably.

The binary system has a time-dependent quadrupole moment that, after half orbit, returns to the same value, so we expect the emission of waves to occur at twice the frequency of rotation. Applying the definition of quadrupole moment to a system of two rotating point-like masses, we have

$$D_{ij} = \frac{1}{2} a^2 \mu \begin{pmatrix} cos2\omega t & sin2\omega t & 0 \\ sin2\omega t & -cos2\omega t & 0 \\ 0 & 0 & 0 \end{pmatrix} \tag{7.41}$$

Then, referring to Eq. 7.30, we compute the gravitational luminosity of the system:

$$L_{GW} = \frac{32}{5} \frac{G}{c^5} \mu^2 a^4 \omega^6 \tag{7.42}$$

In the assumption that the orbital parameters, a and ω, do not change significantly in the time interval of the observation, a condition known as *adiabatic approximation*, we write the third Kepler law in the form

$$\omega^2 a^3 = GM_{tot} \tag{7.43}$$

Substituting in (7.42), the luminosity, i.e. the rate of energy loss of the system, takes the form:

$$L_{GW} = -\frac{dE}{dt} = \frac{32}{5}\frac{G^4}{c^5}(m_1m_2)^2(m_1+m_2)\frac{1}{a^5} \qquad (7.44)$$

We can also differentiate the orbital (kinetic + potential) energy of the system $E = -(Gm_1m_2)/(2a)$ and, comparing, find that the diminishing distance $a(t)$ of the two bodies follows the law:

$$\frac{da}{dt} = -\frac{64}{5}\frac{G^3}{c^5}(m_1m_2)(m_1+m_2)\frac{1}{a^3} \qquad (7.45)$$

Integration of this relation with respect to time yields $a(t)$

$$a^4(t) = a_o{}^4 - \frac{256}{5}\frac{G^3}{c^5}(m_1m_2)(m_1+m_2)\,t$$

that we can rewrite as

$$a(t) = a_o\left[1 - \frac{t}{\tau_{coal}}\right]^{\frac{1}{4}} \qquad (7.46)$$

where a_o is the distance at the initial time and τ_{coal} the characteristic time to coalescence of the phenomenon

$$\tau_{coal} = \frac{5}{256}\frac{c^5}{G^3}\frac{a_o{}^4}{(m_1m_2)(m_1+m_2)} \qquad (7.47)$$

From Eq. 7.45 and differentiating Eq. 7.43 we also see how the angular velocity (or the rotation period) of the system varies over time:

$$\frac{d\omega}{dt} = -\frac{96}{5}\left[\frac{G^3}{c^2}\frac{m_1+m_2}{a}\right]^{\frac{3}{2}}\frac{G^2}{c^2}\frac{m_1m_2}{a^4} = \frac{3}{2}\omega_0\frac{L_{GW}}{E} \qquad (7.48)$$

Integrating Eq. 7.43 for ω and substituting Eq. 7.46 for $a(t)$, we derive the orbiting angular frequency:

$$\omega = \omega_o\left[1 - \frac{t}{\tau_{coal}}\right]^{-\frac{3}{8}} \qquad (7.49)$$

where the initial angular frequency is:

$$\omega_o = \sqrt{\frac{G\,M_{tot}}{a_o^3}} \qquad (7.50)$$

The relation (7.48) explains, although only qualitatively, the radio observations for the PSR 1913+16 system. In the case of a elliptic orbit, the calculations are less straightforward, but lead to similar results. The brightness increases rapidly with eccentricity and the emission is more intense near the periastrum, where the effect of reaction to the emission of radiation is more significant, as well as the tendency to make the orbit circular.

In the case of PSR 1913+16, we are dealing with a gravitational wave source which has an orbital period of 7 h and 45 min, the orbit is highly eccentric ($e \sim 0.6$) and emits g.w. with a luminosity $L_{GW} \simeq 10^{25}$ W. The Fourier spectrum of gravitational radiation is composed of various harmonics of the orbital frequency.[7]

Let us consider a couple of examples of the luminosity associated with existing systems. The β Minoris LMi binary system is 30 pc from Earth, has a 2.9-year orbital period and a brightness of $L_{GW} \sim 10^{21}$ W. The corresponding value at Earth for metric perturbation is $h \sim 10^{-17}$. The signal falls in a frequency band $\sim$22 nHz, well outside that of the interferometric gravitational detectors both on Earth and Space.

The binary system with the shortest period currently known is HM Cancri, o RX J0806.3+1527: it is composed of a pair of white dwarfs, 0.5 $M_\odot$ each, with $a_0 \sim 80000$ km apart, and it has an orbital period T = 321.5 s. This means an emitted g.w. power $L_{GW} = 4 \cdot 10^{28}$ W; the two stars will coalesce in $t_{coal} \sim 10^5$ years. Being at a distance $r = 490$ pc, according to Eqs. 7.31, 7.32, we expect a signal on Earth

$$ h = \frac{T}{4\pi} \sqrt{\frac{4GL_{GW}}{c^3 r^2}} \simeq 10^{-21} $$

The signal frequency $2/T = 6$ mHz, is still too low for terrestrial detectors, but well in the band of the future LISA observatory, that will operate in the frequency range $10^{-4} - 10^{-1}$ Hz.

There is also a large population of binary systems in the galaxy that emit at frequencies between $10^{-4} - 10^{-7}$ Hz. For the sources closest to us the intensity is such as to produce waves with $h \sim 10^{-21}$. The number of sources is so high to produce a random overlap of signals as the dominant effect. This results in a background noise of a gravitational origin analogous to the cosmological background noise.

7.6   Transient Signals

7.6.1   The Coalescence of Binary Systems

We have mentioned in the previous paragraph the formula of luminosity for a binary system in a circular orbit (see Eq. 7.42).

The loss by radiation reduces the orbital radius and causes the stars to move closer until they collide. The intensity of the gravitational ground signal, averaged on all possible detector orientations with respect to the source, is:

---

[7] In the case of the PSR 1913+16, the most intense Fourier component is the eighth harmonic.

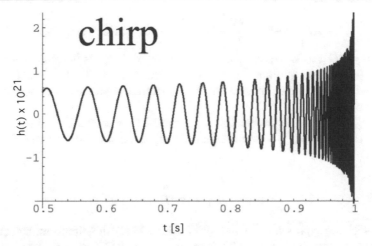

Fig. 7.5 The *chirp*, i.e. a gravitational wave signal emitted by the coalescence of a double neutron star system in the absence of spin and matter transfer

$$h \simeq 10^{-23} \, \mu \left(\frac{M_{tot}}{M_\odot}\right)^{\frac{2}{3}} \left(\frac{\nu}{100\,\text{Hz}}\right)^{\frac{2}{3}} \left(\frac{100\,\text{Mpc}}{r}\right)$$

The time scale for reduction of the orbital radius is expressed by the relation:

$$\tau = \frac{\nu}{\dot{\nu}} \simeq 8 \, \frac{1}{\mu} \left(\frac{M_{tot}}{M_\odot}\right)^{-\frac{2}{3}} \left(\frac{\nu}{100\,\text{Hz}}\right)^{-\frac{8}{3}} \, s$$

The wave emitted by the system appears similar to a sinusoidal function in which both frequency and amplitude increase over time up to the moment when the stars merge (last stable orbit). This waveform is called *chirp* (Fig. 7.5).

To describe the type and frequency evolution of this signal, it is customary to introduce the *chirp-mass*:

$$\mathcal{M} = \left[\frac{(m_1 m_2)^3}{m_1 + m_2}\right]^{1/5} = \left[(m_1 m_2)^2 \mu\right]^{1/5} \tag{7.51}$$

As we shall see, this combination of the masses m_1 and m_2 of the binary system, largely determines how rapidly the frequency of the signal evolves over time. The two components of the metric perturbation h are found, from the relations of the previous section, with a lengthy but straightforward calculation (we refer, for instance, to the book of Maggiore 2018):

$$h_+(t) = 2\frac{(G\mathcal{M})^{\frac{5}{3}}}{c^4} \frac{(\pi \nu(t))^{2/3}}{r} (1 + \cos^2 i) \cos(2\Phi(t) + \Phi_0)$$

$$h_\times(t) = -4\frac{(G\mathcal{M})^{\frac{5}{3}}}{c^4}\frac{(\pi\nu(t))^{2/3}}{r}\cos i\,\sin(2\Phi(t) + \Phi_0) \qquad (7.52)$$

where $\nu = 2/T_{orbit}$ is the frequency of emission of g.w., i is the angle between the orbital angular momentum and the observer's line of sight, $\Phi(t) \equiv 2\pi\nu(t - r/c)$ is the orbital phase of the equivalent one-body system around the center of mass of the binary system.

The Eq. 7.52 contain the instantaneous orbital frequency $\nu(t)$ and terms dependent on the phase $\Phi(t)$, whose dominant components oscillate at twice the orbital frequency:

$$\Phi(t) - \Phi_0 = 2\pi \int_0^t 2\nu(t')dt' \qquad (7.53)$$

In the adiabatic approximation and for circular orbits, using the formulas related to the evolution over time of the orbital frequency Eqs. 7.47, 7.49, we obtain

$$\Phi(t) - \Phi_0 = -2\left[\frac{c^3(\tau_{coal} - t)}{5G\mathcal{M}}\right]^{\frac{5}{8}} \qquad (7.54)$$

This relation shows how the measurement of the wave phase allows to obtain direct information on the chirp mass $\mathcal{M}$.

In a more rigorous approach we should also consider higher order amplitude corrections that contain other harmonics (terms containing phase of the type $n\Phi(t)$, being n a positive integer). Furthermore, the expressions (7.52) are valid for a system on an almost circular orbit, an assumption that is not entirely realistic. A detailed post-Newtonian calculation must take into account spin-orbit and spin-spin couplings between the two components of the binary system; they produce a characteristic modulation of the emitted gravitational signal.

The final frequency of the gravitational signal, at the end of the spiralling, when the two massive objects collide, increases up to (Thorne 1987):

$$f^* \sim 5\left(\frac{M_{tot}}{M_\odot}\right)^{-1} \text{kHz} \qquad (7.55)$$

The relation (7.55) only holds if the system is composed of compact objects. When the two objects collide (merging) and in the following dynamic evolution of the new object, the masses are moving at relativistic speeds in a regime of extreme gravitational fields. Merging implies a particularly violent dynamics that can lead to the formation of a black hole, with a significant release of energy in the form of gravitational radiation. The post-Newtonian approximation is certainly no longer valid and numerical simulations need to be developed to predict the emission. The characteristic time interval of the merging phase is very short: from a few milliseconds in the case of stellar mass black holes to a few seconds in the case of large mass black holes. During this phase, if a lighter star falls toward a heavier and more compact one, a significant amount of matter will likely have a high angular momentum such as to counteract the infall. In the most accredited models the process of matter transfer determines the formation of an accretion disk around the black hole, which can feed an intense jet of gamma rays along the rotation axis (*Gamma Ray Burst, GRB*).

After the merger phase of the two progenitors stars (neutron stars and/or black holes), the new compact object evolves toward an equilibrium state. In this final phase it vibrates at its fundamental modes, emitting gravitational radiation at the frequencies of those oscillation modes that have a quadrupole symmetry (*ringdown* phase). The gravitational wave flux emitted after merging is calculated by considering the overlap of quasi-normal modes of the final object. The *ringdown* duration depends on the mass of the final object: in practice it can consist of two or three cycles only.

The coalescence phase of the binary systems is now well modelled and the amplitudes of the two components of the wave emitted are expressed by the relations (7.52). Since $\Phi(t)$, the phase of the signal is known by the post-Newtonian calculation, it is possible to measure $\mathcal{M}$, ν, t_0 and Φ_0 by properly analyzing the data. The remaining unknown parameters of those equations can be inferred thanks to the different response of the antennas, depending on the polarization and the direction of wave propagation. Thus, from the observations of a network of three non-co-located detectors, we extract three independent combinations for the polarization state and two time delays of the observed signal.

The quantities ν, h and $\frac{\nu}{\dot{\nu}}$ are measurable, and Schutz (1986) noted that in the product $h\frac{\nu}{\dot{\nu}}$ all masses cancel out, and we are left with an equation relating just frequency ν and distance r. It is then possible to derive r independently from the values of the stellar masses. This ability to measure distances is unique in Astronomy. From the knowledge of these parameters and by adequately considering the response function of the detectors, it is possible to determine the distance between the source and the detectors with an accuracy of $\Delta r/r \sim 1 - 10\%$.

The scale of the astronomic distance is based on the measurement of the Doppler effect of electromagnetic signals, i.e. on the *redshift* of the source $z = \nu_{(em)}/\nu_{(oss)} - 1$. The astronomic redshift is measured rather easily, because the emission and absorption spectra of the various atoms are very distinct and very well known. The *redshift* value is not obtained directly from the gravitational signals associated with the coalescence of binary systems. This depends on the cancellation of the simultaneous contribution of z parameter which affects a in different way the quantities concurring to define the amplitude $h(t)$. In fact, when the source is at *redshift* z, the Eq. 7.52 must be rewritten mapping $\mathcal{M} \rightarrow \mathcal{M}_z = (1+z)\mathcal{M}$ and $\nu \rightarrow \nu_{(em)}/(1+z)$. The phase $\Phi(t)$, being by definition the time integral of the pulsation $2\pi\nu(t)$, exhibits a similar cancellation effect and is independent of z.

Therefore, to study the evolution rate of the universe, z must be extracted by detecting an electromagnetic counterpart of the gravitational signal, for example, by identifying the host galaxy of the event. Knowing the *redshift* of the electromagnetic counterpart and deriving r from the relative gravitational signal, it is possible to measure the Hubble constant, through the relation $H_0 = c\,z/r$, and also to study the function of evolution of the star formation rate. Combining e.m. observations (the redshift of galaxy NGC 4993) and GW data from the coalescence of two neutron stars (Abbott et al. 2017a) a value $H_0 = 70^{+12}_{-8}$ km s^{-1}Mpc^{-1} has been determined, in discrete agreement with CMB and Supernovae observations.

As discussed above, modelling the merger phase of collapse is difficult. In the case of a pair of stars with significantly different mass, we expect the smaller body, once a given distance from the other star is reached, to trigger mass transfer due to tidal interactions. Angular momentum is also transferred from one star to the other and in practice the secondary star tends to turn into a thick axis-symmetrical disk that orbits around the primary star.

The time scale of the evolution of the system depends on the transport mechanisms of matter and the dissipation of angular momentum from the disk: it is longer when the process is entirely dominated by matter viscosity. In the case where the initial mass ratio is large, the mass transfer rate would approach a limit due to the emission of gravitational radiation. However, this scenario appears unlikely, due to the enormous value of matter flow.

We have already observed that the steady state of mass transfer is typical of a double system with a large mass ratio. We also need to point out that the case of a binary system of two neutron stars with a large mass ratio is not likely. The transfer process leads to the interesting possibility that the smaller star, losing mass, is reduced to a minimum value below which free expansion is unstable and therefore this mass could explode before colliding with the more massive and voluminous companion. Most of the energy released in the final collapse is converted in production of neutrinos, which are very difficult to detect due to the distance of the event from Earth. Thus, one can hypothesize a scenario in which the gravitational radiation is the only observable signal produced by these phenomena.

Indeed, in the coalescence or collapse of compact objects, GW emission is roughly simultaneous with the release of neutrinos and high energy photons in the explosion phase. Measuring the arrival time of GW and of the gamma burst (GRB) events has allowed us to measure the velocity of gravitational waves with an accuracy limited only by the intrinsic delays to the collapse mechanism, as discussed in Sect. 7.3.

In the Fig. 7.6 sketches a summary of the potential evolutionary scenarios of a binary system. Depending on the masses and the state of rotation of the components of the binary system, the merger can evolve in multiple directions (Bartos et al. 2013). Each of them is associated with a different type of both gravitational and gamma ray. In the case of systems formed by two neutron stars (NS), the various evolutionary paths are selected by comparing the total mass of the system $M_{tot} = m_1 + m_2$ with the threshold value of 3 $M_\odot$, also considering the difference $|m_1 - m_2|$ between the masses of the individual components. Another discriminating element is the comparison of M_{tot} with the maximum mass value of a non rotating neutron star, $M_{NS}^{(static)}$. In the case of a neutron star-black hole system (NS-BH), the evolutionary scheme differs on the basis of the comparison between the radius of the innermost stable circular orbit relative to the coalescence phase, R_{ISCO} with the typical distance where the neutron star begins to be taken apart by the intense tidal forces (tidal disruption—TD) R_{TD}: this value depends on the star equation of state. Note that mass transfer can begin long before reaching the critical distance R_{TD}. The formation of the accretion disk around the black hole seems essential to create the conditions

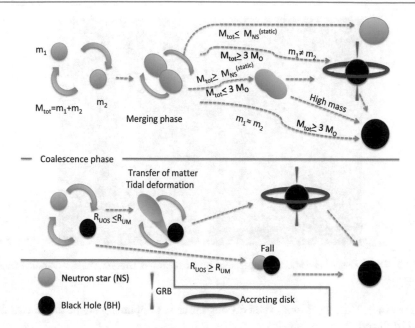

Fig. 7.6 Possible evolutionary scenarios of a binary system. *Credits* Bartos et al. (2013)

for GRB emission. Its formation depends on the masses and on the state of rotation of the objects. If the neutron star is disrupted by tidal forces before falling into the black hole, the mass of the accretion disk that is formed depends on the spin of BH, on the initial alignment of the spin of the two objects and on the ratio m_{BH}/m_{NS}.

If the detector has sufficient bandwidth, the signal-to-noise ratio for the chirps can be drastically improved by filtering the data with an appropriate template, as discussed further on. The assumed shape of the signal plays a fundamental role in the design of such templates.

7.6.2   The Supernovae

Supernova explosions are spectacular processes where the formation of a collapsed body formation is expected. Supernovae are classified into two categories: type I and II.

- The reference event for a type I supernova is the nuclear detonation of a white dwarf, occurring after the accretion from a companion. The most optimistic models suggest that a significant fraction of the white dwarf mass is involved in forming a neutron star. In any case, formation of a black hole is inhibited by the insufficient mass of the progenitor. Furthermore, the presence of surrounding matter at the time of collapse can be the cause of the limited optical emission associated with

Fig. 7.7 The sky before and after the supernova 1987a explosion. *Credits* Anglo-Australian Observatory

the event. Thus, to classify such events, the term of optically silent supernova was introduced.

- The type II supernova is associated with the collapse of the central area of a star and the shock wave determines the ejection of the optically luminous outer mantle.

In the millennium just completed, a dozen of supernovae were optically identified in our galaxy.

In this category fall the most famous supernova examples: the SN1054 was recorded by Asian astronomers in the year 1054 and modern observations have identified the pulsar PSR 0532 in the nebula of the crab as the neutron star born in this collapse.

SN 1987A marks a supernova explosion located in the great Magellanic cloud, at 51 kpc. On Feb 23, 1987, three detectors placed in three different underground laboratories on Earth revealed neutrinos produced by the inverse β decay process. This was absolutely the first direct observation of a collapse and marks the beginning of a new way of doing Astronomy based on the observation of weak interaction processes (Fig. 7.7).

If the mass involved in the collapse process is sufficiently large, it is reasonable to assume that some type II supernovae will form black holes. In this case the shock wave that causes the explosive rebound of the surrounding material should be weak. The outer mantle recedes and the emission of neutrinos and their diffusion through the outer mantle material tend to cause an extension of the time scale of the process from milliseconds to seconds. However, in the final stage the collapse occurs on a time scale typical of free fall and the emission of pulses of gravitational wave should take place on milliseconds scales (see for a complete review Piran et al. 1991).The amount of radiation emitted depends on the amount of kinetic energy associated with the non-spherical motion. The fundamental quantity that can play an important role in limiting the sphericity of collapse is the rotation of the star. If its angular momentum is initially high, the central part involved in the collapse should increase its velocity. Depending on the initial conditions, the system can evolve breaking up into two or more parts that, quickly spiralling, would tend to merge again, with emission of gravitational radiation. If the critical mass value is exceeded (between 1.4 and 3 solar masses $M_\odot$) the collapse continues until a black hole is formed.

Once completed its formation, the new object can still emit gravitational radiation at the frequencies of its own normal oscillation modes, excited by the infall of matter. The frequency range of interest is very large: from 0.1 to 10 kHz. The low frequencies will be excited during the initial phase of collapse, while the high frequencies will show up in the later phase, mainly due to the excitation of the normal modes of the collapsed star. Different authors have derived the characteristics of the gravitational signal; we report here a rough estimate of the value of h and of the emission frequency ν, as given by Thorne (1987):

$$\nu \approx \frac{c^3}{5\pi GM} \approx 1.3 \cdot 10^4 \frac{M_\odot}{M} \text{Hz} \tag{7.56}$$

$$h \approx \sqrt{\frac{15\,\epsilon}{2\pi} \frac{G}{c^2} \frac{M}{r}} \tag{7.57}$$

The value of the conversion efficiency ϵ of the rest mass into gravitational radiation is uncertain. Theoretical evaluations are in disagreement and have gone from the initial, optimistic estimates of $\epsilon \sim 10^{-2}$ to the more modern ones (Dimmelmeier et al. 2001) $\epsilon \sim 10^{-7}$. This is due to the fact that the collapse models have evolved over time and these estimates also vary according to the values of the dynamic quantities characterizing the initial state, such as the angular momentum of the body.

Assuming that the signal is emitted at a frequency of 1 kHz from a body of 13 $M_\odot$ with a conversion efficiency $\epsilon = 10^{-2}$, we would have $h \simeq 10^{-17}$ in the case of a source at the galactic center (10 kpc). This event would have been detectable even by the resonant detectors that have continuously operated till a few years ago. The rate of formation of massive black holes is, however, probably very low. It has then been necessary to push the sensitivity of the detectors to a level that can detect events produced in the Virgo cluster of galaxies ($\sim$2500 galaxies at 10 Mpc from the Earth, $h \simeq 10^{-20}$): within that population, we expect such an event to occur with reasonable probability within the lifetime of gravitational wave detectors.

According to the models based on perturbative calculations, for slow rotations the efficiency ϵ depends on the fourth power of the angular momentum. However, increasing the speed of rotation conditions arise that could lead to disruption. In this case the efficiency of emission by the non-ax-symmetrical residual parts could be much higher. Saenz and Shapiro (1979, 1981) have developed a model of collapse starting from an initial geometry of homogeneous spheroid which rebounds on itself until it leads to the formation of a neutron star. An effort has been made to derive better evaluations for ϵ (Muller 1984) using the quadrupole formula and a more realistic hydrodynamic model of a rotating star.

Full simulations imply solving a fluid dynamics problem coupled to the equations of General Relativity, where neutrino physics plays a crucial role. Results obtained to date, using the largest computing centers available to scientists, well cover two-dimensional problems. 3D simulation, much more costly in terms of computing time, have recently appeared (Takiwaki et al. 2016). The authors used an adaptive lattice step, with tightest mesh where more precision is required, down to a record step of 87 m.

If we limit our considerations to the results of the calculations made for the formation of a black hole (Piran and Stark 1986), we can observe that the emission of gravitational radiation is weak during the hydrodynamic phase. However, a strong signal is expected in the final moments of collapse. This radiation is due to the excitation of the quasi-normal modes of oscillation at low frequency of the black hole. The waveform of the final burst of the gravitational wave is independent of the details of the collapse process and is mainly polarized in the h_+ state.

A complete simulation of the process is a challenging problem. It requires to develop to numerically solve the relativistic equations in the non-perturbative case coupled with the hydrodynamics equations. The numerical code must be particularly accurate. High precision is necessary because the gravitational wave emission is only a secondary effect of the general phenomenon. Multi-dimensional simulations of CCSNe, using the most power computers available nowdays, are currently at the frontier of research regarding the two main explosion paradigms: the *neutrino-driven* mechanism, thought to be active for slowly rotating progenitors and responsible for the most common SNe, and the *magneto-rotational mechanism*, active only for fast rotating-progenitors and responsible for rare but highly energetic events, like hypernovae and long GRBs.

Several groups are currently tackling this problem with two- and three-dimensional simulations using the world's most powerful supercomputers. Multiple challenges arise during the numerical modelling: (i) accurate solution of the neutrino transport equations during the evolution; (ii) folding into the model the complete interactions of electron, muon and tau neutrinos and their anti-particles with matter; (iii) use of high resolution to numerically resolve fine structure features in the convective and turbulent flow around the proto-neutron star; this is of special importance for the development of magneto-rotational instabilities in fast-rotating progenitors; (iv) accurate (general relativistic) description of gravity; (v) use of sophisticated equations of state to describe the behaviour of matter at high densities. The different groups studying the problem use different approaches to handle each of these challenges and, to this point, no one has carried out a definitive three-dimensional simulation including all the physical ingredients and with sufficiently high resolution to provide the world-wide community with results that can be accepted with confidence. Despite the complexity of the problem, these calculations give acceptable remnant neutron-star masses and already predict few distinct signatures of GW signals in both time and frequency domains. The core-bounce signal is the part of the waveform which is best understood (Dimmelmeier et al. 2001) and it can be directly related to the rotational properties of the core (Dimmelmeier et al. 2008). However, fast-rotating progenitors are not common and their bounce signal will probably be difficult to observe in typical galactic events, due to its high frequency and low amplitude.

In addition, during the post-bounce evolution of the newly formed proto-neutron star (PNS), the convection determines the excitation of highly damped modes in the PNS by accreting material and instabilities,[8] with a peculiar GW emission. In this case the GW waveforms last for about 200−500 ms until the supernova explodes while, in the case of black hole formation, the typical duration is 1 s or above. The peculiarity is that the signal frequencies raise with time due to the contraction of the PNS, whose mass steadily increases. This peculiarity can drive the filtering procedure to extract signal embedded in the detector noise ad increase the distance at which we can observe the event. In addition characteristic frequencies of the PNS can be as low as ~ 100 Hz, especially those related to *g-modes*,[9] and SASI, which make them a target for ground-based interferometers, that have the highest sensitivity at those frequencies.

7.7   The Stochastic Gravitational Wave Background

The Stochastic Gravitational Wave Background (SGWB) is a stationary gravitational-wave noise of cosmological or astrophysical origin. It covers a large range of frequencies: at the lower end around 10^{-16} Hz we can infer its properties measuring the B-modes of the Cosmic Microwave Background (CMB), while at frequencies between 10^{-10} and 10^{-6} Hz the Pulsar Timing Arrays (PTA) are suited to detect these signals. LISA will explore the range $10^{-4} - 10^{-1}$ Hz, while higher frequencies $(1 - 10^3)$ are accessible with the ground-based advanced interferometers, as LIGO, Virgo and KAGRA.

The SGWB is generally detectable by looking at a correlated contribution of the noise output in two (or more) outputs of GW detectors operating on the same frequency bandwidth or in a combined Michelson-Sagnac interferometer.

Several potential sources for the background are hypothesised across various frequency bands of interest, with each source producing background with different statistical properties.

Since we are dealing with a random process, SGWB is determined by its statistical properties such as mean, variance and power spectral density. The spectrum of the SGWB is usually described by an energy density ρ_{GW}; with this energy density, as usual in cosmology, we build the dimensionless quantity $\Omega_{GW}(\nu)$,

$$\Omega_{GW}(\nu) = \frac{1}{\rho_c} \frac{d \rho_{GW}}{d \log(\nu)} = \frac{\nu}{\rho_c} \frac{d}{d\nu} \rho_{GW} \tag{7.58}$$

---

[8] Standing Accretion and Shock Instability—SASI.

[9] *g-modes* are a particular class of vibrational modes that have no classical counterpart.

which is the ratio of the gravitational-wave energy density per unit logarithmic frequency, to the critical energy density, i.e. the value of density needed for a closed universe:

$$\rho_c = \frac{3\,c^2 H_o{}^2}{8\pi G}$$

where H_o is the present value of the Hubble–Lemaître constant.

The sources of the stochastic background can be broadly divided into two categories:

- The physical processes that occurred at the earliest moments of the universe certainly created a stochastic background that survives, to some extent, until today. This is analogous to the cosmic microwave background, which is an electromagnetic record of the early universe.
- SGWB is created also by the superposition of a large number of independent sources as binary black hole and binary neutron star merging during the history of the universe. In other words, the incoherent sum of numerous unresolved gravitational wave signals should result in a stochastic background of gravitational waves.

7.7.1   The Cosmic GW Background

Cosmological backgrounds arise from early universe sources, suggested by various inflationary mechanisms like cosmic strings and other defects. While these sources are rather hypothetical, a detection of a background originated by them would be a major discovery of new physics. The detection of such an primordial background would have a profound impact on early-universe cosmology and on high-energy physics.

GWs arise from primordial fluctuations in the spacetime metric which have been parametrically amplified by inflation. The frequency spectrum of SGWB is expected to be extremely wide, from 10^{-16} to 10^{10} Hz (Allen 1988). The lower end corresponds to the inverse Hubble scale when the universe became matter dominated. Below this frequency, the spectrum grows as ν^{-2}. At the upper end, the spectrum is cut off by the (short but finite) time scale at which inflation ends and the universe enters the radiation-dominated phase. In the observation band of LIGO and Virgo some theoretical models predict for the SGWB spectrum a power law behaviour, which assumes, for the fractional energy density in gravitational waves, the form

$$\Omega_{GW}(\nu) = \Omega_\alpha \left(\frac{\nu}{\nu_{ref}}\right)^\alpha \tag{7.59}$$

where α is the spectral index, ν_{ref} and Ω_α are two theory dependent parameters.

Some papers, inspired by superstring theory, suggest that cosmic strings may have been produced in the early universe and then expanded to cosmic sizes (see, for example, Polchinski (2004) for an overview).

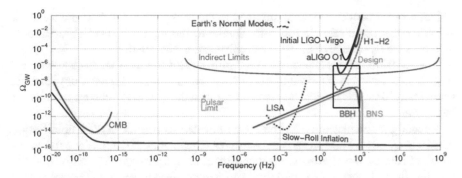

Fig. 7.8 Constraints on the SGWB and the projected design sensitivity of the interferometer network LIGO-Virgo assuming two years of coincident data. We report also the constraints resulting from other kind of measurements: CMB measurements at low multipole moments, indirect limits from the Cosmic Microwave Background (CMB) and Big-Bang Nucleosynthesis, pulsar timing and from the ringing of Earth's normal modes. Note the ν^{-2} slope at frequencies below 10^{-16} Hz, predicted by some inflationary models. *Credits* Christensen (2017)

In the Fig. 7.8 we show the limit set on SGWB using the LIGO-Virgo data collected during the first observation run of the advanced detectors, as well as constraints from previous analyses, theoretical predictions, the expected limits at design sensitivity for Advanced LIGO and Advanced Virgo, and the expected sensitivity of the proposed Laser Interferometer Space Antenna (LISA).

Since the resulting cosmic strings only interact gravitationally, their effects are scaled by the product of Newton's constant and the string tension, $G\mu$. Because the energy stored in the strings is eventually converted into gravitational waves, strings would produce a large stochastic gravitational wave background. There are several possibilities for the geometry of the compact dimensions of string theory, including localised branes, and these warped extra dimensions allow the string tension to be much lower than the Planck scale, anything above the TeV scale.

An ensemble of such objects could produce a background accessible to the interferometers on the Earth and/or in space.

7.7.2   The Astrophysical GW Background

The astrophysical background is produced by an huge number of weak, independent, and unresolved astrophysical sources. Many different astrophysical sources contribute to the Astrophysical Gravitational Wave Background (AGWB): e.g., stellar-mass black hole and neutron star binaries, merging supermassive black hole binaries, rotating neutron stars and even stellar core collapse. The nature of these sources depends on the sensitive frequency band of the signal. For example, in the acoustic bandwidth (Hz to kHz) the most likely source of the stochastic background is an astrophysical background from binary neutron-star and from stellar mass binary black-hole mergers. This prediction results from an estimate of the merger rate based on the first observation of the NS-NS merger (the GW170817 event observed by the

LIGO-Virgo network). This rate, summed to the background rate from binary black hole (a much more reliable estimate, thanks to the numerous GW observations) increases the amplitude of the total astrophysical background.

The AGWB nature is expected to be significantly different from its cosmological counterpart, which is expected to be, at least for inflation, stationary, unpolarised, statistically almost Gaussian, and isotropic, by analogy with the cosmic microwave background.

Recently, particular attention has been devoted to predict SGWB polarisation (Cusin et al. 2019a) and AGWB anisotropies (Cusin et al. 2019b). Since the astrophysical sources are located in cosmic structures and they are not uniformly distributed, the GW energy flux from them (both of resolved or unresolved case) should not be constant across the sky and it should depends on the direction of observation. The fact that the source are not uniformly distributed is just one of the causes of the anisotropy; the second one is related to the deflection effect due to massive structures present along the GW propagation path. For a background of cosmological origin, lensing by large scale structures is the main source of anisotropy and actually, GW and CMB anisotropies, are correlated.

Note that, when a flux of unpolarised radiation coming from a given direction impinges on a massive object, the outgoing radiation can be polarised, due to the dependence of the absorption cross section on the polarisation. Thus, to characterise the polarisation change induced by the diffusion by massive structures, in analogy with the electromagnetic case, we will refer to the change in the GW Stokes parameters, which are defined as

$$I = |h_+|^2 + |h_\times|^2 \quad Q = |h_+|^2 - |h_\times|^2 \tag{7.60}$$

$$U = 2\mathrm{Re}(h_+^* h_\times) \quad V = 2\mathrm{Im}(h_+^* h_\times) \tag{7.61}$$

where I is proportional to the intensity of the GW, Q to the difference between the intensities of two polarised contributions.

It can be shown that U provides the same information as Q in a frame rotated by $\pi/8$, while the parameter V describes a phase difference between h_+ and $h_\times$, which results in circular polarisation. In general these parameters depend on the propagation direction of the outgoing GW radiation, so that these are basic quantities used to model the angular power spectrum of AGWB, that hopefully will be a real observable for the future GW detectors. For an *astrophysical* background in the PTA and LISA frequency bands, the above cited authors predict that the amount of polarisation generated is suppressed by a factor $10^{-4} - 10^{-5}$ with respect to anisotropies. For a *cosmological* background, predictions are even more discouraging: an additional suppression factor of two order of magnitudes is indeed expected.

As mentioned above, the primary target of the Earth-based interferometric detectors will be the background from unresolved binary mergers of stellar-mass black holes and/or neutron stars throughout the universe. In an optimistic scenario, this detection can take place with the present, second generation of detectors (Advanced LIGO and Advanced Virgo). The detection of this background will provide information about stellar-mass binary-black-hole populations at much larger distances than those accessible for the resolved mergers. It will also complement searches for gravitational-wave backgrounds at much lower frequencies, as the probable detect ions before the end of the decade by PTA, and possibly by CMB searches.

Finally, we mention that a huge number of ultra-compact binaries in our galaxy (millions of sources) will form an unresolved foreground signal in LISA (2013). It will appear as noise, but being modulated in amplitude during the yearly rotation, it will be detectable. The overall strength will tell us about the distribution of the sources in the Galaxy, as the different Galactic components (thin disc, thick disc, halo) contribute differently to the modulation. Their relative amplitudes will be used to set upper limits on the halo population, yet largely unknown.

References

Abbott, B., et al.: Observation of gravitational waves from a binary black hole merger. Phys. Rev. Lett. **116**, 061102 (2016)

Abbott, B., et al.: GW170817: observation of gravitational waves from a binary neutron star inspiral. Phys. Rev. Lett. **119**, 161101 (2017)

Abbott, B., et al.: Gravitational waves and gamma-rays from a binary neutron star merger: GW170817 and GRB 170817A. App. J. Lett. **848**, L13 (2017)

Allen, B.: Stochastic gravity-wave background in inflationary-universe models. Phys. Rev. D **37**, 2078 (1988)

Astone, P., et al.: Evaluation and preliminary measurement of the interaction of a dynamical gravitational near field with a cryogenic gravitational wave antenna. Z. Phys. C **50**, 21 (1991)

Astone, P., et al.: Long term operation of the "EXPLORER" cryogenic gravitational wave detector. Phys. Rev. D **47**, 362–375 (1993)

Astone, P., et al.: Experimental study of the dynamic Newtonian field with a cryogenic gravitational wave antenna. Eur. Phys. J. C **5**, 651 (1998)

Bartos, I., Brady, P., Marka, S.: How gravitational-wave observations can shape the gamma-ray burst paradigm. Class. Quantum Grav. **30**, 123001 (2013)

Bondi, H., Pirani, F.A.E., Robinson, I.: Gravitational Waves in General Relativity. III. Exact Plane Waves. Proc. R. Soc. Lond. A **251**, 519 (1959)

Brans, C., Dicke, R.H.: Mach's principle and a relativistic theory of gravitation. Phys. Rev. **124**, 925 (1961)

Christensen, N.L.: Searching for the stochastic gravitational-wave background with advanced LIGO and advanced virgo. In: Auge, E., Dumarchez, J. (eds.) 52nd Rencontres de Moriond on Gravitation. Jean Tran Thanh Van (2017)

Cusin, G., Durrer, R., Ferreir, P.G.: Polarization of a stochastic gravitational wave background through diffusion by massive structures. Phys. Rev. D **99**, 023534 (2019)

Cusin, G., Dvorkin, I., Pitrou, C.: Uzan J-P Properties of the stochastic astrophysical gravitational wave background: astrophysical sources dependencies. Phys. Rev. D **100**, 063004 (2019)

Dimmelmeier, H., Font, J.A., Möller, E.: Gravitational waves from relativistic rotational core collapse. Astrophys. J. Lett. **560**, L163–L166 (2001)

Dimmelmeier, H., et al.: Gravitational-wave burst signal from core collapse of rotating stars. Phys. Rev. D **88**, 064056 (2008)

Eardley, D.M., Lee, D.L., Lightman, A.P.: Gravitational-wave observations as a tool for testing relativistic gravity. Phys. Rev. D **8**, 3308 (1973)

Einstein, A.: Näherungsweise Integration der Feldgleichungen der Gravitation (Approximate integration of the field equations of gravitation). Königh Pruss. Akad. der Wissenschaften Sitzungsberichte, Erster Halbband (1916), p. 688

Einstein, A.: Gravitationswellen (On Gravitational Waves). Kö nigh Pruss. Akad. der Wissenschaften Sitzungsbericthe, Erster Halbband, 154 (1918)

Gravitational Radiation and Supernova: In: Wheeler, J.C., Piran, T., Weinberg, S. (eds.) Supernovae. World Scientific Publishing, Singapore (1991)

Heaviside O.: Electromagnetic Theory, Appendix 1 and 2. London (1893)

Hulse, R.A., Taylor, J.H.: Discovery of a pulsar in a binary system. Astrophys. J. Lett. **195**, 251 (1975)

Maggiore, M.: Gravitational Waves, vol.1: Theory and Experiments. Oxford University Press, Oxford (2018)

Metzger, B.D., et al.: Electromagnetic counterparts of compact object mergers powered by the radioactive decay of r-process nuclei. MNRAS **406**, 2650 (2010)

Muller, E.: The collapse of rotating stellar cores: the amount of gravitational radiation predicted by various numerical models. In: Bancell, D., Signore, M. (eds.) Problems of Collapse and Numerical Relativity. Springer, Dordrecht (1984)

Penrose, R.: A spinor approach to general relativity. Ann. of Phys. (New York) **10**, 171 (1960)

Penrose, R., Rindler, W.: Spinors and Space-Time. Cambridge University Press, Cambridge (1984)

Piran, T., Stark, R.F.: Numerical relativity, rotating gravitational collapse and gravitational radiation. In: Centrella, J.M. (ed.) Dynamical Spacetime and Numerical Relativity. Cambridge University Press, Cambridge (1986)

Pirani, F.A.E.: On the physical significance of the Riemann tensor. Acta Physica Polonica **15**, 389 (1956)

Poincaré, E.H.: Sur la Dynamique de l'Electron. Rendiconti del Circolo matematico di Palermo **21**, 129–176 (1905)

Polchinski, J.: Cosmic superstrings revisited. AIP Conf. Proc. **743**, 331 (2004)

Saenz, R.A., Shapiro, S.L.: Gravitational and Neutrino Radiation from Stellar Collapse: Improved Ellipsoidal Model Calculations. App. J. **229**, 1107 (1979)

Saenz, R.A., Shapiro, S.L.: Gravitational radiation from stellar collapse: III damped ellipsoidal models. App. J. **244**, 1033 (1981)

Schutz, B.F.: Determining the Hubble constant form gravitational wave observations. Nature **323**, 310 (1986)

Takiwaki, T., Kotake, K., Yudai Suwa, Y.: Three-dimensional simulations of rapidly rotating core-collapse supernovae: finding a neutrino-powered explosion aided by non-axisymmetric flows. MNRAS Lett. **461**, L112–L116 (2016)

The LISA Consortium, The Gravitational Universe. arXiv:1305.5720

Thorne K. *Gravitational radiation* in 300 years of Gravitation, Hawking S.W. and Israel W. ed., Cambridge Univ. Press (1987)

Tsubono, K.: Detection of continuous waves. In: Blair, D.G. (ed.) The Detection of Gravitational Waves. Cambridge University Press, Cambridge (1991)

Wagoner, R.V.: Gravitational radiation from accreting neutron stars. Astrophys. J. **278**, 345 (1984)

Weber, J.: Detection and generation of gravitational waves. Phys. Rev. **117**, 306 (1960)

Weinberg, S.: Photons and gravitons in perturbation theory: derivation of maxwell's and Einstein's equations. Phys. Rev. **138** B 988 (1965)

Weinberg, S.: Gravitation and Cosmology: Principles and Applications of the General Theory of Relativity. Wiley, New York (1972)

Further Reading

Sathyaprakash, B.S., Schutz, B.F.: Physics, Astrophysics and Cosmology with Gravitational Waves. Living Rev. Relat. **12**, 2 (2009)

The Resonant Detector

8

8.1 Introduction

For several decades, roughly between 1960 and 2005, the experimental search for gravitational waves was mainly carried on by *resonant detectors* or "Weber bars". These detectors were eventually phased out when long-baseline interferometers came to maturity and outperformed them both in sensitivity and bandwidth. Nevertheless, besides the historical relevance, there is an actual interest in studying this sophisticated and ingenious apparata, where many senior GW scientists initiated their formation, due to the many technologies, still used these days, that were conceived and developed for these detectors. Moreover, resonant detector methods and strategies have been recently implemented (Goryachev et al. 2021) in searches for GW in the MHz frequency region. We refer to Aguiar (2010) for a readable and thorough historical review of the many experimental projects that have involved resonant GW detectors, with an extensive bibliography.

As mentioned in Sect. 7.1, the idea of an elastic detector of GW was suggested by F. Pirani in 1956. Joseph Weber (and J. A. Wheeler) immediately considered the *elastic detector* and worked out a detailed analysis by 1960. Weber then went on to build several of these detectors and was followed by many groups around the world: we count three generations of Weber bars (room temperature, cryogenic and ultra-cryogenic), which we briefly describe here.

A useful, although misleading shortcut in dealing with the effect of GW on free (or bound) masses is the following: GW cause, as we saw in Chap. 7, relative motion of two test masses at distance ℓ: $x(t) = \frac{1}{2}\ell h(t)$. In classical mechanics, force is what causes motion of masses, according to $\vec{F} = m\ddot{\vec{x}}$. If we take the second time derivative of the first relation, and compare it with one-dimensional Newton's second law, we can define a *GW pseudo-force*

© The Author(s), under exclusive license to Springer Nature Switzerland AG 2022
F. Ricci and M. Bassan, *Experimental Gravitation*, Lecture Notes in Physics 998,
https://doi.org/10.1007/978-3-030-95596-0_8

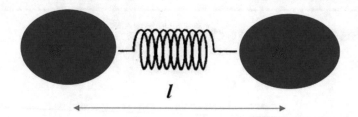

Fig. 8.1 Schematic of Weber's oscillator composed of two point masses connected by a spring

$$F_{GW} = \frac{1}{2}m \; \ell \ddot{h}(t) \tag{8.1}$$

We shall use this relation to derive the antenna response to GW. It is however clear from it that the acceleration produced by GW grows with the linear dimension ℓ of the detector, but does not depend on its mass m, in accordance with the Equivalence Principle. Large mass detectors were developed not to increase the signal, but, as we shall show, to reduce the thermal noise.

8.2   The Resonant Detector

8.2.1   Principle of Operation

A gravitational wave is a tidal effect. As such, any detector needs to extend over a finite region of space to feel such effect. In the case of e.m. beam detectors, test masses are set far from each other (km, for Earth-based interferometers, Gm for spaceborne detectors) in an almost free falling condition and the space-time is probed by light bouncing between them.

J. Weber in 1960 modified the equation of the geodesic deviation to include a term of elastic interaction between two freely gravitating masses[1] and derived a solution.

Consider a simple harmonic oscillator, as shown in Fig. 8.1: two point masses under the driving pseudo-force of a GW $h(t)$. For simplicity, we take the wave as impinging normal to the line (along an x-axis) joining the two test masses. The equation of motion for either test mass of such a system is

$$\ddot{x} + \frac{2}{\tau}\dot{x} + \omega_o^2 x = \frac{1}{2}l \; \ddot{h}(t) \tag{8.2}$$

---

[1] The spring schematizes the restoring effect to the state of stable equilibrium of the system; this effect is essentially due to the internal electromagnetic forces among the microscopic components of the elastic body. It could therefore be argued that the changing metric of space-time also influences these binding forces. This effect would be relevant only if the gravitational energy density (including the rest energy) were comparable with the electromagnetic energy density of the system. We deduce that in a first approximation this effect is negligible.

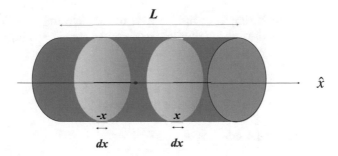

Fig. 8.2 Extension of the oscillator to a macroscopic, continuous resonator. The gravitational wave acts on the whole cylinder: it pushes and pulls each section of the bar. On each pair of sections, symmetric with respect to the centre, the effect is computed in the same way as in the case of Fig. 8.1

where we have introduced the dissipation factor $1/\tau$ and the restoring force, characterized by the angular resonance frequency ω_o.

We solve this differential equation in the frequency domain, introducing the Fourier transforms $H(\omega)$ for the forcing term and $X(\omega)$ for the variable x (Fig. 8.2):

$$X(\omega) = \frac{l}{2} \frac{\omega^2 H(\omega)}{\omega_o^2 - \omega^2 + \frac{i2\omega}{\tau}} \tag{8.3}$$

8.2.2   The Weber Bar

To move beyond the conceptual experiment, and design a truly feasible device, we need to replace the idealized harmonic oscillator with an elastic, extended body. Weber's choice was a metal cylinder, a strategy replicated in most successive experiments. A detailed treatment of the detector implies modifying the elasticity equation for a solid, dissipative body by introducing a *tidal* term due to the impinging gravitational wave. Here we limit our considerations reporting the results for a thin rod of mass M and length L, of a material with sound velocity v_s, vibrating along its axis (the x-axis, see Fig. 8.2) in its first longitudinal vibration mode.

In the case of a plane GW propagating perpendicular to the cylinder axis, the extremes of the rod behave as a simple harmonic oscillator (Weber 1961) with equivalent rest length ℓ and equivalent mass M_{eq}[2] given by

$$\ell = \frac{4}{\pi^2} L \quad M_{eq} = \frac{M}{2} \tag{8.4}$$

---

[2] The equivalent mass of an oscillating body depends on the vibrational mode considered (not all elements of the extended solid oscillate with the same amplitude) and on the point where the vibration is measured: the equivalence is based on the kinetic energy: $\frac{1}{2} M_{eq} \dot{\ell}^2 = \frac{1}{2} \int_{body} \rho \dot{x}^2 dV$.

This equivalence, which simplifies the analysis of the detector, is valid near the resonance frequency of the first longitudinal mode of the bar $\omega_o = \pi v_s / L$. Suppose the antenna is excited by a pulse of the form

$$h(t) = h_o \cos \omega_o t \qquad \text{for } 0 < t < \tau_g$$
$$= 0 \qquad \text{outside this interval}$$

For such a pulse, the detector response is

$$x(t) = -\frac{L}{\pi^2} h_o \tau_g \omega_o e^{-\frac{t}{\tau}} \sin \omega_o t \tag{8.5}$$

In the case of excitation due to a monochromatic wave at the resonant frequency of the bar, we have

$$x(t) = -\frac{2L}{\pi^2} Q h_o \sin \omega_o t \tag{8.6}$$

where $Q = \omega_o \tau / 2$ is the quality factor of the oscillator. This is a first hint of the importance of using bars of materials with low dissipation, i.e. large Q.

In the most sensitive experiments, the rod is a massive metal body weighing few tons. It hangs, through ingenious suspension systems that isolate it from external mechanical noise, in a vacuum chamber, to reduce air friction. The detectors of second and third generation had metal cylinders, $L \simeq 3$ m and $M \simeq 2300$ kg, cooled at low (4 K) and ultra-low (0.1 K) temperature to reduce the thermal noise and increase the bar Q: in those cases, the vacuum chamber ("experimental vacuum") was part of a complex cryostat with many (six, in NAUTILUS Astone et al. 1991) nested shells.

The choice of material fell, for most cryogenic detectors, on aluminium alloy Al 5056, characterized by low acoustic dissipation, $Q \sim 10^7$, at liquid helium temperature. The group at University of Western Australia chose instead to cool a cylinder of $\sim$1000 kg of Niobium, a very expensive material that exhibits, at low temperature, $Q \sim 2 \cdot 10^8$. The cylindrical shape of the antenna was the most common choice for these detectors.

However, as early as in 1976, Wagoner and Paik (1977) had introduced the idea of a spherical antenna: by exploiting its five quadrupolar oscillation modes, this type of detector had the advantage of omni-directionality, while cylinders have a peaked $sin^4(\theta)$ antenna pattern.[3] Spherical antennas, made of an Al-Cu alloy, 60 cm in diameter and cooled to low (4 K) and very low temperature (10–100 mK) were built in Brazil (Aguiar et al. 2006) and in the Netherlands (Gottardi et al. 2007). Pictures of these detectors are shown in Fig. 8.3.

---

[3] See Sect. 9.6.2 for a definition of antenna pattern.

Fig. 8.3 Left: a diagram of the Minigrail spherical antenna, developed at the University of Leiden (NL). The large pumping system serves the specially designed dilution refrigerator, capable of cooling the sphere to 65 mK. Right: photo of the Brazilian antenna Mario Schenberg. Its design is similar to Minigrail, without the ultra-low temperature cooling stage. Courtesy of the Schenberg and Minigrail groups

8.2.3   The Transducer

Technically speaking, a transducer is any device that converts a physical quantity into another, like a loudspeaker (electric signals $\rightarrow$ sound waves) or a Hall probe (magnetic field $\rightarrow$ electric voltage).

The vibrations of the resonant detectors must be monitored by a motion transducer, usually an electromechanical one: a device capable of converting the amplitude of vibration into an electromagnetic signal. The working principle consists in storing a large e.m. energy $\mathcal{E}$ in a small gap: vibration of one surface bounding this gap perturbs the energy distribution and produces an electrical signal. Many commercial devices exist (to begin with microphones and strain gauges), but none is sensitive enough for these purposes, so researchers had to develop their own: all designs studied and developed fall into three general categories.

- Virtually all room-temperature detectors used piezoelectric ceramics, positioned at the central plane of the bar: although the detailed operations of piezoelectric materials require complex tensorial relations, for these purposes it is enough to say that they respond to applied strain with a voltage signal, through a simple relation $\delta V = \alpha \delta x$, with $\alpha \sim 10^7$ V/m. Their main drawback was the destructive effect on the antenna quality factor.
- Passive, or linear, transducers that store a permanent DC field; they are charged once and then virtually isolated from the outside. They preserve a linear phase and amplitude relation between input and output and are relatively simple devices. Linear transducers, that were employed by most detectors who actually had long observation runs, come in two varieties:
  - (i) Capacitive: An electric field close to breakdown ($\sim 2 \cdot 10^7$ V/m) is stored in the gap of a large capacitor. Capacitive transducers were used on the Italian detectors Explorer, NAUTILUS, AURIGA.

- (ii) Inductive: A superconducting current flowing in a flat coil generates a magnetic field close to the critical value ($\sim$0.1T for pure Nb) at the surface of a vibrating ground plane. Inductive transducers equipped the US antennas in Stanford Univ. and Allegro at LSU.
- Active or parametric transducers, in which an AC bias field at high *pump* frequency ω_p is constantly fed into the gap by an external generator. The mechanical vibration phase-modulates this signal and produces sidebands that contain the motion information. These devices have high potential for superior performance because the energy conversion from the pump frequency to the signal frequency takes place with a parametric gain of a factor ω_p/ω_o but are undoubtedly more complex than linear transducers and need, just like an interferometer, amplitude and phase stabilization for the pump signal. Active transducers were realized, both at microwave (Niobe at UWA) and optical pump frequencies, but it was difficult to achieve stability comparable with linear devices.

While many features common to parametric devices can be found in the Appendix C about the Fabry-Perot resonator, we focus here on linear transducers and, for a specific example, on the operating principle of the capacitive, electrostatic transducer. A parallel plate capacitor charged to a constant charge $q_0 = V_0 C_0$ responds to vibrations $x(t)$ of one of its plates, generating a voltage signal $v(t)$ given by

$$\frac{\delta v(t)}{V_0} = -\frac{\delta C(t)}{C_0} \qquad \rightarrow v(t) = \alpha x(t)$$

The transduction coefficient[4] $\alpha = \partial^2 \mathcal{E}/\partial x \partial q$ in this particular case is given by the stored electric field V_0/d and is limited by the breakdown field in the gap d: it can be as large as $5 \cdot 10^7$ V/m. Unfortunately, being a reciprocal device, a linear transducer also performs the inverse conversion with exactly the same efficiency α:

$$f(t) = \alpha q(t)$$

This means that any noise current $i_n = dq_n/dt$ in the input circuit is transformed into a fluctuating force f_n acting on the antenna: this effect is called *back action* and is negligible always but in the most sensitive applications, like indeed GW detectors. A different, more general description of a transducer is given in terms of a linear two-port network, with two mechanical variables, the force $f(t)$ and the velocity $\dot{x}(t)$, and two electrical ones, the circulating current $i(t)$ and the voltage $v(t)$ across the output electrical impedance. These quantities are related, just as in an all electric two-port, by the following relations:

$$f(t) = Z_{11}\dot{x}(t) + Z_{12}i(t) \qquad\qquad (8.7)$$
$$v(t) = Z_{21}\dot{x}(t) + Z_{22}i(t)$$

---

[4] In an alternative derivation, we could write the Lagrangian of the electric + mechanical system: the transduction effect would be found in the interaction part of the Lagrangian.

The Z_{ij} terms of the so-called transduction matrix are, in general, linear differential operators. Clearly, Z_{22} is the output impedance of the electrical circuit, and Z_{11} is the mechanical impedance of the oscillator. Z_{21} and Z_{12} are the direct and reverse transduction coefficients, respectively. They are not independent: $Z_{21} = -Z_{21}^*$ and in a lossless transducer, where the real part is negligible, they are equal. We recover the previous description with the relation $\alpha = i\omega\, Z_{21}$, but the impedance matrix is richer in details. For our example, in the capacitive transducer, we have

$$Z_{12} = Z_{21} = \frac{V_0}{i\omega\, d}; \qquad Z_{11} = \frac{\dot{X}(\omega)}{F(\omega)} = M_{eq}\omega_o; \qquad Z_{22} = \frac{1}{i\omega C_0}$$

The output electric impedance Z_{22} is part of a more complex impedance Z_{el}, a circuit that connects the transducer to the amplifier.

A figure of merit for transducers is the *energy coupling coefficient* β, which represents the fraction of vibrational energy that the device can transduce into electrical in one cycle of oscillation. We give its expression both in the Z_{ij} and in the α description:

$$\beta = \frac{|Z_{12}|^2}{M_{eq}\omega_o |Z_{el}|} = \frac{\alpha^2}{M_{eq}\omega_o^3\, |Z_{el}|} \tag{8.8}$$

The Resonant transducer Note the antenna mass M_{eq} at the denominator: noise considerations, discussed in the next section, require M_{eq} to be as large as possible (1–5 tonnes), forcing β to values $\sim 10^{-6}$. A brilliant breakthrough (Paik 1976) allowed a big improvement in transducer sensitivity by equipping the antenna with a light mass resonator tuned to the same frequency and coupled to the transducer (in our example, one vibrating plate of the capacitor, with a mass $m \sim 300$ g). In this way the two oscillators exchange energy and the light one vibrates by a larger amplitude, thus producing a larger signal in the transducer. In other terms, Eq. 8.8 now has m, rather than M_{eq}, in the denominator, and the coupling can increase to $\beta \sim 10^{-2}$.

The dynamics of the mechanical system results more complex but it can still be analysed in terms of two normal modes of vibration: the splitting between the eigen-frequencies is set by the ratio of the effective[5] masses: $\omega_\pm = \omega_o(1 \pm \sqrt{m/M_{eq}})$.

8.2.4   The Amplifier

The transducer output, typically in the pV range ($v \sim \alpha \cdot hL \cdot m/M_{eq} \sim 10^7 \cdot 10^{-22} \cdot 10^3$ V when a resonant transducer is used), is connected to an amplifier that must make this signal manageable while adding as little noise as possible. Gain is not a relevant parameter, as long as the amplified signal is large enough to neglect the noise of the following amplification stages: it amplifies by the same factor both signal and noise, thus leaving the SNR unchanged. We model a real-world amplifier

---

[5] For the transducers shown in Fig. 8.4, we have $m_{eq} \simeq m$.

Fig. 8.4 Left: a capacitive transducer, with two large surfaces (150 mm diameter) facing a plate resonating at $\omega_o/2\pi = 930\,\text{Hz}$. Center: an inductance-modulation transducer, with the flat "pancake" superconducting coils. Right: a parametric transducer composed of two highly reentrant Niobium cavities, resonating at 480 MHz, facing the two sides of a vibrating plate

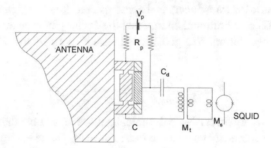

Fig. 8.5 Block diagram of the readout for a resonant antenna equipped with a resonant capacitive transducer, biased to a d.c. voltage V_p and a SQUID amplifier. The total impedance Z_{el} is the proper composition of the transducer capacitance ($Z_{22} = 1/i\omega C_0$), the decoupling capacitor C_d, the matching transformer M_t, and the SQUID input transformer M_s ($Z_{in} = i\omega M_s$)

as an ideal one (noiseless, with open input circuit), plus two sources of noise: current and voltage noise. Due to fundamental considerations, both sources are invariably present in any amplifier. Besides, the input impedance of a real voltage amplifier is not infinite but has a large value Z_{in} in parallel to the input.[6] The input impedance Z_{in} contributes, with the transducer impedance Z_{22} and any possible matching circuit, to the total circuit impedance Z_{el}. We show in Fig. 8.5, as an example, the complex readout of Explorer, NAUTILUS, AURIGA, where a superconducting transformer was required to impedance match the high impedance of the capacitive transducer to the low input impedance of a SQUID amplifier. So, Z_{el} turned into a resonant LC circuit.

An amplifier has a voltage noise[7] source $v_n = \sqrt{S_v(\nu)}$ measured in $[V/\sqrt{Hz}]$, and a current noise source $i_n = \sqrt{S_i(\nu)}\,[A/\sqrt{Hz}]$: one of these is added to the output and the other generates noise currents that circulate in the input circuit, giving rise

---

[6] For a current amplifier we would have: the input impedance of a real current amplifier is not null but has a small value Z_{in} in series with the input.

[7] We assume the reader to be familiar with the basic notions of noise and its statistical description in terms of frequency spectra. A quick recap of these concepts is given in the first sections of Chap. 10.

Fig. 8.6 Schematic of a real voltage amplifier: an ideal (noiseless, open input) device plus a finite, although large, input impedance and generators of current and voltage noise

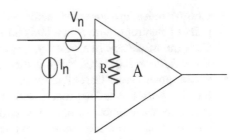

to the back-action effect mentioned above. These noise spectra are assumed white, i.e. with an amplitude distribution independent of frequency, in the small band of interest around the mechanical resonance ω_o. We make the assumption, not always verified but which simplifies the representation, that the two random variables, v_n and i_n are statistically independent.

We shall focus here on a voltage amplifier model, as shown in Fig. 8.6, although a dual model for current amplifier is also encountered.[8] The voltage source represents the *output noise*: it entirely drops on the (almost) infinite input impedance and appears unmodified (but for the gain) at the output. The current source gives rise to input noise, i.e. noise currents circulating in the input circuit.

As an equivalent description, the two following parameters are often used:

$$T_n = \frac{i_n v_n}{k_B}; \qquad \lambda = \frac{v_n}{i_n |Z_{el}|} \qquad (8.9)$$

where k_B is the Boltzmann constant and T_n is called the *amplifier noise temperature*: $k_B T_n$ is the minimum signal energy detectable, with SNR = 1. λ is a matching parameter: it measures how far the actual impedance Z_{el} is from the *noise match impedance* v_n/i_n; the optimum value is $\lambda \sim 1$.

8.2.5 The d.c. SQUID

The direct current Superconducting QUantum Interference Devices (d.c. SQUIDs) were used on low-temperature antennas, because of their unrivalled low noise in the kHz region: they are superconducting devices working on the interplay of magnetic flux quantization and phase locking of Josephson junctions. Their linearized behaviour is well understood, and they have exhibited sensitivity close to the quantum limit, although real-world devices, tightly coupled to an input circuit, like those needed in a GW detector, have higher noise levels.

To outline the working principle of a SQUID biased by a d.c. current (there is also a r.f. variety) we need to recall a few "pills" of superconductivity.

---

[8] The voltage amplifier model is suitable even when dealing with SQUIDs that are typically described as current amplifier.

- **Cooper pairs**: In a superconductor (SC) the carrier of electric current are *Cooper pairs* of electrons, with spin 0 and charge $q = 2e$.
- **Flux quantization**: In a SC ring, magnetic flux is quantized: $\Phi(B) = n\Phi_o$, with $\Phi_o = h/2e = 2.01 \cdot 10^{-15}$ Wb is the flux quantum.
- **Order parameter**: The collective motion of SC pairs can be described, following the Ginzburg-Landau approach, with a macroscopic *order parameter* (a sort of collective wave function) $\psi = |\psi| exp(i\, s(r))$ such that $|\psi|^2 = n_s$ represents the carrier density. The assumption is made that only the phase depends on position. The current density is then $\vec{J_s} = \frac{2e\hbar}{m}|\psi|^2\vec{\nabla}s(\vec{r})$.
- **Minimum coupling**: In the presence of an e.m. field, the prescription of quantum mechanics is to apply the *minimum substitution* $\vec{p} \rightarrow \vec{p} - q\vec{A}$ or $\vec{\nabla} \rightarrow \vec{\nabla} - \frac{2ie}{m}\vec{A}$, where $\vec{A}$ is the e.m. vector potential.
- **Josephson effect**: Consider a SC interrupted by a thin layer of normal conductor or insulator: a *Josephson Junction* (JJ). ψ can experience tunnel effect and maintain continuity, but there is a voltage drop across the junction (typically some 10^4 V), and the phase oscillates. The phenomenon is described by two equations:

$$I = I_c \sin s \tag{8.10}$$

$$\frac{ds}{dt} = \frac{2e\,V}{\hbar} \tag{8.11}$$

where I_c is the critical current, i.e. the maximum current that JJ can stand before superconductivity is disrupted. Equation 8.11 shows that the phase across the JJ oscillates at a very high frequency:

$$\omega_J = \frac{2eV}{\hbar} = 2\pi \cdot 486 \ \text{MHz/}\mu\text{V}$$

Consider now a ring with two JJs as shown in Fig. 8.7: we require the phase s to be single-valued around the loop: $s = \oint_\ell ds = s_1 - s_2 = 2n\pi$. This phase relation implies a phase-locking between the two JJs accomplished, thanks to a current circulating in the ring $I_{circ} = I_2 - I_1$, that is superimposed to the bias current $I_T = I_2 + I_1$. If we now shine a magnetic field onto the ring we have

$$s = s_1 - s_2 - \frac{2e}{h} \oint \vec{A} \cdot d\vec{\ell} = 2n\pi \tag{8.12}$$

The added integral is easily identified with the magnetic flux $\Phi(B)$ in the ring. So, a changing applied flux will modify the phase relations and the voltage across the ring. Equation 8.10 shows that this effect is periodic, with flux periodicity Φ_o.

A tight analogy exists between this device and a Young, two slit interferometer: the two rapidly oscillating light field in the slits are replaced by the oscillating phases across the two JJs; the superposition of the two fields gives rise to a d.c. periodic response: proportional to $\sin(2\pi\, \delta x/\lambda)$ in one case, to $\sin(2\pi\, \Phi_{ext}/\Phi_o)$ in the other.

Besides this analogy, we suggest the following physical interpretation: suppose to linearly increase an external magnetic flux Φ_{ext}: a *shield* current is generated in

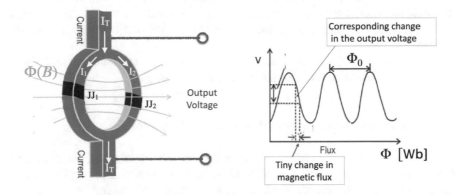

Fig. 8.7 On the left the conceptual scheme of DC SQUID: dimensions of the Josephson junctions are exaggerated. On the right the V versus Φ_{ext}, showing the periodic response, typical of interference

the SC ring to keep the magnetic flux $\Phi_{ext} + \Phi_{circ} = n\Phi_o$ constant inside it; when the shield flux exceeds the value $\Phi_o/2$, the JJs go normal, and let one flux quantum in and we restart with a smaller current, a more energetically favourable condition.

Although the dynamics of a d.c. SQUID is quite complex (we should also account for capacitive and resistive shunts across the JJs, the inductance of the ring, mutual inductance to the outside world, noise sources...) and it must be solved numerically, the response to small signals can be well modelled with an electric two-port (see Sect. 8.2.3) having as input variables the applied flux Φ_{ext} and the bias current I_T; the output variables are the voltage drop V and the circulating current I_{circ}, the latter giving rise to the back action. So, strictly speaking, SQUIDS are transducers, as they convert changes in magnetic flux into voltage signals. SQUIDs can be used as very sensitive current amplifiers, simply by coupling the input current to the ring via a mutual inductance, as schematically shown in Fig. 8.5.

The SQUID sensitivity, expressed in units of the quantum of magnetic flux $\Phi_o = \pi \hbar/e = 2.068 \cdot 10^{-15}$ Wb, can be as low as $10^{-7} \Phi_o/\sqrt{Hz}$. Inserting the SQUID in a feedback loop, with the error signal provided by an additional inductive coupling, substantially extends the useful linear response regime.

A great benefit of d.c. SQUIDs is their extremely low input noise: it has been measured only in a special, suitably degraded device. Therefore, back action is negligible for these devices. For a detailed presentation of SQUIDs and their applications, we refer to Clarke (2010).

8.3   Noise in Resonant Detectors

We shall distinguish between intrinsic and external noise sources. In the second class we list the disturbances due to environmental noise, which is mainly due to seismic and acoustic vibrations, as well as to the boil-off of refrigerating liquids

(Astone 1992): they propagate to the mechanical oscillator and excite its vibrations. To mitigate these disturbances, the antenna is suspended in a vacuum by a system of mechanical filters that provide attenuation of the order of 250 dB at the resonant frequency.

The intrinsic noise sources, that represent the fundamental limitation to the sensitivity of the detectors, are of thermal and electronic origin.

8.3.1   Thermal Noise in the Mechanical Oscillators

Thermal or Nyquist noise is due to the Brownian motion of the atoms of the bar and is the mechanical equivalent of Johnson noise in a resistor in thermal equilibrium at temperature T; it consists of a random force with white monolateral spectrum

$$S_f = 4k_B \, T \frac{M_{eq} \, \omega_o}{Q} \tag{8.13}$$

with $Q = \omega_o \tau / 2$ is the quality factor. It causes a displacement of the bar extremes that is filtered by the resonant transfer function of the antenna

$$S_x = \frac{S_f}{M_{eq}^2} \frac{1}{(\omega_o^2 - \omega^2)^2 + 4(\frac{\omega}{\tau})^2} \tag{8.14}$$

a Lorentzian shape, with a very narrow width if the decay time τ is large. In the time-domain language, this is equivalent to saying that the amplitude of vibration of the oscillator changes very slowly with time, i.e. is highly correlated. Equation 8.14 shows why a large oscillator mass M_{eq} is important.

The displacement autocorrelation function is deduced from the inverse Fourier transform of Eq. 8.14:

$$R_{xx}(\Delta t) = \frac{S_f \, \tau}{4M_{eq}^2 \omega_o^2} e^{\frac{-\Delta t}{\tau}} \cos(\omega_o t) \tag{8.15}$$

For $\Delta t = 0$ we get the variance of the displacement

$$\sigma_x^{\,2} = \frac{k_B T}{M_{eq}\omega_o^2} \tag{8.16}$$

that is just a fancy way to express a basic concept: the mean kinetic energy of the vibration mode is provided by the equipartition theorem: $\frac{1}{2}M_{eq}v^2 = \frac{1}{2}k_B T$.

At the output of a linear transducer, we have the voltage v_{th} due to thermal noise, with variance

$$\sigma_{V th}^{\,2} = \alpha^2 \sigma_x^{\,2} \tag{8.17}$$

However, if a resonant transducer is employed, an additional thermal noise source must be added. This term is relevant, due to the small mass of the second oscillator;

Fig. 8.8 The NAUTILUS antenna, a bar of three meters in length and 2.3 tons of aluminium, was cooled for the first time at CERN, reaching a temperature of 95 mK. The antenna was transferred to the INFN laboratories in Frascati where it was in operation for more than 20 years. Courtesy of the ROG group

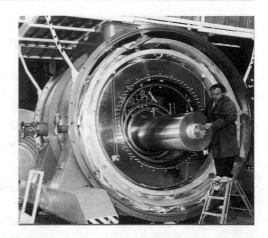

however, the gain in transduction efficiency makes its use worthwhile. An optimum mass for the second oscillator exists, balancing these two competing effects.

The reduction of thermal noise is pursued by reducing the antenna temperature. In the last generation of resonant detectors (AURIGA and NAUTILUS) the Al body of $M = 2300$ kg was cooled to temperatures close to 100 mK, at that time the lowest temperature ever achieved by an object of mass $\sim 10^3$ kg (Astone et al. 1991) (Fig. 8.8).

8.3.2   Amplifier Noises

- **Amplifier wideband (output) noise:** The wideband noise voltage v_n does not take part in the dynamics of the detector but is added to its output. It has a white spectrum (at least in the small bandwidth of interest) and its effect is then proportional to the bandwidth used.
- **Amplifier input (resonant) noise—Back Action:** It is the noise current i_n, generated by the amplifier or by lossy elements in the readout circuit. This noise source is usually negligibly small, but GW resonant detectors are extremely sensitive devices and the input noise appears in a twofold way. First, the noise current, circulating in the circuit impedance Z_{el}, generates an additional noise voltage:

$$S_0 = S_v + |Z_{el}|^2 S_i = S_v(1 + \lambda^{-2}) \tag{8.18}$$

The second effect is the back action: As shown by Eq. 8.7, the noise current that flows into the transducer produces, by reverse conversion, a noise force acting on the mechanical oscillators: $S_f^{(b.a.)} = |Z_{12}|^2 S_i$. Its spectral characteristics are hardly distinguishable from the thermal noise, and its variance is much smaller, so its effect is often accounted just with a correction to thermal noise, raising the thermodynamic temperature T to a slightly higher value T_e. From Eq. 8.13 we have

$$T_e = T + \frac{|Z_{12}|^2 \tau}{4 M_{eq} k_B} S_i \tag{8.19}$$

Although in the cryogenic antennas, equipped with SQUID, the back-action effect is negligible, this contribution to the linear readout measurement cannot be eliminated.

A linear transducer is conceived to continuously monitor the position (or the momentum) of the oscillator, providing a signal at the output of the system. This mechanism of back action is the classical equivalent of the influence of the measurement on a quantum state (see, for example, Caves 1983; Braginsky et al. 2003). In fact, the sensitivity achieved with linear measurement schemes has the ultimate limit related to the quantum nature of the oscillator and it is due to the Heisenberg principle: this is the *Standard Quantum Limit*.

This barrier can be circumvented by applying the *quantum non-demolition—QND* measurement strategies. In the equivalent classical case, i.e. when the system is affected by the *back action* noise, it has been experimentally demonstrated that, with a back-action evading (BAE) system (Cinquegrana et al. 1993) based on a specially designed parametric transducer, the limit can be surpassed.

Summarizing, the total noise voltage spectrum in the detector is the sum of *white* amplifier noise S_0, thus proportional to the measurement bandwidth $\Delta \nu = \Delta \omega / 2\pi$, and a *resonant* term, due to thermal and back-action noise, that has the Lorentzian shape of Eq. 8.14:

$$S_n(\omega) = \alpha^2 S_x(T_e, \omega) + S_0 \tag{8.20}$$

or, integrating over frequency:

$$\sigma_{V_{tot}}^2 = \alpha^2 \sigma_x^2(T_e) + S_0 \Delta \nu \tag{8.21}$$

where the explicit dependence on T_e reminds us that the back action must be included in the resonant noise.

8.4   Burst Sensitivity

8.4.1   Detection Bandwidth

An antenna affected only by thermal noise would have, in principle, infinite bandwidth: indeed, thermal noise and GW signal act on the detector in the same way, and appear at the output through the same transfer function (i.e. the antenna resonance). Therefore their ratio (SNR) is a constant independent of frequency. The presence of amplifier noise, which has a white spectrum, modifies this framework, as shown in Fig. 8.9: the response to a GW (and to thermal noise) is visible only near resonance, where it can peak above the white amplifier noise.

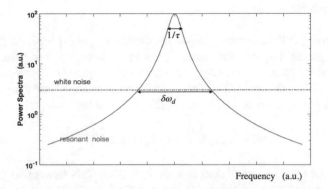

Fig. 8.9 The interplay between resonant thermal noise and wideband amplifier noise determines the detection bandwidth that can be much wider than the resonant bandwidth

The width of this region is determined by the relative strength of these two noise sources and is determined by the two frequencies where the noise spectrum Eq. 8.20 is reduced to half the maximum value

$$\delta\omega_d = \frac{2}{\tau}\sqrt{\frac{1+\Gamma}{\Gamma}} \qquad \text{with} \qquad \Gamma \equiv \frac{S_0/\tau}{\sigma_{V_{th}}^2} \tag{8.22}$$

Γ is the ratio between the wideband noise in the resonance bandwidth $1/\tau$ and the integrated resonant noise of Eq. 8.17.

In cryogenic detectors we had $\Gamma \sim 10^{-6}$ so that the SNR bandwidth was several Hz wide, much larger than the resonance bandwidth $2/\tau \sim$ mHz.

As $\Gamma \ll 1$, we shall approximate in the following expressions $1 + \Gamma \sim 1$. So we are led to distinguish between two very different time (or bandwidth) scales:

(a) *Mechanical bandwidth* of the resonant detector, $\Delta\omega = 2/\tau$ (resonance width), of the order of few mHz
(b) *Detection bandwidth* $\delta\omega_d \simeq \Delta\omega/\sqrt{\Gamma}$, $O(10\,\text{Hz})$, obtained by calculating the spectral trend of SNR.

The use of a resonant transducer does not modify this picture, although it improves it quantitatively.

Nevertheless, a band of tens of Hertz around the mechanical resonance of the detector (1 kHz in the case of the cryogenic cylindrical bars) is still insufficient to extract any information on the spectral shape of a GW signal. Thus, most of the searches focused on looking for sudden energy innovations, i.e. featureless δ-like signals.

8.4.2   Data Filtering

The mean energy in the antenna is proportional to the $\sigma_{V_{tot}}^2$ of Eq. 8.21. Proper filtering of data can greatly improve the sensitivity of these detectors. The basic idea is to look for a sudden change in the energy of the detector, against a slowly changing (as shown by Eq. 8.15) thermal noise. This is achieved by taking a time derivative of the output or, in first approximation, a finite difference at a time interval Δt. We have

$$\Delta \mathcal{E} = k_B T_e \frac{\Delta t}{\tau} + k_B T_n (1 + \lambda^{-2}) \frac{\omega_0 \beta}{\Delta t} \tag{8.23}$$

where β is the coupling coefficient defined in Eq. 8.8. This shows that there exists an optimum sampling time, balancing the contributions of thermal *innovation* and amplifier noise. Recalling that Nyquist's sampling theorem commands, for a desired bandwidth $\Delta \nu$, a minimum sampling time $\Delta t = \frac{1}{2\Delta \nu}$, the optimum sampling time recovers the detection bandwidth $\delta \omega_d$ of Eq. 8.22.

$\Delta \mathcal{E}_{min}$, often expressed as $k_B T_{eff}$, is a measure of the sensitivity integrated into frequency over the detector bandwidth.

In a more sophisticated approach, we can further improve the sensitivity by applying the Wiener-Kolmogorov, that is described in some detail in Sect. 10.11. In this case, the filtered output responds to an δ-like excitation of the detector as (Bonifazi 1978)

$$s(t) = \frac{V_s}{2\sqrt{\Gamma}} \exp\left(-\frac{|t|}{t_{wk}}\right) \tag{8.24}$$

where V_s is the maximum amplitude at the filter output that needs to be calibrated, and

$$t_{wk} = \tau \sqrt{\Gamma} = \frac{1}{\delta \omega_d} \tag{8.25}$$

The filtered signal $s(t)$ has a double exponential shape as it depends on $|t|$, and peaks at $t = 0$: we conclude that the filter response is neither advanced nor delayed with respect to the incoming signal. The Fourier transform of $s(t)$ is a complex function with square modulus and phase:

$$S^2(\omega) \propto \frac{1}{\frac{1}{t_{wk}^2} + \omega^2} \qquad \phi = \arctan(-\omega t_{wk}) \tag{8.26}$$

It is straightforward to verify that $\delta \omega_d = t_{wk}^{-1}$ has the meaning of detector bandwidth.

Following this procedure an expression of the minimum observable energy change at the detector output is derived:

$$\Delta \mathcal{E}_{min} = 4 k_B T_e \sqrt{\Gamma} \tag{8.27}$$

Equation 8.27 shows that the sensitivity gain for a δ-like signal achieved by the filter depends on the spectral ratio Γ, which must therefore be as low as possible: electronic noise in the detector bandwidth must be small compared to thermal noise.

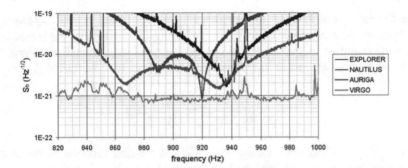

Fig. 8.10 An example of achieved spectral sensitivities of the resonant antennas NAUTILUS, AURIGA and EXPLORER. In this case the spectral strain sensitivity is about $10^{-21}/\sqrt{Hz}$. The sensitivities are compared to what Virgo had achieved in 2005. Credits: Acernese et al. (2008)

More refined methods, operating in the frequency domain, were developed over the years, leading to bursts sensitivities of the order of $T_{eff} \simeq 1\,mK$: a gain of 4000 over the thermodynamic, Brownian, mean energy.

8.4.3 Spectral Sensitivity

In order to compare with the typical sensitivity curves of an interferometer, we need to compute the quantity $S_h(\omega)$, i.e. the spectral sensitivity of the detector referred to its input. This is performed dividing Eq. 8.20 by the antenna transfer function: a simple Lorentzian for a single-mode antenna (as discussed so far), and a more complex function, with two resonance peaks, when a resonant transducer is employed.

Once referred to the input, the shape of the filter response $S(\omega)$ is overturned giving rise to the typical curves of Fig. 8.10, where we show the spectral sensitives achieved during a joint data taking of EXPLORER, NAUTILUS and AURIGA. In this case EXPLORER was running with the resonant transducer well tuned to the antenna resonance, while the NAUTILUS sensitivity was maximized on one frequency, where the pulsar remnant of Supernova Sn1987a was supposed to emit. AURIGA was instead optimized for maximum bandwidth. The picture shows how, in 2005, the performance of resonant bars was surpassed, both in sensitivity and bandwidth, by the interferometer Virgo: this led to the eventual dismissal of the resonant detectors, while Virgo further improved its sensitivity by two orders of magnitude in the following decade.

8.5 Resonant Detectors and Cosmic Rays

A particular source of noise is the acoustic effect produced by cosmic rays that affect the mechanical oscillator. This noise is both external and fundamental, i.e. unavoidable, although it could be reduced by locating the detector in an underground laboratory. Indeed, both extensive air showers and very energetic single particles have been shown capable to excite the antenna through a mechanism well explained by the so-called thermo acoustic model: loss of energy in the bulk of the massive bar → local heating → thermal expansion → excitation of longitudinal vibrations. The model predicts an amount of vibrational energy $\mathcal{E}$ given by (Coccia 1995)

$$\mathcal{E} = \frac{4}{9\pi} \frac{\gamma^2}{\rho L v_s^2} \left[\frac{dW}{dx} \right]^2 \cdot f(\vec{r}) \tag{8.28}$$

where $\frac{dW}{dx}$ is the ionization energy loss, $f(\vec{r})$ is a scalar function of the cosmic ray trajectory ($0 \le f \le 1$), ρ is the bar density and γ is the Gruneisen parameter: $\gamma = V(dP/d\mathcal{E})_V$ describes the coupling between changing volume and vibrational excitation in a solid. One of the thermodynamical definitions of Gruneisen parameter is $\gamma = \alpha_{th} v_s^2 / C_p$, where α_{th} is thermal expansion coefficient and C_p the heat capacity at constant pressure.

This disturbance became significant when the sensitivity reached the level $h \sim 1 \cdot 10^{-19}$. For this reason the most advanced resonant detectors were equipped with a cosmic ray telescope to veto the events induced by high energy particles hitting the bar (Astone et al. 2000). Analysis of these events allowed us to discover that some very energetic particles caused a local transition from the superconducting to the normal state of aluminium (cooled to 0.14 K), producing extremely large signals in the detector (Astone et al. 2008).

8.6   The Search of Periodic Signals with Resonant Detectors

Resonant antennas have also been used to search for periodic signals, such as those emitted by pulsars. In the 1970s, the Japanese group of Tokyo University and KEK studied different geometries in order to maximize the antenna response. Their goal was to select a vibrational mode with quadrupole symmetry resonating at a frequency close to 60 Hz, i.e. twice the rotation rate of the pulsar in the Crab nebula (*Crab pulsar*). They first developed a square-shaped detector with cuts, obtaining a sort of four-leaf clover (see Fig. 8.11). A second resonator, dumbell-shaped and cooled to 4 K, was later built, whose first torsional resonance mode was again tuned to 60 Hz (Tsubono 1991).

If the expected signal frequency is close to resonance, the detection strategy is relatively simple. We can assume that the dominant noise is due to thermal noise (Brownian and back-action noise), which corresponds to the $\Gamma \ll 1$ condition. This can always be achieved, as shown in Eq. 8.21, by extending the integration time t_{obs}, i.e. narrowing the bandwidth around the frequency of interest: that is because the power of electronic noise is proportional to the measurement bandwidth.

We compute the explicit expression of the SNR for a sinusoidal signal with amplitude h_0 and frequency $\omega \ne \omega_o$ by expanding the noise spectrum of Eq. 8.20 and comparing it with the signal of Eq. 8.6, but off-resonance:

Fig. 8.11 Quadrupole vibration mode of the cloverleaf antenna of the University of Tokyo. The depth of the lateral cuts tunes the resonance to the desired value, close to 60 Hz. The slab is suspended on the four nodal points of the vibrational mode, marked by black dots, to minimize noise input and ensure high Q. Credits: Tsubono (1991)

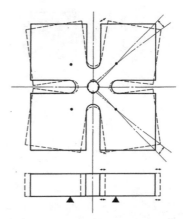

$$SNR(\omega) = \left(\frac{Lh_o}{\pi^2}\right)^2 \frac{\omega^4 M_{eq} Q \, t_{obs}}{\omega_o k_b T_e} \cdot \frac{1}{1 + \Gamma \left[Q^2(1 - \frac{\omega^2}{\omega_o^2})^2 + \frac{\omega^2}{\omega_o^2}\right]} \qquad (8.29)$$

This same relation, setting $SNR(\omega) = 1$ and $t_{obs} = 1$ s (unitary bandwidth) and solving for h_0, yields the spectral sensitivity curve of Sect. 8.4.3.

The detection band, measured at full width of half maximum (FWHM) height of Eq. 8.29 is

$$\Delta\omega_d = \frac{\omega_o}{Q} \frac{1}{\sqrt{\Gamma}} \qquad (8.30)$$

in agreement with Eq. 8.22: with the parameter values quoted above ($\omega_o = 2\pi \, 900$ Hz; $Q \sim 10^6$; $\Gamma \sim 10^{-4} - 10^{-7}$) this allows a useful band of the order of a few tens of Hertz.

8.6.1 The Cold Damping

The sensitivity to monochromatic signals, as shown in Eq. 8.29, is maximized when the antenna resonance is tuned to the signal. This is achieved in various ways:

– Modifying the detector geometry by machining it, raising or lowering the resonant frequency
– Adding small masses at appropriate points of the oscillating body: this can only lower the mechanical frequency
– By exerting static forces on the antenna, installing electric or magnetic actuators, or taking advantage of the *electric stiffness* of the transducers themselves. Magnetic transducers have a positive magnetic stiffness (raise the resonant frequency), while electric devices have it negative (they lower ω_o)

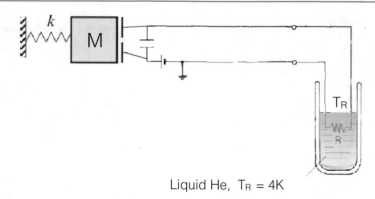

Liquid He, $T_R = 4K$

Fig. 8.12 The cold damping method for widening the detector band without increasing the thermal noise through a 4 K cooled resistance

In a long-term search for periodic signals, the detector must be able to track the frequency of the signal, if this changes. Indeed, the frequency of the gravitational wave signal can change over time due to various reasons:

- The Doppler effect caused by the orbital and the diurnal motion of the Earth, of the order of ± 0.03 Hz.
- A pulsar generally shows a small slowing of the rotation frequency (*spin down*) of the order of ~ 0.01 Hz/year.
- The signal can exhibit *glitches*, i.e. sudden jumps in the rotation frequency (Chap. 12 has more details on pulsar frequency and its tracking).
- Furthermore, the resonance frequency of the antenna not only changes with temperature, mainly due to thermal dilation, but also due to changes in the speed of sound. The magnitude of the effect depends on the vibration mode considered and on the antenna material, as well as on the operating temperature. At cryogenic temperatures, this effect is reduced, but slow variations are observed, related to the level of the cryogenic liquids or to changes in the electric or magnetic field present in the transducer, that change its stiffness.

If the antenna Q is too high, the signal can easily fall out of the useful bandwidth $\Delta \omega_d$ during the search. On the other hand, a low Q factor degrades the sensitivity of the detector, and should be avoided.

To overcome these difficulties, researchers at Tokyo University first used, in their search for a 60 Hz signal from the Crab pulsar, a very ingenious system to obtain a widening of the detection band without altering the signal to noise ratio (Hirakawa et al. 1977). Consider the scheme shown in Fig. 8.12. The harmonic oscillator represents the antenna, at room temperature ($T \simeq 280$ K), equipped with an electrostatic transducer to which the resistance R is connected. The equations that characterize the electromechanical system are

$$M\left(\frac{d^2x}{dt^2} + \frac{\omega_o}{Q_A}\frac{dx}{dt} + \omega_o^2 x\right) + E_o q = 0$$

$$R\frac{dq}{dt} + \frac{1}{C}q + E_o x = 0 \tag{8.31}$$

where ω_o, Q_A, M are, as usual, the antenna resonant (angular) frequency, quality factor and mass, respectively; E_o is the average electric field in the transducer capacitor C, q its charge and R is the readout resistor.

Combining these two equations, with some approximations, we get

$$\frac{d^2x}{dt^2} + \omega_o\left\{\frac{1}{Q_A} + \beta\frac{RC\omega_o}{1 + (RC\omega_o)^2}\right\}\frac{dx}{dt} +$$

$$+ \omega_o^2\left\{1 - \beta\frac{1}{1 + (RC\omega_o)^2}\right\}x = 0 \tag{8.32}$$

where β is the electromechanical coupling factor of the capacitive transducer:

$$\beta = \frac{CE_o^2}{M\omega_o^2} \tag{8.33}$$

Equation 8.32 describes the motion of an oscillator with a modified quality factor:

$$\frac{1}{Q^*} = \frac{1}{Q_A} + \frac{1}{Q_R} \quad \text{with} \quad \frac{1}{Q_R} = \beta\frac{RC\omega_o}{1 + (RC\omega_o)^2} \tag{8.34}$$

We now focus on the thermal fluctuations of this system.

If we lower the temperature of the resistance R by plunging it in a liquid helium bath, $T_R = 4$ K, a new stationary status is established, with heat exchanged between the antenna and the resistor. Antenna and resistance exchange the respective fractions of dissipated thermal power and, at equilibrium, we have

$$\frac{\omega_o}{Q_A}k_B(T^* - T_A) = \frac{\omega_o}{Q_R}k_B(T_R - T^*) \tag{8.35}$$

It follows that

$$\frac{T^*}{Q^*} = \frac{T_A}{Q_A} + \frac{T_R}{Q_R} \tag{8.36}$$

Combining Eqs. 8.34, and 8.36 we get

$$T^* = \frac{\left(1 + (RC\omega_o)^2\right)T_A + \beta Q_a RC\omega_o T_R}{\left(1 + (RC\omega_o)^2\right) + \beta Q_A RC\omega_o} \tag{8.37}$$

In the case $\beta Q_A >> 1$, we can simplify this relation to get the equivalent temperature

$$T^* = T_R + 2T_A \frac{1}{\beta Q_A} \tag{8.38}$$

that can be substantially lower than T_A. The thermal fluctuations of the antenna vibration x now take place at the equivalent temperature T^*:

$$\sigma_x^2 = \frac{k_B T^*}{M\omega_o^2} \tag{8.39}$$

Such lowering of the equivalent temperature of the system is called *cold damping*. Inevitably, this comes with a lowering of the detector's Q. Therefore the signal decreases along with the noise and the SNR does not change. This method makes it possible to extend the detection bandwidth, thus making it easier to track the periodic signal in frequency.

References

Aguiar, O.D.: The past, present and future of the resonant-mass gravitational wave detectors (2010). arxiv:1009.1138

Aguiar, O.D., et al.: The Brazilian gravitational wave detector Mario Schenberg: status report. Class. Quantum Grav. **23**, S239 (2006)

Astone, P., et al.: First cooling below 0.1 K of the new gravitational-wave antenna "Nautilus" of the rome group. Europhys. Lett. **16**, 231 (1991)

Astone, P., et al.: Noise behaviour of the Explorer gravitational wave antenna during the λ transition to the superfluid phase. Cryogenics **32**, 668 (1992)

Astone, P., et al.: Cosmic rays observed by the resonant gravitational wave detector NAUTILUS. Phys. Rev. Lett. **84**, 14 (2000)

Astone, P., et al.: Detection of high energy cosmic rays with the resonant gravitational wave detectors NAUTILUS and EXPLORER. Astroparticle Physiscs **30**, 200 (2008) and references therein

Bonifazi, P., Ferrari, V., Frasca, S., Pallottino, G.V., Pizzella, G.: Data analysis algorithms for gravitational-wave experiments. Nuovo Cimento **1C**, 465 (1978)

Braginsky, V.B., et al.: The noise in gravitational wave detectors and other classical force measurements is not influenced by test mass quantization. Phys. Rev. D **67**, 082001 (2003)

Caves, C.M.: Quantum non-Demolition Measurements. In: Meystre, P., Scully, M. (eds.) Quantum Optics, Experimental Gravity, and Measurement Theory. Springer, Boston (1983)

Cinquegrana, C., Majorana, E., Rapagnani, P., Ricci, F.: Back-action-evading transducing scheme for cryogenic gravitational wave antennas. Phys. Rev. D **48**, 448 (1993)

Goryachev, M., et al.: Rare events detected with a bulk acoustic wave high frequency gravitational wave antenna. Phys. Rev. Lett. **127**, 071102 (2021)

Coccia, E., Marini, A., Mazzitelli, G., Modestino, G., Ricci, F., Ronga, F., Votano, L.: A cosmic-ray veto system for the gravitational wave detector NAUTILUS. Nucl. Inst. & Meth. Phys. Res. A **355**, 624 (1995)

Clarke, J.: SQUIDs: then and now. Int. J. Mod. Phys. B **24**, 3999–4038 (2010)

Acernese, F., et al.: First joint gravitational wave search by the AURIGA-EXPLORER-NAUTILUS-virgo collaborations. Class. Quantum Grav. **25**, 205007 (2008)

Gottardi, L., et al.: Sensitivity of the spherical gravitational wave detector MiniGRAIL operating at 5 K. Phys. Rev. D **76**, 102005 (2007)

Hirakawa, H., Hiramatsu, S., Ogawa, Y.: Damping of Brownian motion by cold load. Phys. Lett. A **63**, 199 (1977)

Paik, H.J.: Superconducting tunable diaphragm transducer for sensitive acceleration measurements. J. Appl. Phys. **47**, 1168–1178 (1976)

Ricci, F.: Mechanical noise and low temperature physics aspects of the gravitational wave experiment. In: Posada, E., Violini, G. (eds.) The Search of Gravitational Waves, p. 157. Word Scientific, Singapore (1982)

Tsubono, K.: Detection of continuous waves. In: Blair, D.G. (ed.) The Detection of Gravitational Waves. Cambridge University Press, Cambridge (1991)

Wagoner, R.V., Paik, H.J.: Multi-mode Detection of Gravitational Waves by a Sphere. B. Bertotti ed. Accademia Nazionale dei Lincei, Roma (1977), p. 257

Weber, J.: General relativity and gravitational waves. Interscience Tracts on Physics and Astronomy. Interscience Publishers Inc, New York (1961)

Interferometric Detectors of Gravitational Waves

9

9.1 Introduction

The pioneering idea of using a light signal that travels between two freely gravitating masses and characterizes the metric of space is already present in Piran's work (Pirani 1956).

J. Weber, the scientist who devised and built the first resonant detectors (Chap. 8), together with R. Forward, first considered the possibility of replacing the spring with a beam of laser light. Forward, with L. Miller and G. Moss, performed an experiment of this type at the *Hughes* research lab (Moss et al. 1971; Forward 1978).

Several classes of detectors can be traced back to this principle *spacecraft Doppler tracking*, a method based on the monitoring of Doppler signals for tracking artificial satellites, *planetary ranging*, or the measurements of planetary orbitals, the combined observation of the arrival times of the radio signals emitted by the pulsars, and finally the long-baseline Michelson interferometers.

In each of these techniques the effect of the gravitational wave is that of influencing the propagation of the signal e.m. and the test masses. In the case of *pulsar timing* the signals are simply received at Earth (the first gravitating mass) and accurately compared with a phase reference derived from the best available (in terms of stability) atomic clocks. In the other case, the signal originates from a mass and is then reflected backwards (or re-transmitted while maintaining phase coherence) from the second mass. Since the seismic noise is absent, these techniques can be used to detect signals in the frequency band from 10^{-8} to 1 Hz.

Finally, it should be emphasized that the idea of the interferometric detection of gravitational waves was taken up and literally re-invented in 1972 by R. Weiss at M.I.T. (Weiss 1972). In fact, he carried out a real feasibility study of the experimental configuration, laying the foundations of the current Virgo and LIGO detectors.

The original version of this chapter was revised: Chapter have been updated with the correction. The correction to this chapter can be found at https://doi.org/10.1007/978-3-030-95596-0_15
© The Author(s), under exclusive license to Springer Nature Switzerland AG 2022, corrected publication 2023
F. Ricci and M. Bassan, *Experimental Gravitation*, Lecture Notes in Physics 998, https://doi.org/10.1007/978-3-030-95596-0_9

9.2   The Photon Detector

We analyse the effect of a gravitational wave on a system of free masses initially at rest, making the obvious choice to use a coordinate system where the TT gauge holds: we call it the TT reference frame. Under this hypothesis we can show that, while the time-space structure around them changes, the coordinates of the test masses do not change over time: an accelerometer fixed on each mass would not feel any effect, while if a photon travels between two of these masses, the transit time follows the metric change of the space-time.

We consider two test masses, TM1 and TM2, in free fall along two geodesics, whose distance is deduced on the basis of the equation of geodesics deviation (5.9). It can be easily shown that the masses initially at rest remain in this condition, in the TT frame, despite the effect of the plane wave described . This follows from the fact that the time components of the Christoffel symbols of the second type (Γ^r_{jt}) are also null in the perturbed metric. Therefore, the *coordinate* distance between the two masses remains unperturbed (say L) even when the wave passes by. In other words, in the TT gauge, it is the metric itself that gets distorted by the wave, in such a way that the positions of the test masses initially at rest do not change.

So, what changes is the geometry of the space hit by the wave, that we assume to be a plane wave, polarized $h_{yy} = -h_+$, propagating along the z-axis.

$$h_+(\vec{r}, t) \rightarrow h_0 \, cos(\omega_g t - kz + \phi) \tag{9.1}$$

As a result, the corresponding ds^2 is

$$ds^2 = g_{\mu\nu}dx^\mu dx^\nu = c^2 dt^2 - (1 + h_+(z,t))dx^2 - (1 - h_+(z,t))dy^2 \tag{9.2}$$

If we limit our considerations to e.m. radiation of wavelength much shorter than that of the gravitational wave (i.e. in the limit of geometric optics), the trajectories described by the light rays are still of the geodesic null type. In this way, the integration of the photon geodesic in the perturbed space-time still expresses the apparent distance (the optical path) covered by the photon in a return trip between the two freely gravitating masses. We can quantitatively express this by calculating the propagation (or delay) time t_r, corresponding to the forward and backward travel of $2L$ along one of the axes (we'll take the x-axis) in the plane $z = 0$. We start from the null geodesic:

$$ds^2 = 0 \quad \rightarrow \quad c^2 dt^2 = [1 + h_+(\vec{r}, t)]dx^2$$

For the round trip of each photon, leaving TM1 at the time t_0, bouncing off TM2 and returning to TM1 at t_r, we can write

$$2L = \int_0^L dx + \int_L^0 d(-x) = c \int_{t_0}^{t_r} \frac{dt}{\sqrt{1 + h_+(t)}} \tag{9.3}$$

We now series expand $(1 + h_+)^{-1/2}$ and integrate, assuming for h(t) a waveform like Eq. 9.1:

$$2L \simeq c(t_r - t_0) - \frac{h_0}{2\omega_g}[sin(\omega_g t_r) - sin(\omega_g t_0)] \tag{9.4}$$

As we are limiting our solution to first order in h_0, we can take $t_r \simeq t_0 + 2L/c$, in the argument of the sine function in Eq. 9.4; so we get for the round trip time

$$t_r - t_0 = \frac{2L}{c} \pm h_0 \frac{L}{c} \frac{\sin \eta}{\eta} \cos(\omega_g t + \eta) \tag{9.5}$$

where $\eta = \frac{\omega_g L}{c}$ and the "-" sign holds for photons travelling along the x-axis (and "+" for the y-axis).[1]

Clearly, the round trip travel time is, in the unperturbed ($h_+ = 0$) case, $(t_r - t_0)^{(0)} = 2L/c$; also, if the distance L is short enough, and $h_+(t)$ changes slowly enough to be considered constant during the travel time, we simply have $(t_r - t_0)^{(1)} = (2 + h_+(t))L/c$ so that the GW produce a perturbation $\delta(T_{rt}) = \frac{L}{c} h_0$. This is also the simple, intuitive result we would get by computing the effect in a *detector-based reference frame*. Note also that the travel time depends on time, via the phase of the gravitational wave.

We now compute the effect of the GW on the electromagnetic field of a monochromatic light wave, of frequency $\omega_L/2\pi$, travelling between TM1 and TM2: Call $A(t_0) = A_0 e^{-i\omega_L t_0}$ the (complex) amplitude at the beginning of the round trip, and $B(t_0) = A(t_r)$ its value at the end. To evaluate B(t) we substitute in the exponential t_r as given by Eq. 9.5 and we expand to first order in h_0.

To simplify the notation, from now on we drop the index "0", for a generic initial time, and drop "+" from h_+ and we introduce the function[2] $sinc(x) = \frac{sin(x)}{x}$

$$B(t) = A(t_r) = A_0 e^{2i\omega_L L/c} \cdot$$
$$\cdot [e^{-i\omega_L t} + \frac{ih_0}{2} \frac{\omega_L L}{c} sinc(\eta)(e^{i\eta} e^{-i(\omega_L + \omega_g)t} + e^{-i\eta} e^{-i(\omega_L - \omega_g)t})] \tag{9.6}$$

We notice that, due to the action of GW, besides the field at frequency ω_L, we now have two sidebands at frequencies $\omega_L \pm \omega_g$. In an interferometer, the light beam can bounce back and forth between the mirrors, feeling several times the effect of GW: therefore we shall use, as incoming e.m. field in the detector, a combination of these three frequencies:

$$A(t) = \left(A_o + \frac{1}{2}he^{i\omega_g t} A_1 + \frac{1}{2}he^{-i\omega_g t} A_2\right)e^{-i\omega_L t} \tag{9.7}$$

---

[1] We are still in a very simplified case, as we consider only h_+ and we assume a very lucky direction of propagation $\vec{k}$, perpendicular to L for the plane wave of Eq. 9.1. For a more general approach, see Vinet (1988).

[2] This definition of sinc(x) is the most common, but it is not unique. Theorists will recognize the function $j_0(x)$, spherical Bessel function of zeroth order.

In a detailed discussion of a complex optical detector, it is useful to introduce an operator D that describes the propagation along the round trip path of length $2L$. This is achieved by iterating the procedure described above: substituting t, as defined in expression Eq. 9.5, into Eq. 9.7, and expanding to first order in h_0. Defining also $\xi = \omega_L L/c$, we obtain

$$
\begin{aligned}
B(t) = e^{i2\xi}\Bigg[A_o e^{-i\omega_L t} + \\
\frac{1}{2} h_o e^{-i(\omega_L - \omega_g t)}\left(A_1 e^{-2i\eta} \pm i\xi sinc(\eta)e^{-i\eta}A_0 \right) + \\
\frac{1}{2} h_o e^{-i(\omega_L + \omega_g t)}\left(A_2 e^{2i\eta} \mp i\xi sinc(\eta)e^{i\eta}A_0 \right)\Bigg]
\end{aligned}
\tag{9.8}
$$

We can rewrite this using a matrix form:

$$
\begin{pmatrix} B_o \\ B_1 \\ B_2 \end{pmatrix} = D \begin{pmatrix} A_o \\ A_1 \\ A_2 \end{pmatrix}
\tag{9.9}
$$

where

$$
D = e^{2i\xi} \begin{pmatrix} 1 & 0 & 0 \\ i \pm \xi \frac{sin(\eta)}{\eta}e^{-i\eta} & e^{-2i\eta} & 0 \\ i \pm \xi \frac{sin(\eta)}{\eta}e^{i\eta} & 0 & e^{2i\eta} \end{pmatrix}
\tag{9.10}
$$

The matrix formalism is a powerful tool to analyse complex optical system; Eq. 9.10 gives us the opportunity to treat also the interaction GW-travelling light within this formalism: each of the optical components of the system is described by a matrix, element of a non-commutative algebra in which we define a global operator O that represents the detector as a whole (Vinet 2020).

9.3   The Michelson Interferometer

The optical configuration of a Michelson interferometer consists of a laser beam split into two equal parts by a beamsplitter, i.e. a mirror with a partial reflecting surface: the two resulting beams (one originating by partial reflection and the other by partial transmission) travel towards two "end" mirrors, M1 and M2, then reflected back to the beamsplitter where they recombine before hitting the photodetector. The typical configuration used in the ground-based GW detectors is characterized by setting the beamsplitter at $45°$ with respect to the incoming beam: the two split beams then travel in orthogonal paths. If the two end mirrors, M1 and M2, are positioned at equal distance ($L1 = L2$), see Fig. 9.1, one end mirror overlaps with the virtual image of the other generated by the beamsplitter.

Although this is the optimal configuration for GW detection, other configurations can be considered: in particular, the LISA triangle can be treated as a combination of three Michelson interferometers, each of them forming an angle of $60°$.

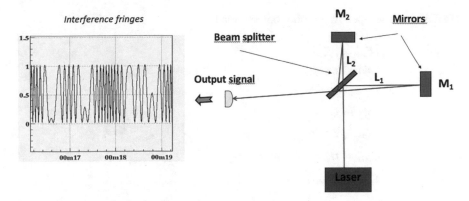

Fig. 9.1 The simplest scheme of a Michelson interferometer

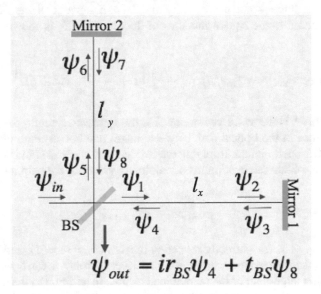

Fig. 9.2 The e.m. fields in the different points inside the Michelson interferometer

A more general analysis of the Michelson interferometer can be done using the optical matrix formalism. Here we refer to the fields as defined in Fig. 9.2. The two mirrors have reflectivities, r_1 and r_2, that can be different; r_{bs} and t_{bs} indicate the reflectivity and transmittivity of the beamsplitter, respectively; ψ_i, $i = 1...n$ is the e.m. field at the different point inside the interferometer; and ψ_{in} and ψ_{out} are the e.m fields at the input and output port of the Michelson set-up. Using the matrix formalism developed in the previous sections

$$P_{in} \propto |\psi_{in}|^2$$

$$\psi_1 = t_{bs}\, \psi_{in} \qquad\qquad\qquad \psi_5 = r_{bs}\, \psi_{in}$$

$$\psi_2 = e^{-ikL_x}\psi_1 \qquad\qquad\quad\; \psi_6 = e^{-ikL_y}\psi_5$$

$$\psi_3 = ir_1\, \psi_2 \qquad\qquad\qquad\;\; \psi_7 = ir_2\, \psi_6$$

$$\psi_4 = e^{-ikL_x}\psi_3 \qquad\qquad\quad\; \psi_8 = e^{-ikL_y}\psi_7$$

$$\psi_{out} = ir_{bs}\psi_7 + t_{bs}\psi_8$$

$$P_{out} \propto |\psi_{out}|^2$$

Carrying through the products we finally find the e.m. power collected at the output port: proportional to the square modulus of the output field, is shown in the same figure

$$P_{out} \propto |\psi_{out}|^2 = P_{in}r_{bs}^2 t_{bs}^2 (r_1^2 + r_2^2)\left[1 + \frac{2r_1 r_2}{r_1^2 + r_2^2}\cos 2k\delta L\right] \qquad (9.11)$$

where $k = 2\pi/\lambda$ is the wave vector and λ is the light wavelength, $\delta L = L_x - L_y$ is the difference of the optical path between mirror and beamsplitter along the two directions. The coefficient multiplying the cosine is the contrast C: it is measured by changing δl to obtain the maximum and minimum values of the output power, being

$$C = \frac{2r_1 r_2}{r_1^2 + r_2^2} = \frac{P_{max} - P_{min}}{P_{max} + P_{min}} \qquad (9.12)$$

C varies between 1, for perfectly reflecting mirrors ($r_1 = r_2 = 1$), and 0, when no interference takes place. However, even with perfect mirrors, we can have $C < 1$, due to an imperfect alignment of the recombined beams. In addition to the output beam, directed towards the detector, another beam is sent back from the interferometer towards the laser source. The field of this beam is in phase opposition with respect to that at the output port (just count the reflections, each reflection adds $\pi/2$ to the phase), complying with energy conservation. This implies that when the output power is at its minimum value, the beam reflected back at the input port is at its maximum.

9.4   The Suspended Mirror

We now return to the problem of the interaction of a GW with a light beam: in Sect. 9.2 we had taken the simplifying hypothesis that the two test masses be in free fall. In a real interferometric system installed on the ground, the bodies are suspended mirrors to form pendulums. We now calculate the phase difference measured by this system by applying the equation of geodesic deviation for the three masses with respect to the centre of mass of the whole system. As a result, the action of the gravitational wave can be seen as an acceleration field $a(t, \vec{\xi}_{c.m.s.})$ to which the three suspended masses are subjected:

$$a_j^{(n)}(t, \vec{\xi}_{c.m.s.}^{(n)}) = \frac{1}{2c^2} \frac{\partial^2 h_k^{j(TT)}}{\partial t^2} \xi_{c.m.s.}^{k}{}^{(n)}$$

where

$$\vec{\xi}_{cms}^{(n)} \equiv (x^{(n)} - x_{cms}, y^{(n)} - y_{cms}, z^{(n)} - z_{cms})$$

indicates the position vector of the n-th pendulum (n = 0, 1, 2, see Fig. 9.3). If the origin of our reference is placed on the beam separator (*beamsplitter*), we have the equations of motion for the various masses:

$$\ddot{x}_0 + \tau^{-1}\dot{x}_0 + \omega_p^2 x_0 = -\frac{1}{2}(\ddot{h}_{xx} x_{cms} + \ddot{h}_{xy} y_{c.m.s.})$$

$$\ddot{y}_0 + \tau^{-1}\dot{y}_0 + \omega_p^2 y_0 = -\frac{1}{2}(\ddot{h}_{yx} x_{cms} + \ddot{h}_{yy} y_{c.m.s.})$$

$$\ddot{x}_1 + \tau^{-1}\dot{x}_1 + \omega_p^2 x_1 = -\frac{1}{2}(\ddot{h}_{xx}(L - x_{cms}) + \ddot{h}_{xy} y_{c.m.s.})$$

$$\ddot{y}_2 + \tau^{-1}\dot{y}_2 + \omega_p^2 y_2 = \frac{1}{2}(-\ddot{h}_{yx} x_{cms} + \ddot{h}_{yy}(L - y_{c.m.s.}))$$

For sake of simplicity we have omitted the (TT) symbol; we introduced both the relaxation time τ and the characteristic pulsation ω_p, which we assume, for simplicity, equal for all pendulums. The phase difference measured by the detector is

$$\delta\phi = 4\pi[(\Delta x - \Delta y)] \tag{9.13}$$

where $\Delta x = x_0 - x_1$ and $\Delta y = y_0 - y_2$.

This quantity is deduced by combining the linear differential equations previously written:

$$\ddot{\Delta\phi} + \tau^{-1}\dot{\Delta\phi} + \omega_p \Delta\phi = \frac{4\pi}{\lambda}\ddot{h}_{xx} L \tag{9.14}$$

The solution of this equation depends on the form of $\ddot{h}_{xx}$.

If the metric perturbation $h(t)$ is a burst of amplitude h_o and duration $\Delta t \ll 1/\omega_p$, we have

$$\delta\phi(t) \simeq \frac{4\pi}{\lambda} L h(t) + \frac{4\pi}{\lambda} L h_o [\omega_p \Delta t \, sin(\omega_p t) e^{-\frac{t}{\tau}}] \tag{9.15}$$

which holds for large values of the oscillator quality factor $Q \equiv \omega_p \tau/2 \gg 1$.

This equation shows how the detector output is a measure of $h(t)$; the second term represents, for times $t > \Delta t$, the memory effect on the pendulum system of the occurred interaction with the gravitational wave. This term, being weighted by the factor $\omega_p \Delta t << 1$, can usually be neglected.

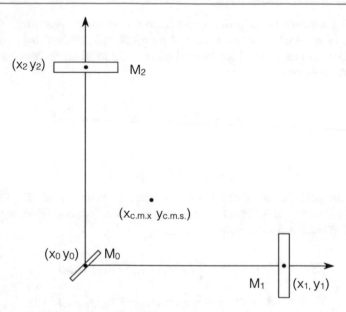

Fig. 9.3 Michelson interferometer scheme for the calculation of Sect. 9.4

In the case of a sinusoidal gravitational wave

$$h(t) = h_o e^{-i\omega_g t}$$

we have for $\omega_g \gg \omega_p$ and $Q \gg 1$

$$\delta\phi = \frac{4\pi L}{\lambda} h_o e^{i\omega_g t} \tag{9.16}$$

This means that the interferometer can detect sinusoidal signals at frequencies higher than the resonance of the pendulum suspension.

The approach followed here is equivalent to the matrix treatment, where the mirrors are considered in free fall, as long as the frequency of the mirror suspension is lower than those characteristics of the signal (Fig. 9.3).

In this section, we have consistently neglected the factor $sinc(\omega_g L/c)$, see, for example, Eq. 9.8, accounting for the signal reduction when the light travel time inside the interferometer arm is comparable with half of the GW period. Although this is certainly acceptable for a simple Michelson,[3] we shall recall this correcting factor in Sect. 9.5 where we discuss interferometers with optically extended arms.

---

[3] Even at the highest frequency of 5 kHz, the GW wavelength is $\lambda_{gw} = 60$ km, and the biggest interferometer, LIGO, has arms 4 km long.

Fig. 9.4 Two possible
configurations for a
Michelson interferometer
with "extended arms". The
FP configuration is shown at
the top, and in the lower part
of the figure, the optical
delay line. The colour
drawing highlights the need
for the trajectories of the
bouncing light beam to be
well separated

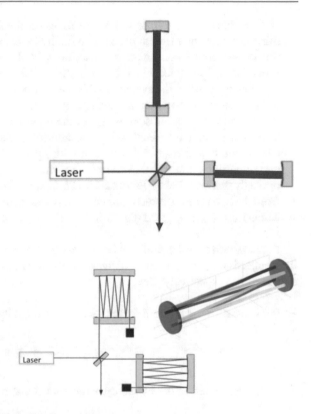

9.5   The Experimental Configuration

In the previous section we have shown how the sensitivity of an interferometric
detector depends on the length of the optical path of light in the arms. The light path
can be increased by inserting an auxiliary mirror positioned in the vicinity of the
beam separator into each of the Michelson interferometer arms. The light passes the
first mirror and then bounces back and forth several times along the Michelson arm
before exiting (*being extracted*) and interfering with the light from the other arm.
The total length of the optical path is chosen so as to optimize the response function
of the interferometer for a given frequency of the gravitational wave: one chooses the
time of entrapment of light in the arm equal to half of the period of the target wave
to be detected. For example, if one wants to optimize the interferometer to detect
signals at 1 kHz, the optical path should be of the order of 150 km.

There are two possible optical configurations that allow us to obtain a higher
trapping time than the simple Michelson interferometer: (Fig. 9.4)

–   The *multi-pass* Michelson, which includes in each arm an optical delay line (*Delay
    Line*—DL). DL consists of two spherical mirrors arranged in a quasi-confocal
    configuration (Herriott et al. 1964). The light beam enters through a hole made
    in the first (*input*) mirror and travels back and forth for N round trips, reflecting

N times off the *end* mirror and $N - 1$ times off the *input* one. The beam hits the mirrors always in different points, and finally exits from another hole in the input mirror and recombines on the beamsplitter with the other beam coming out of the second arm. The recombined beam impinges on a photodiode that monitors the interference state of the system. In this way, we extend the *effective path length* to $2NL$, at the cost of attenuating the light power by a factor $\sim (r_i r_e)^N$, where r_i, r_e are the reflectivities (close to unity) of the input and end mirror, respectively.

- The Michelson–Fabry-Perot (FP): the interferometer is obtained by placing, in each of the two arms, besides the (virtually) perfectly reflecting mirror at the far end, a partially transmitting mirror after the beamsplitter so as to compose an optically resonant Fabry-Perot cavity (Pérot and Fabry 1899). The light returning from the two cavities is recombined in phase opposition, as it is done in the simple Michelson. For a detailed discussion on Fabry-Perot, see Appendix C.

A few parameters are useful to characterize a FP cavity; we first define the amplitude reflectivities r_i, r_e (just as above) and the trasmissivities t_i, t_e of the input and end mirrors, respectively; then

- the finesse, a sort of merit factor for an optical cavity

$$\mathcal{F} = \frac{\pi \sqrt{r_i r_e}}{1 - r_i r_e} \gg 1 \qquad (9.17)$$

- the power stored in the FP cavity, when on resonance, is enhanced by

$$P_{eff} = P_{in} \frac{2\mathcal{F}}{\pi} \qquad (9.18)$$

- the frequency spacing of the resonance lines, *free spectral range—FSR*,

$$\Delta \nu_{FSR} = \frac{c}{2L} \qquad (9.19)$$

- the width of the resonance lines, *full-width half-maximum—FWHM*,

$$\Delta \nu_{FWHM} = \frac{\Delta \nu_{FSR}}{\mathcal{F}} \qquad (9.20)$$

- the time that a light wavefront (or packet) remains in the cavity, the *storage time* τ_s

$$\tau_s^{FP} = \frac{2L}{c} \frac{(r_i r_e)^{\frac{1}{2}}}{1 - r_i r_e} = \frac{2L}{\pi c} \mathcal{F} \qquad (9.21)$$

A detailed comparison of the two systems is a cumbersome task, and we refer to the complete analysis of these detectors (Vinet et al. 1988) based on the algebra of optical matrices. Here we just report the expressions of the phase change observed at the output of the detector, caused by a monochromatic gravitational wave of angular frequency ω_g in the two cases:

$$\delta\phi_{DL} = h_o\frac{2\pi NL}{\lambda}sinc(\omega_g NL/c) \qquad (9.22)$$

$$\delta\phi_{FP} \simeq h_o\frac{8\mathcal{F}L}{\lambda}\frac{1}{\sqrt{(1+\omega_g^2\tau_s^2)}}sinc(\omega_g L/c) \qquad (9.23)$$

We note two main differences with respect to the phase change $\delta\phi_{Mich} = h_0\frac{4\pi L}{\lambda}$ in a single-pass Michelson (see Eq. 9.16):

– The optical path length (2L) is now extended to 2NL or $8\mathcal{F}L/\pi$, respectively.
– The response is no longer independent of frequency: the factors $sinc(\omega_g NL/c)$ for the DL configuration, and

$(1+\omega_g^2\tau_s^2)^{-1/2}$ for FP, produce a slow reduction of the phase change signal when the frequency increases.

The factor $sinc(\omega_g L/c)$ for the FP, which we have reintroduced for completeness, is on the other hand irrelevant: its value is virtually 1 for physical arm lengths of a few kilometres and frequencies below 10 kHz.

In the relations above, and in particular in Eq. 9.23, we have neglected the possible losses of the mirrors.

If we compare the performance at equal storage time, under the reasonable hypothesis $\omega_g\tau_s \leqslant \pi$, it turns out that the FP configuration is only a factor of two more sensitive than DL.

However, the choice between these configurations also requires an analysis of the difficulties encountered in carrying out each one. In the case of delay lines the main problem is the effect of light scattered by the mirror, which overlaps with the main beam, producing spurious signals. In particular, the condition of avoiding any overlap between the reflection spots of the beam on each mirror of the DL imposes the use of spherical mirrors of large diameter and excellent quality, especially on the edge. In the FP cavity, on the other hand, the beam is also reflected many times, but always from the central section of the mirror, at normal incidence; therefore a smaller diameter is required for the mirrors, as well as for the vacuum pipes connecting them: this means, for a many-kilometre ultra-high vacuum system, a relevant reduction of costs.

The difference between the static length of optical paths in the two arms is a problem common to the two configurations because it makes them sensitive to fluctuations in the laser frequency. This difference is due, in both cases, to asymmetries in the practical implementation of the two arms: unequal curvature radii for the DL, unequal finesse (that depends on both the curvature radius and the reflectivity of the mirrors) in the case of the FP cavities. This asymmetry places even more demanding conditions on the reduction of frequency noise of laser light.

In conclusion, the choice of the FP system compared to the DL is due to two main reasons:

- The FP configuration needs mirrors with diameters five times smaller than the DL device.
- The light inside the cavity FP travels back and forth along the same path so that the problem of the scattered light is greatly reduced.

For this reason, LIGO, Virgo and KAGRA detectors preferred to adopt the FP configuration.

The complexity of the configuration is then further increased by the addition of a *power recycling* mirror at the input, to increase the laser intensity circulating in the arms, of a *signal recycling* mirror at the output, to enhance the sensitivity at a particular frequency range, and of a *light squeezing* system at the output, to achieve a shot noise reduction through the use of suitably prepared light wavepackets. All these developments are briefly described in the following sections.

9.6   Interferometer Signal and Noises

9.6.1   The Signal

As a first step in the derivation, we shall focus on a simple Michelson interferometer ($N = 1$), with virtually perfect alignment and parameters: $r_1 = r_2 = 1, r_{bs} = t_{bs} = 1/2$, but leaving the contrast C in the equations, as it also depends on the quality of the alignment. We shall however maintain a possible difference in the length of the arms that produces a static phase difference $\alpha = \frac{4\pi}{\lambda}(L_x - L_y)$. In the presence of a gravitational wave that generates an additional phase difference $\phi_{gw} = \frac{4\pi L}{\lambda}h(t)$, the light power P_{out} at the interferometer output (Eq. 9.11) reduces to

$$P_{out} = \frac{P_{in}}{2}\left[1 + C\cos(\alpha + \phi_{gw})\right] \tag{9.24}$$

where P_{in} is the input power. Equation 9.24 allows us to deduce the sensitivity of the instrument to a gravitational signal:

$$\delta P_{gw} = \frac{\partial P_{out}}{\partial \phi_{gw}}\delta\phi_{gw} = \frac{P_{in}}{2}C\sin(\alpha)\,\delta\phi_{gw} \tag{9.25}$$

Thus, the signal is maximized by the choice $\alpha = (k + 1)\pi/2$ with k integer. This would naively suggest to adjust the interferometer so that the output light is halfway between the conditions of constructive (maximum light power, bright fringe) and destructive (minimum light, dark fringe) interference: this is called condition of half fringe, or grey fringe.

But the correct approach is, rather than maximizing the signal, to configure the interferometer in order to optimize the signal to noise ratio (SNR). Although many noise sources must be included in this optimization, it is instructive here to evaluate the SNR considering only the *optical, or readout* noise components, which limits the sensitivity in the higher range of the detector bandwidth. Other noises, and their role in determining the sensitivity, will be mentioned at the end of this chapter.

So, we must compare the signal δP_{gw} of Eq. 9.25 with the fluctuations δP_{shot} in the output power caused by a fundamental noise source: granular noise (or *shot noise*) on the output photodiode.

9.6.2 The Antenna Pattern

Gravitational waves of the same amplitude and frequency but arriving from different directions will produce different length changes in the interferometer. For example, a GW incident perpendicular to the interferometer plane will produce strain on both arms and the effect at the interferometer output GW is summed while, if the same GW is incident in-line with one arm does not produce a length change in this arm (GW in GR are transverse) but only on the other so that the signal at the interferometer output is half of the previous case. Additionally, if a GW is incident from a non-orthogonal angle each arm will see the projection of the GW onto its axis and therefore measure a reduced amount of strain. Thus, the detector sensitivity depends also on the propagation direction of the GW with respect to the interferometer; besides, this effect on the detector is different from one polarization to another. In analogy with the electromagnetic antennas, we define the GW antenna patterns of a Michelson interferometer: we assume a coordinate system determined by the detector and by the gravitational wave source location. The detector arms lie along the x and y axes with the beamsplitter at the origin O. The sky locations of the source is specified by two polar angles ϕ and θ as defined in Fig. 9.5 In general, the gravitational radiation has an arbitrary polarization. We introduce an additional coordinate system for a plane gravitational radiation: X and Y lie on the wavefront plane and Z is the propagation direction. The polarization angle ψ is defined between the X-axis and the line of nodes, which is the intersection of the wavefront plane and Earth equatorial plane. The dashed line in Fig. 9.5 is thus the North-South rotation axis of the Earth.

To proceed further, it is better to separately analyse the two states of polarization of the radiation. Consider first the $+$ polarization tensor, when the polarization angle is $\psi = 0$

$$
e^{+}_{\alpha,\beta} = \begin{pmatrix} 0 & 0 & 0 & 0 \\ 0 & 1 & 0 & 0 \\ 0 & 0 & -1 & 0 \\ 0 & 0 & 0 & 0 \end{pmatrix} \tag{9.26}
$$

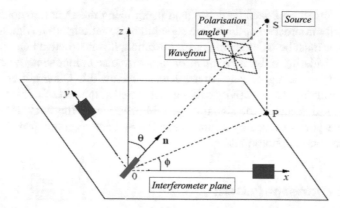

Fig. 9.5 The coordinate systems used to compute the GW antenna pattern of a Michelson interferometer: the origin is set at the beamsplitter, the arms lie along the x and y axes, the source is identified by two polar angles θ and ϕ

Fig. 9.6 Wavefront plane and definition of the polarization angle

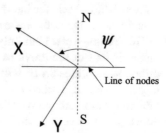

We then convert it in the detector coordinate frame, using the rotation matrix

$$R_\alpha^\beta = \begin{pmatrix} 1 & 0 & 0 & 0 \\ 0 & cos\phi & sin\phi & 0 \\ 0 & -cos\theta sin\phi & cos\theta cos\phi & sin\theta \\ 0 & sin\theta sin\phi & -sin\theta cos\phi & cos\theta \end{pmatrix} \tag{9.27}$$

The explicit computation yields

$$e_{\alpha,\beta}^{+\ d} = (R^{-1})_\alpha^\gamma\, e_{\gamma,\delta}^+ R_\beta^\delta = \tag{9.28}$$

$$\begin{pmatrix} 0 & 0 & 0 & 0 \\ 0 & cos^2\phi - cos^2\theta sin^2\phi & \frac{sin2\theta}{2}sin\phi & 0 \\ 0 & (1+cos^2\theta)\frac{sin2\phi}{2} & sin^2\phi - cos^2\theta cos^2\phi & -\frac{sin2\theta}{2}cos\phi \\ 0 & \frac{sin2\theta}{2}sin\phi & -\frac{sin2\theta}{2}cos\phi & -sin^2\theta \end{pmatrix}$$

where the suffix d reminds us that the polarization tensor is now expressed in the detector frame.

We now start from the simplest and optimal case, already discussed: the antenna response to a gravitational wave of amplitude h_0, polarization $+$, when the directions of the axes (X, Y, Z) coincide with (x, y, z), modulo rotations of $\pi/2$ around the Z-axis, which transform the wave into itself.

Focus now on the optical path difference[4] $\Delta\ell = L_x - L_y$: it can be rewritten in a convenient form by introducing the detector response matrix $A^{\alpha,\beta}$

$$A^{\alpha\beta} = L \begin{pmatrix} 0 & 0 & 0 & 0 \\ 0 & 1 & 0 & 0 \\ 0 & 0 & -1 & 0 \\ 0 & 0 & 0 & 0 \end{pmatrix} \tag{9.29}$$

where L is the arm length of the interferometer. Using this matrix we can write

$$\Delta\ell = \frac{1}{2}(h_{xx} - h_{yy})L = \frac{1}{2}h_0 \, e_{\alpha\beta}^+ \, A^{\alpha\beta} = h_0 L \tag{9.30}$$

Extend now this result to obtain the antenna response for a $+$ polarized wave incident along a generic direction: just replace in Eq. 9.30 $e_{\alpha\beta}^+$ with $e_{\alpha\beta}^{+\,d}$:

$$\Delta\ell_+ = \frac{1}{2}h_0 \, e_{\alpha\beta}^{+\,d} \, A^{\alpha\beta} = \frac{1}{2}h_0 \, L \, (1 + cos^2\theta)cos2\phi \tag{9.31}$$

We can compute, following the same logic path, the antenna response for GW with polarization $\times$, i.e. $\psi = \pi/4$

$$e_{\alpha\beta}^{\times} = \begin{pmatrix} 0 & 0 & 0 & 0 \\ 0 & 0 & 1 & 0 \\ 0 & 1 & 0 & 0 \\ 0 & 0 & 0 & 0 \end{pmatrix} \tag{9.32}$$

In the detector coordinates we have

$$e_{\alpha\beta}^{\times\,d} = \begin{pmatrix} 0 & 0 & 0 & 0 \\ 0 & -sin2\phi cos\theta & cos2\phi cos\theta & cos\phi sin\phi \\ 0 & cos2\phi cos\theta & sin2\phi cos\theta & sin\phi sin\theta \\ 0 & cos\phi sin\theta & sin\phi sin\theta & 0 \end{pmatrix} \tag{9.33}$$

The antenna response for the polarization $\times$ is

$$\Delta\ell_\times = \frac{1}{2}h_0 \, e_{\alpha\beta}^{\times\,d} \, A^{\alpha\beta} = -h_0 \, L \, cos\theta sin2\phi \tag{9.34}$$

We can now compute the general case, i.e. the superposition of two polarized waves of different amplitudes h_+ and $h_\times$, with polarization angle ψ.

---

[4] We consider here, for clarity sake, a simple Michelson interferometer, with $\Delta\ell = \left(\frac{\lambda}{2\pi}\right)\Delta\phi_{gw}$, neglecting the signal enhancement factors due to multi-pass configurations discussed in Sect. 9.5.

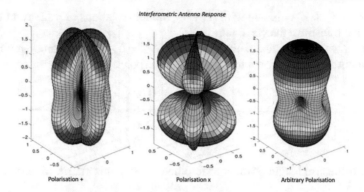

Fig. 9.7 The antenna pattern of a Michelson interferometer for the two polarizations $+$ and $\times$. The third figure is the total combined antenna response averaged over two polarization states

We rotate the wave matrix

$$h_{\alpha\beta} = \begin{pmatrix} h_+ & h_\times \\ h_\times & -h_+ \end{pmatrix} \tag{9.35}$$

by the angle ψ on the wavefront plane (see Fig. 9.6):
$h^d = [R(\psi)]^{-1} \, h \, R(\psi)$ so that, in the detector frame

$$\begin{aligned} h^d_+ &= \cos 2\psi \, h_+ + \sin 2\psi \, h_\times \\ h^d_\times &= -\sin 2\psi \, h_+ + \cos 2\psi \, h_\times \end{aligned} \tag{9.36}$$

Putting all pieces together, we get the detector response as

$$\frac{\Delta \ell}{L} = F^+ h^d_+ + F^\times h^d_\times \tag{9.37}$$

where the general form factors of the antenna for the two polarizations are

$$F^+(\theta, \phi, \psi) = \frac{1}{2}(1 + \cos^2\theta)\cos2\phi\cos2\psi - \cos\theta\sin2\phi\sin2\psi \tag{9.38}$$

$$F^\times(\theta, \phi, \psi) = \frac{1}{2}(1 + \cos^2\theta)\cos2\phi\sin2\psi + \cos\theta\sin2\phi\cos2\psi$$

In Fig. 9.7 we show the antenna pattern of a Michelson interferometer for the two polarizations $+$ and $\times$, as well as the total combined antenna response averaged over the two polarization states. The distance from the origin of a point on the surface is proportional to the amplitude of the detector response to waves coming from that direction: it is largest in the directions normal to the plane of the interferometer. The secondary response characteristics of the antenna are along the four tangential lobes aligned with the antenna arms.

9.6.3 The Shot Noise Limit of the Michelson Interferometer

The shot noise of a laser light is derived by the fluctuations in the number of detected photons (photon counting) and it is a standard example of a measurement of a random variable following the Poisson statistics. The average output power is $\bar{P}_{out} = \bar{n}\hbar\omega_L$, defining $\bar{n}$ the average rate of detected photons. The probability to measure N photons in the time interval τ is

$$p(N) = \frac{\bar{N}^N}{N!}e^{-\bar{N}} \tag{9.39}$$

The expectation value is $\bar{N} = \bar{n}\tau$ and the variance $\sigma_N^2 = \bar{N}$: its square root σ_N measures the quantum fluctuation of the e.m. radiation, so we can see that the relative error on the light power measurement, in the given time interval

$$\frac{\sigma_N}{\bar{N}} = \frac{\sqrt{\bar{n}\tau}}{\bar{n}\tau} = \frac{1}{\sqrt{\bar{n}\tau}}. \tag{9.40}$$

decreases using more laser power ($\bar{n}$) and a longer integration time. The power fluctuations in the time interval τ are then characterized by

$$\delta P_{shot} = \hbar\omega_L\frac{\sigma_N}{\tau} = \hbar\omega_L\sqrt{\frac{P_{out}}{\hbar\omega_L\tau}} = \sqrt{\frac{P_{in}}{2\tau}(1 + C\cos\alpha)\hbar\omega_L} \tag{9.41}$$

Note that the amount of power fluctuations depends on the integration time τ, that defines the instrument bandwidth ($\tau = 1/2\Delta f$). Therefore we can write again Eq. 9.41 as

$$(\delta P_{shot})^2 = 2P_{out}\hbar\omega_L\Delta f \tag{9.42}$$

This noise power P_{shot} does not depend on the observation frequency, but is proportional to the bandwidth Δf chosen: it is an example of *white noise*, and we define the *unilateral* shot noise power spectral density (PSD) (see Chap. 10)

$$S_P^{(shot)} \equiv \frac{\delta P_{shot}^2}{\Delta f} = 2\hbar\omega_L P_{out} \qquad [\text{W}^2/\text{Hz}]$$

We can now evaluate the signal to noise ratio, comparing Eqs. 9.25 and 9.41

$$\sqrt{SNR} = \frac{\delta P_{gw}}{\delta P_{shot}} = \frac{1}{2}\sqrt{\frac{P_{in}}{\hbar\omega_L\Delta f}}\frac{C\sin\alpha}{\sqrt{1 + C\cos\alpha}}\frac{2\pi L}{\lambda}h(t) \tag{9.43}$$

SNR has a maximum for

$$\cos(\alpha_{opt}) = \frac{1}{C}(-1 + \sqrt{1 - C^2}) \approx -1 + \sqrt{2(1 - C)} \tag{9.44}$$

value very close, for $C \simeq 1$, to the condition of darkness ($cos\ \alpha_{opt} = -1$, dark fringe). Besides, operation on the dark fringe is required when a radiofrequency modulation is added to the laser light, as will be discussed in Appendix A.

Nevertheless, for sake of simplicity, we shall evaluate the sensitivity of the Michelson interferometer with the naive assumption of operating at the maximum signal response ($sin\alpha = 1$). The minimum detectable GW amplitude is simply found by setting $SNR = 1$ in Eq. 9.43:

$$\sigma_h = \frac{\lambda}{4\pi L}\sqrt{\frac{2\hbar\omega_L}{\eta P_{in}\tau}} = \frac{1}{L}\sqrt{\frac{\hbar c\lambda}{4\pi\eta P_{in}\tau}} \tag{9.45}$$

where we have introduced, for completeness, the efficiency η of the detection photodiode.

From this (using again $\tau = 1/2\Delta f$) we can define a shot noise spectrum *referred to the input*, i.e. measured in h units:

$$S_h^{(shot)}(\nu) = \frac{1}{L^2}\frac{\hbar c\lambda}{2\pi\eta P_{in}} \tag{9.46}$$

We find the obvious confirmation that sensitivity increases with the arm length L, with observation time τ (for periodic or long-lasting signals) and with the input power P_{in}. It is customary to express in units of h, as done here, all noise sources, acting in different points of the apparatus: this is useful as it allows to compare noises, to evaluate their effect on the sensitivity of the instrument and to add them up to obtain a comprehensive noise spectrum.

We report here the following formula, useful to estimate the order of magnitude of the shot noise contribution to limit the measurement of the GW signal:

$$\sigma_h = 5.2 \cdot 10^{-20}\left(\frac{1000\ m}{L}\right)\sqrt{\frac{\lambda}{1.064\ \mu m}}\sqrt{\frac{10\ W}{P_{in}}}\sqrt{\frac{1\ ms}{\tau}}. \tag{9.47}$$

With a geometrical distance of $L = 1$ km between the beamsplitter and the mirrors of the Michelson interferometer, the optical path can be increased by a factor ~ 100 by means of a Fabry-Perot. This allows us to achieve $\sigma_h \sim 10^{-22}$ that still represents a major limitation on the detector sensitivity.

For the DL and FP configurations the calculation are more complex (Meers 1988; Maggiore 2018). Here we report the results of these algebraic developments:

$$\sigma_h^{(DL)} = \frac{1}{2\,NL}\left(\frac{\hbar c\lambda\Delta f}{\pi\eta P_{in}r^{2N-1}}\right)^{\frac{1}{2}} \cdot \frac{1}{sinc(\frac{\omega_g\tau_s}{2})} \tag{9.48}$$

where we assumed, for simplicity, $r_i = r_e \equiv r$.

Using Eq. 9.23, a similar limit is obtained in the case of the FP cavities with negligible losses:

$$\sigma_h^{(FP)} = \frac{1}{4\mathcal{F}L}\left(\frac{\pi\hbar\lambda c\Delta f}{\eta P_{in}}\right)^{\frac{1}{2}}\frac{1}{\sqrt{1+(\omega_g\tau_s)^2}} \cdot \frac{1}{sinc(\omega_g L/c)} \tag{9.49}$$

Unlike the simple Michelson, for both configurations the *shot* limit is a function of the angular frequency of the gravitational signal ω_g and in particular for $\omega_g > (2\pi\tau_s)^{-1}$ the sensitivity decreases.

In practical interferometers, to limit the contribution of the very low frequency noise that characterizes the radiation emitted by the laser, the light frequency is modulated at a frequency of the order of MHz. High efficiency and high speed modulators are realized exploiting the electro-optical and acousto-optical effects in crystals, whose refractive index can be changed by applying an electric field or mechanical stress. The modulators based on these effects realize the phase modulation (direct or through polarization variations) or frequency (see the dedicated Appendix A on modulation techniques). The modulated signals of the interferometer are then used to control the interferometer itself. Under modulation conditions, the gravitational signal appears in one of the side bands of the light signal spectrum and the maximum of the SNR ratio occurs again when the signal on the interferometer output photodiode is at the extinction (destructive interference, or dark fringe). In this condition the light reflected backwards from the interferometer towards the laser (second output) is maximum.

9.6.4   Light Recycling

Power Recycling.   R. W. P. Drever contributed many brilliant ideas and solutions that are pillars of modern interferometers. Among these, he proposed (Drever 1983) to exploit the fact that, when the interferometer is operated in the dark fringe, all the light exits from the second output, towards the laser. The smart idea, to further lower the limit due to the shot noise, consists of re-injecting into the interferometer the light fraction reflected back towards the laser source. This allows to increase the light power in the arms and is achieved by adding another element to the optical scheme in order to realize an additional resonant optical cavity. A semi-reflecting mirror, the *recycling mirror*, is placed between the laser and the beamsplitter in a position such that the recycled light is in phase with the laser light that is freshly injected into the system, and this verifies a new optical resonance condition. The general theory of light recycling was developed by Vinet et al. (2020), Vinet (1988), Meers (1988). In practice, the recycling scheme can be seen as a peculiar FP cavity, composed of a real mirror and a virtual one, made by the Michelson detector. With the cavity kept in resonance, the light power impinging on the beamsplitter is increased, in agreement with the formula Eq. 9.18, by a recycling factor K ($P_{in} = K \cdot P_{laser}$), which in practice depends on the light storage time in this cavity (similar to that of Eq. 9.21) that in turn depends on the chosen reflectivity of the recycling mirror.

The net result is that the shot-noise-limited sensitivity is enhanced by the square root of this factor K.

An exhaustive discussion of the power recycling system should include optical losses. Indeed, the increase in power inside the recycling interferometer is limited by the losses, mainly due to the absorption and scattering of the cavity mirrors. It can be shown (Meers 1988) that the configuration gain offered by the recycling of light, with respect to the *non-recycled* case, depends on the storage time inside the recycling cavity that, in the end, is limited by the losses.

Signal Recycling The search for signals in a given frequency region can be optimized by the *signal recycling* configuration (Pegoraro et al. 1978; Drever 1983), which consists in improving the sensitivity in a narrower (few hundreds Hz or less) frequency band, centred around a frequency of interest ν_g, typically 300–500 Hz. We recall here that, when the interferometer is operated on the dark fringe, the light at the laser frequency is not present at the photodiode output, while the sidebands containing the GW signal information are completely transmitted. It follows that, in principle, we can send back into the interferometer the signal deprived of the *carrier*, i.e. the laser frequency component of the light. In other words, the strategy consists of storing the light carrying the GW information in one arm for half of the gravity wave cycle π/ω_g. Then, instead of extracting it towards the photodiode, we feed it back into the other arm. During this second half of the wave cycle, since in this second arm both the phase of the signal and the sign of the response are inverted, the phase shift of the light adds up, in the same direction of the first half cycle. The signal recycling with this modality can be repeated many times until the losses of the mirror prevail.

In practice, to achieve this recycling effect on the signal, an additional mirror is used, the sixth of the *dual recycling Fabry-Perot configuration* (not counting the beamsplitter). This last *signal recycling (SR)* mirror is placed between the beamsplitter and the detection photodiode and constitutes, together with the power recycling interferometer considered as a whole, a cavity that is designed to resonate at a frequency ν_{sr}, on or near the frequency ν_L of the laser light (Fig. 9.8).

This additional SR cavity will affect the phase of the carrier circulating in the interferometer by a factor (Vajente 2014)

$$\phi = \frac{2\pi}{\lambda} L_{(SRC)} + \frac{\pi}{4} \tag{9.50}$$

where

$$L_{(SRC)} = l_{SR} + (l_x + l_y)/2$$

is the length of the SR cavity: the distance l_{SR} between the SR mirror and the beamsplitter, plus the mean value of the Michelson arm lengths (see Fig. 9.8), which is typically of the order of few tens of metres.

The computation of the response of a dual (power + signal) recycled interferometer is conceptually simple, but algebraically cumbersome: however, we can consider the power-recycled interferometer as a single box and then treat it as a virtual mirror with a frequency-dependent reflectivity in front of the SR.

The optimization of SR configuration, based only on the *shot noise* reduction criterion, is performed by choosing the reflectivity r_{sr} and the position l_{SR} of the SR mirror.

A lengthy calculation (Strain and Meers 1990; Meers 1989) takes us to the conclusion that, by tuning the choice of ϕ to different values, we can significantly change the shape of the response curve of the detector. With reference to Fig. 9.9, we can

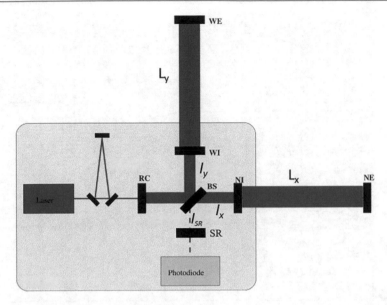

Fig. 9.8 A diagram of a Michelson interferometer with both power recycling mirror (PRM) and signal recycling mirror (SRM). The input mode cleaner is an additional optical cavity, working as a filter, needed to improve the light spectral purity. Courtesy of Virgo collaboration

see that, for $\phi = 0$, the detector bandwidth is extended, with respect to the non-SR case, at the price of a reduced response in the bandwidth. This configuration is called *broad-band signal recycling*. For $\phi = \pi/2$ the low-frequency sensitivity is improved, but the bandwidth is considerably narrowed. For intermediate values of ϕ, a broad frequency peak appears where the detector response is increased. These configurations are often called *detuned signal recycling*.

Signal recycling thus offers us a way to operate a trade-off between bandwidth and gain, letting us also choose where in frequency we can exploit this enhanced gain.

9.6.5   Radiation Pressure Noise

We have seen that the quantized nature of light implies an uncertainty in the measurement of the relative position of the mirrors, which we have attributed to the shot noise. However, it is also inevitable that the quantum fluctuations of light result in fluctuations in the radiation pressure of the light beam and in the impulse transferred to the mirrors.

To estimate the effects of this second contribution related to the quantum nature of radiation, we consider the force f exerted by an electromagnetic wave of power P_{eff} ($\simeq \mathcal{F} \cdot P_{in}$ in the case of a FP cavity) on a mirror of mass M:

$$f = \frac{P_{eff}}{c}. \tag{9.51}$$

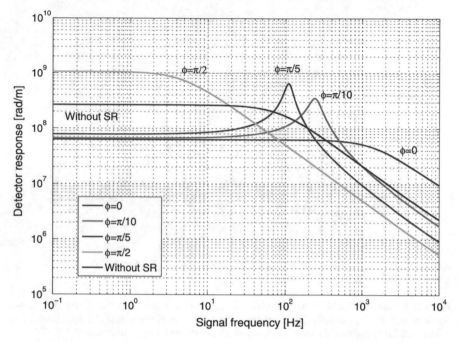

Fig. 9.9 Dual recycling response for different ϕ values (credit Vajente (2014))

This is a static force that causes a constant shift in the equilibrium position of each suspended mirror. But a fluctuation of the number of photons impinging on the mirror implies a fluctuation in the force applied on the mirror surface:

$$\sigma_f = \frac{\sigma_P}{c} = \frac{\hbar \omega_L}{c} \frac{\sigma_N}{\tau} = \sqrt{\frac{\pi \hbar P_{eff}}{c \lambda \tau}} \tag{9.52}$$

The spectral density of this force is then

$$S_f(\nu) = \frac{2\pi \hbar P_{eff}}{c\lambda}. \tag{9.53}$$

The oscillation amplitude of each suspended mirror is

$$x(\nu) = \frac{f(\nu)}{M(2\pi\nu)^2} = \frac{1}{M\nu^2}\sqrt{\frac{\hbar P_{eff}}{8\pi^3 c\lambda}}. \tag{9.54}$$

Since the fluctuation in the two interferometer arms are anticorrelated (a positive fluctuation on one arm corresponds to a negative fluctuation on the other), the change δL is twice $X(\nu)$. As a consequence, the radiation pressure noise referred to the input port and expressed in terms of power spectra density of h is

$$\sqrt{S_{rp}(\nu)} = \frac{2x(\nu)}{L} = \frac{1}{M\nu^2 L}\sqrt{\frac{\hbar P_{eff}}{2\pi^3 c\lambda}}. \tag{9.55}$$

Note that, in order to reduce the effects of this noise, one should increase the mass M of the mirrors and decrease the incident power P_{in}. This last request is opposite to that associated with the reduction of the shot noise. There is therefore an optimum value for the light power that balances the contribution of these two noise sources.

At present, the available light power on the interferometer mirror is still smaller than the optimum value so that the dominant contribution remains that of shot noise. However, the next generation of detectors, for which we foresee a considerable increase in the power of the input laser and of the finesse of the FP cavities, will likely approach that condition.

9.6.6   The Standard Quantum Limit

As we have in the two previous paragraphs, two different types of noise are both related to the quantum nature of electromagnetic radiation, which is the basis of our system of monitoring the dynamic state of the mirrors. We therefore consider together these two sources of noise as a single entity, which are referred to with the name *optical readout noise*:

$$S_{opt}(\nu) = S_{shot}(\nu) + S_{rp}(\nu). \tag{9.56}$$

In the case of a simple Michelson interferometer, we obtain explicit expressions of the spectral density for the shot and the radiation pressure noise, which have different dependence, both on the laser power and on the frequency. In general, at low frequencies, the radiation pressure term is more relevant, while at high frequencies the shot noise dominates.

We also note that, for each frequency value ν_*, there exists the *optimal power* P_{opt} such that $S_{shot}(\nu_*) = S_{rp}(\nu_*)$:

$$P_{opt} = \pi c \lambda M \nu_*^2 \tag{9.57}$$

For example, using a laser light of $\lambda = 1.064\ \mu m$ (Nd-Yag) impinging on a mirror of mass $M = 10$ kg, at the frequency $\nu_* = 10$ Hz, we have $P_{opt} \simeq 1$ MW, which is not too far from the light power stored at present in the interferometer arms of Advanced LIGO and Virgo.

Substituting $P_{in} = P_{opt}$ in S_{opt} we have

$$\sqrt{S_{SQL}(\nu_*)} = \frac{1}{\pi \nu_* l} \sqrt{\frac{\hbar}{M}} \tag{9.58}$$

where SQL is the acronym for *standard quantum limit* of a simple interferometer. Note that this magnitude is not really a specific value of h but rather the locus of points such that $S_{opt}(\nu)$ is minimal in $\nu = \nu_*$ (see Fig. 9.10).

Shot and radiation pressure noises set limits in the capability to monitor the dynamic status of the mirror. Summarizing, we have that

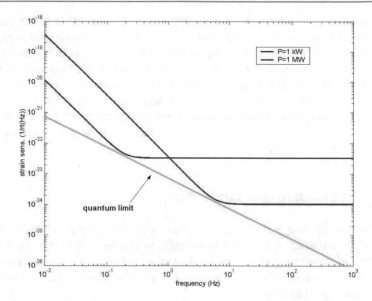

Fig. 9.10 The optical readout noise in a simple Michelson for two values of power stored in the arms. The green line is the locus of points representing the standard quantum limit

- the mirror displacement is measured at a given frequency by the photon counting procedure; its incertitude is given by the shot noise limit as computed in (9.45): $\sigma_{\delta L} = L \cdot \sigma_h$
- the radiation pressure noise induces fluctuations of the mirror momentum; its standard deviation is given by $\sigma_p = \sigma_f \cdot \tau$, with σ_f computed in (9.52).

Not surprisingly, the product of these two noises shows that the ultimate limit in the strategy of a classic monitoring system is given by the Heisenberg uncertainty principle

$$\sigma_p \cdot \sigma_{\delta L} \geqslant \frac{\hbar}{2}$$

9.6.7   The Thermal Noise in the Inteferometer

Particular attention must be paid to a second fundamental limitation to the detector sensitivity: the thermal noise. We mention here some specific aspects of this noise in connection with the sensitivity of interferometric GW detectors.

Thermal noise has two main contributions: normal modes of the mirror and vibration modes of the suspension fibres. Each mirror is a thick disk of fused silica (350 mm in diameter and 200 mm thick, for a total mass of 42 kg in Advanced Virgo). It is suspended by four fused silica fibres of circular cross section: they are thin ($\sim$200 μm in diameter) in the long middle section, and about twice as thick near the two ends, where they are fastened to the mirror and to the previous suspension stage. The fibres are 60 cm long and the resulting pendulum frequency is $\nu_p \sim 0.6$ Hz. The geometry is chosen so as to keep high the other fibre transverse mode frequencies (the first overtone is 500 Hz) and to keep low the vertical stretching mode frequency (9 Hz for the chosen mirror mass).

Each mode of the suspended mirror (both for oscillation and vibration) has an associated fluctuation energy equal to $k_B T$, where k_B is the Boltzmann constant and T is the equilibrium temperature of the mirror. Some of these vibration modes have a displacement field extending over regions of the mirror surface hit by the light beam: The thermal fluctuations of the mirror surface give rise to an ill-defined mirror position, and therefore to a displacement noise. To quantitatively assess this effect, one should also consider that the wavefront of the light mode resonating in the FP cavity (the TEM_{00} mode) has a Gaussian intensity profile: the vibration of a region of the mirror surface is more relevant if hit by a portion of the wavefront with higher intensity. Therefore, the computation of the spectral density of the displacement fluctuations due to the thermal noise must include the convolution of the spatial distributions, on the mirror surface, of the mechanical and electromagnetic modes.

The thermal noise of the mirror pendulum resonances, localized on the suspension wires, is also a source of noise relevant for the sensitivity to GW.

The computation of this contribution is based on the fluctuation-dissipation theorem stating that the unilateral spectral noise densities[5] of the force and displacement fluctuations of a mechanical system at the equilibrium temperature T are, respectively,

$$S_f = 4k_B T \Re[Z(\omega)] \qquad S_x = \frac{4k_B T}{\omega^2} \Re[\frac{1}{Z(\omega)}] \qquad (9.59)$$

where k_B is the Boltzmann constant and $Z(\omega) = F(\omega)/i\omega X(\omega)$ is the impedance of the mechanical system defined via the Fourier transforms of the force and the velocity.[6]

We then assume that each mode $v(t)$ follows the typical equation of the damped harmonic oscillator,

$$M_i \frac{dv}{dt} + \frac{M_i}{\tau_i} v + M_i \omega_i{}^2 \int v(t)dt = f(t) \qquad (9.60)$$

where M_i, ω_i and M_i/τ_i are the effective mass, angular frequency and coefficient of dissipation[7] related to the i-th vibration mode we are considering.

---

[5] The bi-lateral spectrum is defined in the frequency domain $(-\infty, +\infty)$. It is, obviously, half as much as the unilateral one that extends only over positive frequencies, i.e. in the domain $(0, +\infty)$.

[6] Not to be confused with the transfer function (TF), that is the ratio between the Fourier transforms of force and displacement.

[7] τ_i as used here is the *energy* dissipation time constant. Elsewhere, the *amplitude* time constant might be used. As the energy decay is proportional to $A^2(t)$, we have $\tau_{ampl} = 2\tau_{energy}$.

After Fourier transforming Eq. 9.60, we find the mechanical impedance:

$$Z(\omega) = \frac{M_i}{\tau_i} + i M_i (\omega - \frac{\omega_i^2}{\omega})$$

Thus, the bilateral power spectral density of the stochastic force is

$$S_f = \frac{4 k_B T M_i}{\tau_i} \tag{9.61}$$

while the spectral displacement density of the mirror $S_x(\omega_g)$ is

$$S_x(\omega) = \frac{S_f}{|Z(\omega)|^2} = \frac{4 k_B T}{M_i \tau_i} \frac{1}{(\omega^2 - \omega_i^2)^2 + (\frac{\omega}{\tau_i})^2} \tag{9.62}$$

Note the double role played by the coefficient τ_i: it is both the measure of dissipation in the oscillation and the source of the thermal noise force. Hence the name *fluctuation-dissipation* theorem.

Generally, the amount of dissipation is assessed via the *quality factor* of each resonant mode $Q_i = \omega_i \tau_i$.

We can now apply to Eq. 9.62 the usual approximations for frequencies much higher or much lower than the resonant frequency ω_i. Thus, if we consider a GW signal at a frequency $\omega > \omega_p$ higher than the pendulum own frequency ω_p, we can deduce a sensitivity limit in h due to the thermal noise of the pendulum mode of suspension:

$$\sqrt{S_h^{(therm.p.)}} = \frac{1}{L} \frac{1}{\omega^2} \sqrt{\frac{4 k_B T \omega_p}{M Q_p}} \tag{9.63}$$

where the mass of this mode coincides with the mirror mass M and Q_p accounts for several different dissipation mechanisms. This noise becomes less and less relevant as the observation frequency ω is increased, typically above 20 Hz.

On the other hand, the modes of the mirror bulk oscillation are at frequencies above 5 kHz, i.e. higher than the ω of interest and in this case ($\omega < \omega_i$) we have

$$\sqrt{S_h^{(therm.m.)}} = \frac{1}{L} \frac{1}{\omega_i^2} \sqrt{\sum \frac{4 k_B T \omega_i}{M_i Q_i}} \tag{9.64}$$

These limits depend on the nature of the material as well as on the geometry and on the modes of vibration considered.

Assuming $Q_i \simeq 10^6$, the thermal noise contributions can be kept below the threshold value of $S_h^{1/2} \sim 10^{-23}$ Hz$^{-1/2}$ for a mass $M \sim 100$ kg and angular frequency $\omega \sim 2\pi \cdot 160$ Hz.

Measuring the contribution of thermal noise in a gravitational interferometer is a delicate task. The envelope of the two main contributions of thermal noise, that of the pendulum and that of the modes of the mirror, should contribute to defining the spectral region of maximum sensitivity of the instrument. In this case, to reach $S_h^{1/2} \simeq 10^{-23}$ Hz$^{-1/2}$, the Q values of all the modes we are considering should range in the order of $10^6 - 10^8$. This is quite a difficult goal to achieve at room temperature. In particular, the dissipation at the connection points of the suspensions was found to influence the overall dissipation of both the mirror and the suspensions. However, with a careful selection of materials and careful design of the connection points, significant increase in the value of Q values has been achieved. In addition, these values are also enhanced, for crystalline materials, by cooling the system to low temperature. This is one of the reasons to support the use of cryogenic techniques for the KAGRA observatory and for future-generation detectors.

We must notice that the formulas reported above have been computed assuming in Eq. 9.60 a viscous (proportional to velocity) force. In 1990, Saulson (1990) pointed out that these calculations of thermal noise based on velocity-damping models could be in error. In particular, the internal dissipation in materials obey an extension of Hooke's law, in which losses are modelled by a complex spring constant,[8] and including in it a loss angle $\phi(\omega)$, which determines the lag of the response x of the spring when solicited by a sinusoidal force:

$$f_{spring} = -M\omega_o^2[1 + i\phi(\omega)]\Delta x \tag{9.65}$$

By far the most common functional form for the loss angle is $\phi(\omega) = constant$ over a large band of frequencies and its typical values are $\phi << 1$. These results of the acoustic dissipation have been obtained mainly for bulk materials where the loss mechanism is associated with dislocations. Under this assumption the equation of any damped harmonic oscillator, resonating at ω_0, becomes

$$M\frac{d^2x}{dt^2} + M\omega_o^2(1 + i\phi)\Delta x = f(t)$$

and the corresponding mechanical impedance is

$$Z(\omega) = M\phi\frac{\omega_o^2}{\omega} + iM(\omega - \frac{\omega_o^2}{\omega})$$

so that the spectral densities of the force and displacement are

$$S_f = 4k_BTM\phi\frac{\omega_o^2}{\omega} \quad ; \quad S_x(\omega) = \frac{4k_BT\omega_o^2\phi}{M\omega[(\omega^2 - \omega_o^2)^2 + (\omega_o^2\phi)^2]} \tag{9.66}$$

This model yields a frequency dependence significantly different from the viscous case. This is evident far from the resonance, in the two limits discussed above,

---

[8] Analogously to the complex susceptibility in e.m.

$\omega_g \gg \omega_p$ and $\omega_g \ll \omega_i$

$$\sqrt{S_h^{(therm.p.)}} = \frac{1}{L}\sqrt{\frac{4k_B T \omega_p^2 \phi_p}{M}}\frac{1}{\sqrt{\omega^5}} \quad \text{for} \quad \omega_g \gg \omega_p \qquad (9.67)$$

$$\sqrt{S_h^{(therm.m.)}} = \frac{1}{L}\sqrt{\frac{4k_B T \phi_i}{M_i \omega_i^2}}\frac{1}{\sqrt{\omega}} \quad \text{for} \quad \omega_g \ll \omega_i \qquad (9.68)$$

In today's advanced detectors, the dominant noise in the frequency region of highest sensitivity is thermal noise in the mirrors. The main cause is dissipation in the thin film of materials used to produce the reflective coating of the mirrors. Here two mechanisms are related to thermal fluctuations:

- The thermo-elastic effect: the coupling between mechanical and thermal fluctuations produces local deformations in the mirrors and therefore a displacement noise
- The thermo-refractive coefficients of the coating materials. Temperature fluctuations in both the bulk and coating of mirrors produce a change in the refractive index of the materials, and again a displacement noise in mechanical loss in the bulk fused silica is responsible for the substrate Brownian noise term.

For this reason a wide research programme for low-losses coating materials is underway.

9.6.8   Other Sources of Noise

In a real detector the list of sources of noise, limiting the sensitivity to GW, is very long: careful design and construction is required for all components.

Residual gas noise
The whole interferometer is hosted in a huge vacuum enclosure, in order to suppress the noise effects associated with the presence of air along the beam trajectory. Indeed, the presence of any residual gas implies:

- the transmission of acoustic noise to the mirrors
- the increase of thermal noise due to damping of mirror suspensions by air friction
- the excitation of mirror motion by impact with gas molecules
- changes in refractive index due to fluctuations of gas density.

The upper limit for the residual pressure is derived mainly by considering the last noise mechanism listed above. The light beam propagates back and forth in a vacuum enclosure, a tube few kilometres long and ~ 1 m diameter. The density fluctuations of the residual molecules in the tube determines fluctuations of the effective refractive index of the medium crossed by the laser beam. This noise can be modelled by calculating the change in the phase of the Fabry-Perot cavity field as a molecule moves through the beam, and by integrating over the molecular velocity distribution. The noise power spectrum, expressed in h units, due to the residual pressure, is Accadia et al. (2012):

$$S_h^{vacuum}(\omega) = \frac{8\pi}{L^2}\frac{\alpha^2}{v_0}\int_0^L \frac{\rho(z)}{w(z)}exp\left(-\frac{\omega w(z)}{v_o}\right)dz \qquad (9.69)$$

with ρ is the residual density of molecules function of the longitudinal coordinate of the beam tube z, $w(z)$ is the light beam transversal dimension at z, α is the optical polarizability of the molecules while $v_0 = \sqrt{\frac{2k_B T}{m}}$ is the thermal velocity of the molecules, T the temperature and k_B the Boltzmann constant. The goal is to keep the pressure-related noise one order of magnitude below the shot noise: for Advanced Virgo this implies that the noise strain due to pressure fluctuations must be below 10^{-25} Hz$^{-1/2}$ corresponding to a residual gas pressure of about 10^{-7} Pa in the case of hydrogen molecules; for hydrocarbons, the limit is much more stringent, 10^{-11} Pa. In the case of residual gas dominated by higher polarizability gases, the limit pressure should be lowered by one order of magnitude. Reaching such low values of partial pressure is a real challenge: they must be achieved in a tube of a 1 m diameter and of kilometric length. The volume of the vacuum tube determines the pump-out time, but it is the wall surface that determines the limit pressure p_{lim}, that is given by the relation

$$p_{lim} = \frac{S\mathcal{R}}{\Sigma} \qquad (9.70)$$

where (we report here numerical values for Virgo) $S = 25000$ m^2 is the wall surface, Σ is the pumping speed and $\mathcal{R} \sim 10^{-10}$ Pa $\cdot$ 1/cm$^2 \cdot$ s is the degassing rate. Typical values of degassing rate measured on Stainless Steel 304L samples are of the order of 10^{-10} Pa $\cdot$ 1/cm$^2 \cdot$ s.

The required $p_{lim} = 10^{-7}$ Pa then would demand, with a small safety margin, the impressive pumping speed $\Sigma = 10^6$ l/s, i.e. a huge investment in pumping systems, to be distributed along the kilometre-long tube. Therefore, the only way to keep the pumping system at an economically affordable level is to reduce the wall outgassing rate. This can be obtained by *firing* each tube module, 12 m long, before the installation: firing involves heating in air the tube modules to 450 °C,[9] a procedure aimed at increasing the hydrogen atoms mobility. These atoms are trapped inside the metal during the production process at a weight concentration of few part per millions. When cooled to room temperature, the hydrogen outgassing rate is reduced by more than two orders of magnitude. After this procedure, the modules are welded to form the long tube and this huge volume is pumped out while keeping the tube temperature at 150 °C for several days (*bake-out procedure*). This procedure is particularly effective to reduce the residual water vapour adsorbed on the inner surface of tube, and is the standard procedure applied to all ultra-high vacuum systems.

---

[9] The firing temperature is chosen to be well below the brittle temperature for 304L stainless steel.

Moreover, the following technical aspects should be highlighted:

- The mirrors installed in the vacuum chambers are sensitive to any kind of contamination: pumps must be oil-free and must not generate residual particles.
- The pumping stations should produce low acoustic, seismic and electromagnetic noise.

In addition, the residual gas in the test mass vacuum chambers will contribute to the damping of the test mass suspension, potentially increasing the suspension thermal noise. This damping effect is increased by the relatively narrow gap between the test mass and its suspended reaction mass, an effect called proximity-enhanced gas damping. Gas damping noise is most significant in the 10–40 Hz band, and then it falls off with frequency. An approximate expression of the noise spectral density in the relevant region is

$$\sqrt{S_h^{(vac)}} \geqslant 7 \ 10^{-25} \left(\frac{P}{10^{-7} Pa} \right)^{1/2} \tag{9.71}$$

Seismic noise In the low-frequency range, i.e. below 10 Hz, the sensitivity is limited by seismic noise: ground motion drives the structure holding the apparatus, thus coupling a displacement noise to the mirrors. The strategy to fight this noise input uses the property of harmonic oscillators to attenuate all disturbances at frequencies above its resonance. The solution adopted is to suspend each mirror to a chain of several stages in series, each composed of a pendulum, and connected by vertical springs. This technique, called super-attenuator (Ballardin 2001), proposed by A. Giazotto, permits to lower the detector useful bandwidth down to a few Hz.

The super-attenuator is a rather complex device: the loading masses of each pendulum are equipped with vertical springs at the end of which the pendulum wires are fastened (Fig. 9.11).

In this way each pendulum attenuates the horizontal vibrations by a factor $\left(\omega_h / \omega_g \right)^2$, where ω_h is the resonance frequency of the pendulum and ω_g is the frequency of interest. The springs set at the connecting point of the pendulum wire determine the attenuation of vertical vibrations by a similar factor $\left(\omega_v / \omega_g \right)^2$ where ω_v is the resonant frequency of the springs.

With a proper design of the pendulums and of the vertical springs, a total attenuation of the suspension system of the order of 10^{-14} at 10 Hz has been achieved, both for horizontal and vertical vibrations. Each mirror in the Virgo interferometer is suspended by one of these systems.

Laser fluctuations A complex optical layout and several nested feedback loops are used to stabilize the intensity and frequency of the laser and reduce the beam jitter. An important element is a suspended, triangular Fabry-Perot cavity, which cleans up the spatial profile of the laser beam, suppresses input beam jitter, cleans polarization and helps stabilize the laser frequency before sending the beam into the interferometer.

To give an idea of the stabilization performance required, we report the final figure of the frequency stabilization achieved in Virgo: $\sqrt{S_{\delta \nu_L / \nu_L}} \sim 3 \cdot 10^{-21} \, (Hz/Hz)/\sqrt{Hz}$.

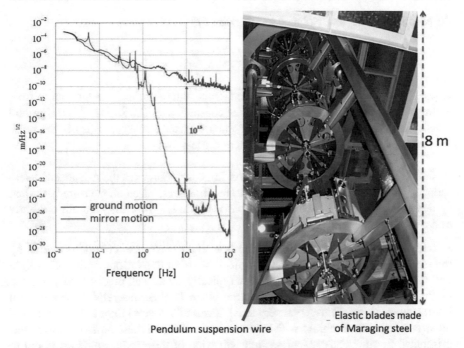

Pendulum suspension wire

Elastic blades made
of Maraging steel

Fig. 9.11 On the right is a picture, taken from below, of the Virgo super-attenuator. The pendulum cables and the Maraging steel blades for the attenuation of the horizontal and vertical seismic motion are indicated. On the left, a measurement of the spectrum of the seismic noise, measured on the ground and on the mass suspended by the super-attenuator. Courtesy of the Virgo Collaboration (Braccini 2010)

Recalling the laser frequency $\nu_L = c/(1.064 \, \mu m) = 282 \, THz$, this means an absolute stability $\delta\nu_L \sim 10^{-6} Hz/\sqrt{Hz}$

The residual contribution of this noise to the overall GW detector sensitivity depends on the static asymmetries present between the two arms of the interferometer. These asymmetries are due to the residual difference in the geometry of the mirrors (curvature radii), and to the different losses in the FP cavities of the two arms. This limit is lowered by improving the cavity symmetry, i.e. matching the two end and two input mirrors with almost the same optical and geometrical characteristics. In addition, a control of their geometry is carried out by a system of thermal compensation of the mirrors.

Newtonian Noise Seismic waves produce density perturbations in the ground next to the test masses, which in turn produce fluctuating gravitational forces on the masses. Thus, fluctuations of the local gravity field around the test masses, caused by ground motion and vibrations of the buildings, chambers, and concrete floor, induce a random motion of the mirrors as force noise (gravity gradient noise).

This seismic gravity-gradient noise is estimated using a representative model for the seismic motion at an observatory site. A good estimate of the coupling to the differential arm length displacement is given by

$$\sqrt{S_h} = \frac{2}{L} \frac{G\beta\rho}{\omega^2} \sqrt{S_{seismic}}$$

where the factor of 2 accounts for the incoherent sum of noises from the four mirrors acting as test masses, G is the gravitational constant, $\rho \simeq 1800 \ kg \ m^{-3}$ is the ground density near the mirror, $\beta \simeq 10$ is a geometric factor and $S_{seismic}(\omega)$ is the noise spectral density of the seismic motion near the test mass.

Gravity gradient noise is one of the limiting noise sources of the Advanced Detectors design in the frequency range 10–20 Hz. Strategies, based on measurements and dedicated algorithms, to subtract this particular noise from the data are currently under study and test in Virgo.

Scattered light noise A fraction of the laser light leaves the volume occupied by the main beam through scattering or reflections from the various optics. This light will hit and scatter from surfaces that are typically not as well mechanically isolated as the suspended optics, picking up large phase fluctuations relative to the main interferometer light. Thus, even very small levels of scattered light can be a relevant noise source if it recombines with the main beam. To limit these noise contributions, originated by the acoustic and seismic vibration of the vacuum pipe anchored to the ground, a *photon killing* system has been developed. The scattering process and the related noise was evaluated via ray tracing and Monte Carlo simulations, to design absorbing elements (baffles) to be distributed along the vacuum pipes. The best known absorbing material is coated glass for welding protection: this solution, however, cannot be adopted because of its fragility. The adopted solution consists of stainless steel diaphragms with a conical profile that reduce the useful diameter from 1.2 to 0.9 m. They have an absorption coefficient close to 50 % and the remaining light is reflected on a wide field.

9.7   Conclusive Notes

Interferometric detectors of GW are now in the so-called second-generation (*advanced detectors*): they are very complex instruments that we could only sketch here. More details can be found in the cited references and in dedicated reviews, like (Vajente 2014).

The performance of the detector at each frequency are characterized in terms of the *amplitude spectral sensitivity* $H(\omega)$: it is simply equal, for SNR=1, to the noise spectral amplitude $\sqrt{S_h(\omega)}$. It is obtained by summing all the power spectral densities of the various noise contributions referred at the instrument input and then taking the square root of this sum and is compared with the break down of all the identified noise sources.

As an example of the performance level of the first-generation detectors, we show in Fig. 9.12 Virgo's sensitivity curve. At frequencies below 60 Hz we notice the steep

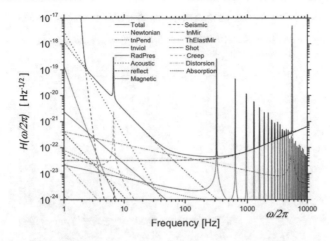

Fig. 9.12 Virgo sensitivity curve achieved in the year 2011. The quantity plotted versus frequency is amplitude spectral sensitivity $H(\omega/2\pi)$. Courtesy of the Virgo Collaboration

increase in noise (i.e. decrease in sensitivity) due to the suspension thermal noise. On the other hand, for at frequencies above 500 Hz the sensitivity is dominated by shot noise that, as we have seen, has white spectrum. Nevertheless, the sensitivity curve slowly decreases because the response to a signal of the FP cavities has a cutoff, as shown in Eq. 9.23.

In Fig. 9.13, we show a more recent noise curve, obtained during the O3 observation run (2019–2020), together with the predicted sensitivity for the run O5, with the adoption of signal recycling.

At the time of writing, the new detector in India, IndIGO is under construction, while KAGRA, Advanced LIGO and Virgo are in an upgrading phase. The installation of new hardware has been almost completed and the detector tuning phase, called in jargon *commissioning*, has started.

More powerful lasers, up to 100 W, have been installed to reduce the shot noise; the optical configuration of Advanced Virgo now includes the signal recycling mirror; new baffles are instrumented with sensors allowing to monitor the scattered light. Both Advanced LIGO and Advanced Virgo are equipped with new optical benches and an additional cavity to pursue a further reduction of the optical noise via *frequency-dependent squeezing* technique.

Squeezing is a method of quantum optics to improve the sensitivity of the detector at high frequencies by further reducing the light phase fluctuation at the interferometer output against the amplitude noise while doing the opposite at low frequencies. In essence, it is based on one form of Heisenberg uncertainty relation: $\Delta N \Delta \phi \geqslant 1/2$, where $N = P/\hbar\omega_L$ is the number of photons in the laser beam, and ϕ is the beam phase. By sophisticated light manipulations (that includes sending proper e.m. "vacuum" field back into the output port of the interferometer), one manages to reduce ΔN, and improve shot noise, at the expenses of $\Delta \phi$, i.e. radiation pressure noise, or vice versa. We refer to Hild (2014) for an introduction to squeezing, and to Danilishin

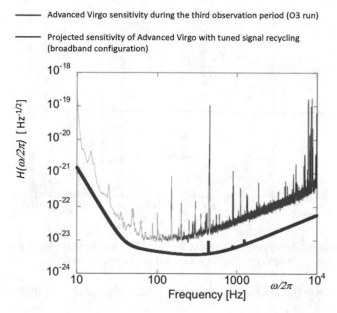

Fig. 9.13 Virgo sensitivity curve achieved in 2019 during the O3 observation run. The lower curve is the projected sensitivity for an upcoming run when the signal recycling will be in place. Courtesy of the Virgo Collaboration

and Khalili (2012) for an in-depth analysis of quantum optics of the GW interferometers.

The alternation of periods dedicated to hardware upgrading with periods of Universe observations will continue for the foreseeable future, pushing these instruments to their ultimate limits, both in terms of durability and performance, to the point where no further significant sensitivity improvements will be possible. Thus, a new generation of detectors will be needed: a third generation (3G), counting Virgo-LIGO as the first and aLIGO-Advanced Virgo as the second.

The design of such complex detectors, planned for the mid-2030s, has already begun: the Einstein Telescope (ET) is the European 3rd generation effort, while the Cosmic Explorer (CE) project is in preparation in the USA. Both will rely on new research infrastructures designed to observe the entire Universe with the GW messenger. ET and CE are conceived following the same basic design of LIGO and Virgo: a Michelson interferometer with Fabry-Perot cavities in the arms plus power and signal recycling. Both new instruments will have longer arms: the present plan calls for 40 km on surface for CE and 10 km underground for ET.

The underground operations will allow to extend the frequency band of the observatory down to a few Hz, thanks to reduced seismic and gravity gradient noise induced by seismic waves.

There is a dichotomy in the optimization of a high-sensitivity interferometer: high light power is needed to reduce the shot noise, dominant at high frequencies. At the same time, due to the inevitable optical losses, high power increases the mirror temperature and the thermal noise, relevant at low frequencies. In ET this issue is solved with a *xylophone*, i.e. by building a pair of complementary interferometers:

- ET-LF, more sensitive at low frequencies, thanks to the use of cryogenic mirrors to reduce the thermal noise
- ET-HF, operated at room temperature and at higher light power stored in the optical cavities for optimum performance at higher frequencies.

Both CE and ET will be designed to accommodate several technology upgrades during the planned 50 years of operation. These new detectors will detect BBH coalescences up to cosmological distances and extend the region of detectable BH masses: sources of several thousands solar masses could be detected up to $z \sim 1 - 5$; while taking advantage of higher low-frequency sensitivity, ET will explore a volume of the universe up to $z \sim 0.5 - 1$ searching for BBH with a total mass of order one solar mass, the value compatible only with the primordial origin of these black holes.

In addition, ET and CE will provide the possibility to detect NS-NS coalescent system to $z \sim 2 - 3$ and boost the multi-messenger astronomy. The 3G detectors will also discover, with all probabilities, new sources: the GWs from supernovae would elucidate the collapse and the explosion mechanism, while the detection of continuous GWs from NSs would provide information on their spin, thermal evolution and magnetic field.

Finally, the 3G detector will search for the stochastic GW backgrounds (SGWB) of cosmological origin. If detected, it would give us the incredible potentiality to look at the earliest moment of the Universe; while the cosmic microwave background observations are revealing how the Universe behaves 380000 years after the Big Bang, and SGWB will allow us to probe the laws of physics up to a fraction of a second from the Big Bang, i.e. at high-energy scales inaccessible even by neutrino observations or by particle accelerators.

3G detectors will probably operate in the same years of the space-based GW antenna LISA, described in Sect. 11: this will constitute a combined observatory with a frequency range covering from 20 μHz to several kHz, with a huge potential for discovery.

References

Accadia, T., et al.: Virgo: a laser interferometer to detect gravitational waves. JInst **7**, P03012 (2012)

Ballardin: Measurement of the VIRGO superattenuator performance for seismic noise suppression. Rev. Sci. Instrum. **72**, 3643 (2001)

Braccini, S., et al.: Measurements of superattenuator seismic isolation by Virgo interferometer. Astropart. Phys. **33**, 182–189 (2010)

Danilishin, S.L., Khalili, F.Y.: Quantum measurement theory in gravitational-wave detectors. Liv. Rev. Relat. **15**, 5 (2012)

Drever, R.W.P.: Interferometric detectors for gravitational radiation. In: Deruelle, N., Piran, T. (eds.) Gravitational Radiation. North Holland (1983)

Forward, R.L.: Wideband laser-interferometer graviational-radiation experiment. Phys. Rev. D **17**, 379 (1978)

Herriott, D., Kogelnik, H., Kompfner, R.: Off-axis paths in spherical mirror interferometers. Appl. Opt. **3**, 523 (1964)

Hild, S.: A basic introduction to quantum noise and quantum-non-demolition techniques. In: Bassan, M. (ed.) Advanced Interferometers and the Search for Gravitational Waves. Springer Int. Publish. (2014)

Maggiore, M.: Gravitational Waves, vol.1: Theory and Experiments. Oxford University Press, Oxford (2018)

Meers, B.: Recycling in laser-interferometric gravitational-wave detectors. Phys. Rev. D **38**, 2317 (1988)

Meers, B.J.: The frequency response of interferometric gravitational wave detectors. Phys. Lett. A **142**, 465 (1989)

Moss, G.E., Miller, L.R., Forward, R.L.: Photon-noise-limited laser transducer for gravitational antenna. Appl. Opt. **10**, 2495 (1971)

Pegoraro, F., Picasso, Radicati, G.: On the operation of a tunable electromagnetic detector for gravitational waves. J. Phys. A **11**, 1949 (1978)

Pérot, A., Fabry, C.: On a new form of interferometer. Ap. J. **9**, 87 (1899)

Pirani, F.A.E.: On the physical significance of the Riemann tensor. Acta Physica Polonica **15**, 389 (1956)

ch9Strainsps1991 Saulson, P.: Thermal noise in mechanical experiments. Phys. Rev. D **42**, 2437 (1990)

Strain, A., Meers, B.J.: Experimental demonstration of dual recycling for interferometric gravitational-wave detectors. Phys. Rev. Lett. **66**, 1391 (1990)

Vajente, G.: Interferometer Configuration. In "Advanced Interferometers and the Search for Gravitational Waves", Bassan M. ed. Springer Int. Publish (2014)

Vinet, J.-Y.: The VIRGO Physics Book, OPTICS and related TOPICS (revision 2020)

Vinet, J.Y., Meers, B., Man, C.N., Brillet, A.: An improved test of the strong equivalence principle with the pulsar in a triple star system. Phys. Rev. D **38**, 433 (1988)

Weiss, R.: Electomagnetically coupled Broadband Gravitational Wave Antenna. Quarterly Progress Report, MIT Research Laboratory of Electronics **105**, 54 (1972). Reprinted as LIGO-P720002-00-R

Further Reading

Saulson, P.R.: Fundamentals of Interferometric Gravitational Wave Detectors. World Scientific, Singapore (1994)

Bond, C., Brown, D., Freise, A., Strain, K.A.: Interferometer techniques for gravitational-wave detection. Living Reviews in Relativity **19**, 3 (2016)

Danilishin, S.L., Khalili, F.Y., Miao, H.: Advanced quantum techniques for future gravitational-wave detectors. Living Reviews in Relativity **22**, 2 (2019)

Data Analysis

10

10.1 Introduction

Several physical processes targeted by the present research studies suffer from large background contamination that needs to be filtered out. Statistical analysis of the data plays a crucial role in boosting the signal-to-noise ratio (SNR) and improve the sensitivity. Scientific methods for data processing and analysis in the area of precision experiments has progressed impressively during the last decades, continuously developing dedicated data analysis strategies: they range from very simple and robust methods to more sophisticated and targeted procedures. The experimental field that most needs these tools, and the one where the most important developments have arisen, is that of the search for GW with ground-based interferometers: we shall specialize our overview to these techniques, bearing in mind that many are suitable for application anywhere a small signal needs to be detected in the presence of large noise. Two excellent references on these topics are the classical books by Papoulis (1965) and Zubakov and Wainstein (1962). For applications to the data analysis of GW detectors, see the book by Schutz (1989).

In this chapter, we summarize basic data analysis tools,[1] starting with few fundamental concepts for the noise characterization. Then, we will review few analysis strategies for different classes of signals, to offer the flavour of the complexity of this important topic of experimental gravitation.

---

[1] The reader is expected to be familiar with the basic concepts and tools of statistics. To fill possible gaps, we suggest, e.g. Ross (2021).

© The Author(s), under exclusive license to Springer Nature Switzerland AG 2022
F. Ricci and M. Bassan, *Experimental Gravitation*, Lecture Notes in Physics 998,
https://doi.org/10.1007/978-3-030-95596-0_10

10.2   Stochastic Processes

A random variable is a function that assigns values to the outcomes of an experiment[2] where a non-deterministic component (noise) is present. By making repeated observations (*realizations*) of the same random phenomenon, different data sets over time are obtained. Noise is the realization of a random phenomenon.

- A stochastic process **X** is a family of *random variables*, whose *realizations* are an ordered set of functions, $x_k(t)$: the index k identifies the element in the family and t represents, in all cases of our interest, the time.

Observing the different realizations at the instant t_0, we are studying the statistics of the random variable $X(t_0)$: these realizations are values that the process assumes at the instant t_0 in the $(k = 1, 2 \ldots)$ replicas of our experiment.

To characterize **X** we make use of the probability density function: focusing on the realizations $x(t)$ taken in the generic instant t, we derive the estimate of the first order probability density function $f_t^{(1)}$ (see Fig. 10.1).

The mean, i.e. the first-order moment of the distribution, is

$$\mu_x(t) \equiv E[X(t)] = \int_{-\infty}^{+\infty} x f_t^{(1)}(x) dx \tag{10.1}$$

and the variance

$$\sigma_x^2(t) \equiv E[\left(X(t) - \mu_x(t)\right)^2] = \int_{-\infty}^{+\infty} (x - \mu_x)^2 f_t^{(1)}(x) dx \tag{10.2}$$

When we consider two time instants, t' and $t"$, we use the *joint* probability density function $f_{t',t"}^{(2)}$ to define the second-order momentum of the distribution, that permits to evaluate the interdependence of the random variable realizations at different times:

$$E[X(t'), X(t")] = \tag{10.3}$$

$$= \int_{-\infty}^{+\infty} \int_{-\infty}^{+\infty} x' x" f_{t',t"}^{(2)}[x', x"] dx' dx"$$

where, with obvious notation, $x' = x(t')$ and $x" = x(t")$.

To generalize this concept, we consider n instants of time and we introduce the probability density function of order n, $f_{t_1 \ldots t_n}^{(n)}$, which will give the more detailed statistical description of the process.

---

[2] Formally, a random variable is a function that can take on either a finite number of values, each with an associated probability, or an infinite number of values, whose probabilities are summarized by a density function.

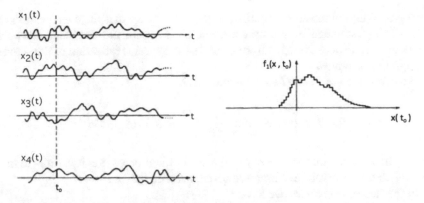

Fig. 10.1 The figure shows four realizations of the stochastic process $x_k(t)$. From many of these realizations, we extract their values at the time parameter t_0. The ensemble of these values constitutes the random variable $X(t_0)$. The histogram on the right is the estimate of the probability density function of the process for the variable $X(t_0)$

10.3 Stationary Processes

In general, the probability densities of all orders change as the points in time are changed.

We define a fundamental property of a class of processes, the **stationarity**: an arbitrary time translation of the entire realizations does not change its statistics.

We distinguish processes that are stationary in a *strict (narrow, strong)* sense or in a *wide (weak)* sense

A process is stationary in the strict sense if its statistics do not change under any time shift, i.e. when:

i- the first-order probability density function is independent of t, and is then the same for each possible realization. It follows from Eq. 10.1 that also the mean does not depend on t;

ii- the second-order density only depends on the difference $\tau = t_2 - t_1$ and not on t_2, t_1 separately;

iii- the n-th order density depends only on the $n - 1$ parameters $\tau_i = t_{i+1} - t_i$, for $i = 1, 2, \ldots, n - 1$.

For a process to be stationary in the wide or weak sense, it is sufficient to meet the first two requirements.

10.4 Ergodic Processes

In *ergodic processes*, observation of a realization of the process *for a sufficiently long time* allows to extract all the statistical properties. *In signal theory, the process is ergodic when the statistical averages converge almost everywhere with the time averages.* With reference to Fig. 10.1, this is like saying that the average value of $x(t_0)$ can be estimated by the time average of any realization x_k. This allows us to calculate statistical averages from a single and generic realization of the process through time averages.

We define the *auto-correlation* function R_{xx},

$$R_{xx}(t, t + \tau) \equiv \lim_{T \to \infty} \frac{1}{2T} \int_{t-T}^{t+T} x(t')x(t' + \tau)dt' \tag{10.4}$$

it gives information on the *memory* of the stochastic process, i.e. it permits to infer how much the observation at time t is correlated with that at $t + \tau$.

In the case of *ergodicity*, we have

$$\mu_x = \lim_{T \to \infty} \frac{1}{2T} \int_{-T}^{+T} x(t)dt \tag{10.5}$$

If the process is ergodic in its statistical auto-correlation, all the realizations of the process have the same auto-correlation function,

$$R_{xx}(\tau) = \lim_{T \to \infty} \frac{1}{2T} \int_{-T}^{+T} x(t)x(t + \tau)dt = E[X(t), X(t + \tau)] \tag{10.6}$$

this is just Eq. 10.3 rewritten as a time average. Moreover, in the case $\mu_x = 0$,

$$\sigma_x^2 = R_{xx}(0) \tag{10.7}$$

We note that both μ_x and σ_x^2 are now time-independent.

We summarize four important considerations related to an ergodic process:

- necessary condition for a process to be ergodic to order n, is the stationarity of the moments up to the same order.

  In particular, we refer to *ergodic average* when the time average and the statistical average coincide; we talk about ergodicity in correlation when statistical and temporal auto-correlations coincide;

- the following fundamental theorem is associated with the names of **Wiener-Khinchin** (Papoulis 1965): *if the process* **X** *is stationary and ergodic at least in its auto-correlation* $R_{xx}(\tau)$, *the power spectral density* S_{xx} *of the process is the Fourier transform of its auto-correlation:*

$$S_{xx}(\omega) = \int_{-\infty}^{+\infty} R_{xx}(\tau) \exp(-j\omega\tau)d\tau \tag{10.8}$$

Conversely, $R_{xx}(\tau)$ can be derived via the inverse Fourier transform of S_{xx}. Note that evaluating at $\tau = 0$ the auto-correlation function of the process with zero mean, yields the variance of the process:

$$\sigma_x{}^2 = R_{xx}(0) = \frac{1}{2\pi} \int_0^\infty S_{xx}(\omega)d\omega \qquad (10.9)$$

showing that the variance is the sum of random contributions at all frequencies.

- When the ergodic process $i(t)$ is the input of a linear system with transfer function $H(\omega)$ (see Appendix B), the output is also an ergodic process $o(t)$ with auto-correlation function $R_{oo}(\tau)$ and spectral density:

$$S_{oo}(\omega) = |H(\omega)|^2 S_{ii}(\omega) \qquad (10.10)$$

- In the presence of two stochastic and ergodic processes $x(t)$ and $y(t)$, we define the cross-correlation function and the cross-spectral density

$$R_{xy}(\tau) = \lim_{T \to \infty} \frac{1}{2T} \int_{-T}^{+T} x(t)y(t+\tau)dt \qquad (10.11)$$

$$S_{xy}(\omega) = \int_{-\infty}^{+\infty} R_{xy}(\tau)\exp(-j\omega\tau)d\tau \qquad (10.12)$$

In previous chapters, we have used the simplified notation $S_x(\omega)$ instead of $S_{xx}(\omega)$. This is often adopted when no cross-spectrum or cross-correlation is involved so that the double subscript is redundant.

10.5 The Matched Filter

The notion of filtering raw data to extract the signal embedded in the noise was originated in the context of the communication theory: N. Wiener in the 1940s broadened the concept interpreting the filter theory on statistical basis. Filter theory is still in evolution in these days, targeting new methods of design of optimum filters. In the following, we will focus on the optimum linear filter,[3] although robust, non-linear methods, like those based on machine learning (Ross 2021), are now being developed and implemented.

The matched filter is the best linear approach to extract a signal of known shape when it is embedded in a stationary Gaussian noise.

We shall apply the following considerations to the gravitational signal $h(t)$ and the noise $n(t)$ from an interferometer, as discussed in Chap. 9, but the analysis, originally derived for e.m. antennas, is very general and is applied to a large class of experiments. Consider a data stream, a function $i(t)$ containing a linear superposition of both the signal $h(t)$ and the noise $n(t)$:

$$i(t) = h(t) + n(t)$$

---

[3] The filter operator $\mathcal{F}$, relates an input $i(t)$ to the output $o(t)$ by $o = \mathcal{F}\{i\}$. $\mathcal{F}$ is linear when the two properties of superposition and scaling are verified: $\mathcal{F}\{ai_1 + bi_2\} = a\mathcal{F}\{i_1\} + b\mathcal{F}\{i_2\}$.

We feed this function $i(t)$ to a linear filter so that the output $o(t)$ is the convolution of $i(t)$ with the filter function $k(t)$:

$$o(t) = \int_{-\infty}^{+\infty} i(t - \tau)k(\tau)d\tau \qquad (10.13)$$

If the input signal is a very short pulse $i(t) \to A\delta(t)$, the output tends to coincide with the filter function: hence the name "filter impulse response" for $k(t)$.

Since we are dealing with a linear combination of noise and signal, we shall have[4]

$$o(t) = \int_{-\infty}^{+\infty} [h(t - \tau) + n(t - \tau)]k(\tau)d\tau = o_h(t) + o_n(t) \qquad (10.14)$$

Our goal is then to find a filter that will maximize the output SNR at a given time t_0. Intuitively, we look for a linear time-invariant filter whose output will be much larger if the signal $h(t)$ is present than when it is absent. To do this, the filter should make the instantaneous power in $o_h(t_0)$, as large as possible compared to the average power $o_n(t)$: this is equivalent to maximizing the standard definition of the SNR.

A well-known property of Fourier transform is the mapping of convolution in the time domain to a simple product in the frequency domain. Therefore, the analysis is simpler in the latter, and we introduce the Fourier transform of the functions $i(t)$, $o(t)$, $k(t)$, denoted by the corresponding capital letters:

$$I(\omega) = \frac{1}{2\pi} \int i(t)e^{-i\omega t} dt \; ; \qquad K(\omega) = \frac{1}{2\pi} \int k(t)e^{-i\omega t} dt$$

$$O(\omega) = \frac{1}{2\pi} \int o(t)e^{-i\omega t} dt \qquad (10.15)$$

Rather than computing the output signal as a time domain convolution

$$o_h(t) = \int_{-\infty}^{+\infty} h(t - \tau)k(\tau)d\tau$$

we derive its Fourier transform, i.e. the product

$$O_h(\omega) = K(\omega)H(\omega)$$

and we end up with the relation

$$o_h(t) = \frac{1}{2\pi} \int_{-\infty}^{+\infty} K(\omega)H(\omega)e^{i\omega t} d\omega \qquad (10.16)$$

---

[4] To avoid ambiguities: in the following, the word *signal* is reserved for the component $h(t)$, not to the entire input $i(t)$; same for the signal output $o_h(t)$.

Being $n(t)$ a stochastic process, the noise contribution at the output of the filters is computed applying the Wiener-Khinchin theorem: the power spectrum of the noise processes at the input $S_n(\omega)$ and the output of the filter S_{o_n} is:

$$S_{o_n}(\omega) = |K(\omega)|^2 S_n(\omega) \tag{10.17}$$

The output noise spectrum S_{on} is easily evaluated when the signal is rare, thus usually absent in the output (like in the detection of GW): in the case S_{on} simply coincide with the spectrum of the filter output S_o.

Recalling the property of the power spectral density of a stationary process with null average, the expected value of the noise variance is

$$E[o_n^2] = \frac{1}{2\pi} \int_{-\infty}^{+\infty} S_{o_n}(\omega) d\omega \tag{10.18}$$

The SNR at the time t_0 is

$$SNR_{t_0} \equiv \frac{|o_h(t_0)|^2}{E[o_n^2]} = \frac{1}{2\pi} \frac{|\int_{-\infty}^{+\infty} K(\omega) H(\omega) e^{i\omega t_0} d\omega|^2}{\int_{-\infty}^{+\infty} |K(\omega)|^2 S_n(\omega) d\omega} \tag{10.19}$$

To find the optimum $K(\omega)$, we proceed by maximizing the SNR, applying the Cauchy-Schwartz inequality, written here for two generic functions $A(\omega)$ and $B(\omega)$:

$$\left| \int_{-\infty}^{+\infty} A(\omega) B(\omega) d\omega \right|^2 \leqslant \int_{-\infty}^{+\infty} |A(\omega)|^2 d\omega \int_{-\infty}^{+\infty} |B(\omega)|^2 d\omega \tag{10.20}$$

By assuming for A and B the following functions:

$$A(\omega) = K(\omega)\sqrt{S_n(\omega)} e^{i\omega t_0} \qquad B(\omega) = \frac{H(\omega)}{\sqrt{S_n(\omega)}}$$

we can rewrite the numerator of Eq. 10.19 as

$$\left| \int_{-\infty}^{+\infty} K(\omega) H(\omega) e^{i\omega t_0} d\omega \right|^2 \leqslant \int_{-\infty}^{+\infty} |K(\omega)|^2 S_n(\omega) d\omega \int_{-\infty}^{+\infty} \frac{|H(\omega)|^2}{S_n(\omega)} d\omega \tag{10.21}$$

Substituting this relation into Eq. 10.19, we obtain an upper bound

$$SNR_{t_0} \leqslant \frac{1}{2\pi} \int_{-\infty}^{+\infty} \frac{|H(\omega)|^2}{S_n(\omega)} d\omega \tag{10.22}$$

Equation 10.21 shows that the upper bound is achieved if we choose as filter the function

$$K(\omega) = C e^{-i\omega t_0} \frac{H^*(\omega)}{S_n(\omega)} \tag{10.23}$$

where C is a constant factor and $e^{-i\omega t_0}$ a phase to be interpreted as a delay effect with respect to the detection time t_0. $H^*(\omega)$ is the complex conjugate of the Fourier transform of the signal $h(t)$.[5] Intuitively, these last two equations make a lot of sense: in the filter of Eq. 10.23 the input is enhanced in the frequency bands where $H(\omega)$, i.e. the signal is large, while it is depressed where the denominator, i.e. the noise dominates. On the other hand, the SNR of Eq. 10.22 shows that, when the signal is properly filtered, the global SNR is obtained by summing over all frequencies the basic SNR density shown in the integrand.

If we factor Eq. 10.23 as:

$$K(\omega) = \left[S_n(\omega)^{-1}\right] \cdot \left[Ce^{-i\omega t_0} H^*(\omega)\right] \tag{10.24}$$

we notice that the first block in square brackets divides the input by the spectrum of the noise. At the output of this first block the noise no longer has any spectral feature, but a constant spectrum: $S_n(\omega) = constant$. In jargon, this operation is called *data whitening*.

10.6   The Case of Signal in White Noise

When we deal with Gaussian noise and white power spectrum, the frequency domain filter for which we have the maximum SNR is simply proportional to the Fourier transform $H(\omega)$ of the signal that we want to detect. The filtering process will pass through only those frequencies that contribute to the signal:

$$K(\omega) = e^{-i\omega t_0} H^*(\omega) \tag{10.25}$$

$h(t)$ is a real function so $H^*(\omega) = H(-\omega)$ and the signal at the filter output is

$$o_h(t) = \frac{1}{2\pi} \int_{-\infty}^{+\infty} H(-\omega)H(\omega)e^{i\omega(t-t_0)} d\omega \tag{10.26}$$

Using the Fourier transform of

$$H(\omega) = \int_{-\infty}^{+\infty} h(t')e^{-i\omega t'} dt' \tag{10.27}$$

reordering the integrals at the filter output, we have

$$o_h(t) = \int_{-\infty}^{+\infty} h(t') \left[\frac{1}{2\pi} \int_{-\infty}^{+\infty} H(-\omega)e^{-i\omega(t'-t+t_0)} d\omega\right] dt' \tag{10.28}$$

---

[5] The matched filter is, indeed, matched to a known, or presumed, form of the input signal $h(t)$.

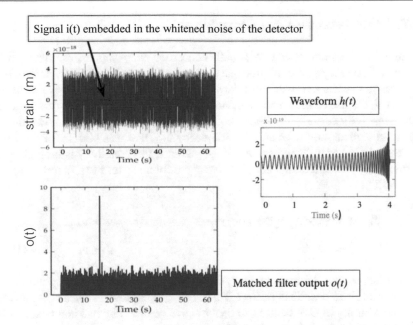

Fig. 10.2 On the top left, the signal embedded in the detector noise in the time domain. On the right, the waveform $h(t)$ used in the matched filter: a chirp, emitted by a binary neutron star system in the final coalescence phase. On the bottom left, the filter output versus time. The signal in the data produces a sharp peak in the matched-filter output at the termination time of the chirp signal

that we rewrite as

$$o_h(\tau) = \int_{-\infty}^{+\infty} h(t')h(t' - \tau)dt' \tag{10.29}$$

where $\tau = (t - t_0)$, shows the delay introduced by the filter. We read in this expression the definition of the auto-correlation function of the signal.

In the presence of the whole input $i(t) = h(t) + n(t)$, the matched filter then acts as

$$o(t) = \int_{-\infty}^{+\infty} i(\tau)h(\tau - t)d\tau \tag{10.30}$$

This result is telling us that the matched filter simply works as a correlator and acts on the input $i(t)$ as by cross-correlating the input with the known signal.

An example of the application of the matched filter to GW data is shown in Fig. 10.2. We convolve the strain data $i(t)$ with the waveform $h(t)$ shown in the figure: he result is a sharp, narrow peak occurring at a time that almost exactly coincides with the termination time of the chirp signal (merging time). This is conventionally taken as the chirp arrival time.

10.7   The Detection of a Chirp Signal

The first direct detections of GW signals concerned *chirp* signals, oscillating functions with frequency and amplitude that grow over time, as it was shown in Fig. 10.2. Thus, let us discuss the matched filtering technique in this case. As we discussed in Chap. 9, the use of the resonant cavities in the interferometer modifies the shape of the sensitivity curve in the high-frequency range of the detector bandwidth. The shot noise contribution rises above a frequency ν_{kn}, that was 400 Hz for the LIGO detectors. At lower frequencies, below $\nu_s \sim 20$ Hz, the sensitivity is limited by the seismic noise so that a crude, but effective approximation for the spectral density of the overall noise is

$$S_n(\nu) = \frac{1}{2}\sigma_n(\nu_{kn})\Big[1 + (\frac{\nu}{\nu_{kn}})^2\Big] \qquad\qquad \nu > \nu_s$$

$$= \infty \qquad\qquad\qquad\qquad\qquad\qquad\quad \nu < \nu_s \qquad (10.31)$$

We then use, as a model for the signal, the waveform generated by two point masses in circular orbit as described in Sect. 7.5.2. Here, we follow a standard assumption of the community of GW hunters, to deal a signals emitted by two identical neutron stars of 1.4 $M_\odot$ (see the example in (Schutz 1991)).

We use a simplified expression of the amplitude of the *chirp* wave (Eq. 7.52) impinging on the detector along the optimal direction and with the most favourable polarization. Note that in this chapter we are focused on GW signal analysis so that ν identifies the signal frequency: for rotating sources, this is twice the frequency of rotation ν of Chap. 7.

$$h(t) = A_h[\nu(t)]\cos\left(2\pi \int_{t_a}^t \nu(\tilde{t})d\tilde{t} + \Phi\right) \qquad (10.32)$$

where t_a is the arrival time of the signal when its instantaneous frequency is equal to ν_a, and Φ the phase at the same time. The amplitude increases slowly over time[6]

$$A_h(\nu) = 2.6\cdot 10^{-23}\Big(\frac{\mathcal{M}}{M_\odot}\Big)^{\frac{5}{3}}\Big(\frac{\nu(t)}{100\text{ Hz}}\Big)^{\frac{2}{3}}\Big(\frac{100\text{ Mpc}}{r}\Big) \qquad (10.33)$$

where we recall the definition of chirp mass $\mathcal{M} = \frac{(m_1 m_2)^{3/5}}{(m_1+m_2)^{1/5}}$. The orbital period is reduced over time and the frequency of the gravitational signal changes at the pace:

$$\frac{d\nu}{dt} = 13\Big(\frac{\mathcal{M}}{M_\odot}\Big)^{\frac{5}{3}}\Big(\frac{\nu}{100\text{ Hz}}\Big)^{\frac{11}{3}} \quad \text{Hz s}^{-1} \qquad (10.34)$$

---

[6] In this equation and in the following of this long paragraph, where numerical reference values appear, the frequencies are normalized in units of 10^2 Hz, the distances r in unit of 10^2 Mpc and the time by 1 s.

From this differential equation, we deduce how the frequency changes with time

$$\frac{v(t)}{100 \text{ Hz}} = \left[(v_a)^{-\frac{8}{3}} - 0.33 \left(\frac{\mathcal{M}}{M_\odot} \right)^{\frac{5}{3}} \left(\frac{t - t_a}{1 \, s} \right) \right]^{-\frac{3}{8}} \tag{10.35}$$

one additional integration yields the expression of the phase of the Eq. 10.32

$$\Phi(t) = 2\pi \int_{t_a}^{t} v(\tilde{t}) d\tilde{t} = \tag{10.36}$$

$$= 3000 \left(\frac{\mathcal{M}}{M_\odot} \right)^{-\frac{5}{3}} \left\{ \left(\frac{v_a}{100 \text{ Hz}} \right)^{-\frac{5}{3}} - \left[\left(\frac{v_a}{100 \text{ Hz}} \right)^{-\frac{8}{3}} - 0.33 \left(\frac{\mathcal{M}}{M_\odot} \right)^{\frac{5}{3}} \left(\frac{t - t_a}{1 \, s} \right) \right]^{\frac{5}{8}} \right\}$$

Using Eq. 10.35, the duration of the chirp signal starting from the moment the frequency is equal to v_a and stopping at the end of the *chirp*: in principle, when $v \to \infty$ but, in practice, when the last stable orbit is reached.

$$T_{coal}(v_a) = 3 \left(\frac{\mathcal{M}}{M_\odot} \right)^{-\frac{5}{3}} \left(\frac{v_a}{100 \text{ Hz}} \right)^{-\frac{8}{3}} s \tag{10.37}$$

Taking advantage that the signal changes slowly in frequency ($T_{coal} \gg 1/v$) an approximate, analytic expression can be derived for the Fourier transform (Dhurandhar and Schutz 1994) of $h(t)$:

$$|H(v)| \simeq 3.7 \cdot 10^{-24} \left(\frac{\mathcal{M}}{M_\odot} \right)^{\frac{5}{6}} \left(\frac{v}{100 \text{ Hz}} \right)^{-\frac{7}{6}} \left(\frac{100 \text{ Mpc}}{r} \right) \text{ Hz}^{-1/2} \tag{10.38}$$

With this, we can evaluate the SNR:

$$SNR = \int_{-\infty}^{\infty} \frac{|H(v)|^2}{S_n(v)} dv \tag{10.39}$$

An interesting event for the detection of the gravitational signal is defined when the filtered variable exceeds an assigned threshold value. This event will be characterized by the arrival time t_{arr} and the amplitude of the signal. The accuracy of the arrival time, which depends in part on the efficiency of the filter, is an essential feature in order to determine delays of the signal seen by a network of gravitational wave detectors and extract information on the direction of arrival.

The time resolution, i.e. the standard deviation of t_{arr}, can be approximatively evaluated, when the signal is narrow band around the value v_o as in (Schutz 1991)

$$\sigma_{t_{arr}} \simeq \frac{1}{2\pi v_o \cdot SNR} \tag{10.40}$$

For example with $SNR = 10$ and $v_o = 100$ Hz, we have a resolution of about 0.16 ms.

The efficiency of the filter is excellent in the statistical sense mentioned above, if the template assumed in the filter coincides with the real signal. The template depends on several parameters: for example in the simplest case in addition to the position in the sky of the emitting system, the polarization angle, the total mass $m_1 + m_2$ and the *chirp mass* $\mathcal{M}$ must be taken into account, determining both the amplitude and the frequency of the signal.

10.8   The Template Bank

The matched filtering technique is based on the hypothesis of our knowledge of the signal to be detected. For example, in the case of the search of GW signals emitted in the coalescence process of a binary black hole system, we can relay on the prediction of general relativity. It means choosing the theoretical waveform of the expected signals to filter the detector data.

The chosen waveform depends on a number of parameters $\{\theta_i\}$, that are not known a priori and constitute the so-called *parameter space*. Each choice of parameters is a point in such space and represents a different waveform, or *template* to be used in the filter. As we shall see, applying to this space the techniques of differential geometry will bring relevant results for the search optimization.

Applying the filter (matched to a given template) we will correspondingly extract from the data stream a number of *events*, i.e. potential candidates of a true signal. Events are selected when the SNR value raises above a threshold a-priori set by the experimentalist: it is usually chosen at few times (5 to 8) the standard deviation of the filtered noise. The signal parameters are then derived by comparing the template to the data following a Bayesian approach (see next Sect. 10.9).

The data are thus filtered again and again using a number of templates, that is known collectively as *template bank*: The template bank consists of a discrete set of points in the parameter space.

The coverage of the parameter space with templates, called *placement*, is an open optimization problem[7]: use of a coarse grid of discrete points leads to an unavoidable loss in the SNR, as we risk to miss the templates that best match the data. Besides, reduction in SNR increases the probability of false dismissal, i.e. not detecting valid candidates. On the other hand, a large number of points rapidly increases the computational cost of carrying out the search. Therefore, the art of template placement for a given type of source lies in maximizing the inter-template separation, with the aim of using the smallest bank size, while covering at the best the template space.

The geometric approach to the placement problem proceeds as follows.

---

[7] The template bank could be seen naively as an uniform grid in the parameter space. However, to appreciate how this is not, in general, a trivial task, just consider the case of placing dots on a sphere at equal distances.

Consider the template, defined by the set of N parameters $\theta_{ref}{}^i$.

$$\vec{\theta}_{ref} \equiv \{\theta_{ref}{}^i\}, \ i = 1...N$$

We begin by calibrating the filter, feeding it with a simulated signal that perfectly matches the template $\vec{\theta}_{ref}$, i.e. a signal with FT $H(\omega; \vec{\theta}_{ref})$, as prescribed by Eq. 10.25. Clearly, this combination, signal and template perfectly matched, gives the maximum possible SNR, but our filter is also sensitive to signals corresponding to nearby parameters, although with a reduced SNR. If we define the acceptable loss in SNR, usually up to 5%, we can compute the volume of the parameter space covered by that template, and the distance to the adjacent ones, in order to efficiently tile the whole space. Consider then one of these nearby templates: $\vec{\theta}_{near} = \vec{\theta}_{ref} + \vec{\Delta\theta}$: the loss in SNR with respect to the optimal case introduces the *ambiguity function*

$$\zeta \equiv \frac{SNR(\theta_{near})}{SNR(\theta_{ref})} \tag{10.41}$$

Applying the filter with $H(\omega; \vec{\theta}_{ref})$ to nearby points, we can express, by power series expansion near its maximum, the reduction in SNR:

$$\zeta(\vec{\theta}_{ref} + \vec{\Delta\theta}) = 1 + \frac{1}{2}\left(\frac{\partial^2\zeta}{\partial\theta^i\partial\theta^j}\right)_{\Delta\theta^k=0} \Delta\theta^i \Delta\theta^j + \cdots \tag{10.42}$$

Using the tools of differential geometry[8], we can interpret the coefficient of the expansion as the metric tensor of the parameter space:

$$g_{ij} = -\frac{1}{2}\left(\frac{\partial^2\zeta}{\partial\theta^i\partial\theta^j}\right)_{\Delta\theta^k=0} \tag{10.43}$$

by means of which we quantify the distance square between two neighbour templates with

$$1 - \zeta = g_{ij}\Delta\theta^i\Delta\theta^j \tag{10.44}$$

Based on the metric 10.43, we map the template placement using the smallest possible number of templates, to fully cover the parameter space (i.e. leave no unexplored regions).

A few main template placement strategies have been developed. We cite here just the construction of a *geometric quasi-regular lattice* of points and the *stochastic* construction built from a set of random proposals.

Under the geometrical approach, a set of coordinates is first identified and then a quasi-regular distribution of patches are placed.

---

[8] We encountered a similar expansion with the tidal tensor, Eq. 1.24.

The stochastic template placement algorithm is built starting from a set of seed points drawn from a uniform statistical distribution over the parameter space. Then, other point are randomly proposed and then accepted only if they lie at a distance $D > \sqrt{1 - \zeta}$ from the others, previously drawn.

The random choice process continues until a pre-set convergence threshold is reached for the full coverage of the parameter space. The stochastic method is robust and can be implemented in higher dimensional curved parameter spaces.

In general, the dimensionality of the parameter space is large and the number of parameters depends on the complexity of the waveform we are assuming. Even in the simplest case of GW signals emitted by a spinless coalescent binary systems, the vector is nine dimensional: two BH masses, inclination angle i, polarization angle ψ, phase at coalescence ϕ_c, right ascension α and declination δ, luminosity distance d_L and time of coalescence t. The dimensionality rises to 15 (adding two spin vectors) for spinning stellar systems, and the parameter number is even larger when more detailed physics is included in the gravitational waveform.

To give an idea of the bank dimension, we refer to the case of the first direct detection of a GW signal from a binary black hole system (Abbott et al. 2016a). At that time, the bank was constructed by targeting compact binary systems with individual masses restricting the total mass up to a maximum of 100 $M_\odot$. The dimensionless aligned-spin magnitude[9] of the individual objects was limited to 0.99: the search was carried on a bank consisting of almost 250000 templates.

10.9   Bayesian Inference

The goal of data analysis is to reconstruct the information, hidden in the experiment output, by comparing the observational data with theoretical waveform of the signal and extracting the parameter values. When dealing with a transient signal emitted by an astrophysical source, the possibility to repeat the identical experiment several times is precluded: we have just a unique event. It follows that the assessment of statistical confidence on the event parameters cannot be based on a strict *frequentist* statistical approach.

Such assessment is instead achieved applying Bayesian inference. The statistical pillar of the inference is the Bayes theorem, which describes the probability of an event, based on prior knowledge of conditions that might be related to the event (Thrane and Talbot 2019).

Let us first introduce the notation.

- θ and i are two variables; for our purposes θ represents the model parameters and i the data of the event.

---

[9] In a black hole, the angular momentum can have a maximum value $|\vec{J}| \leqslant GM^2/c$. It is therefore natural to define a dimensionless spin vector with magnitude $|\vec{j}| = c|\vec{J}|/GM^2 \leqslant 1$.

- $p(\theta)$ and $p(i)$ are the probabilities of observing θ and i, respectively, without any given conditions, i.e. they are the *prior* probabilities.
- $p(\theta|i)$ is the *conditional probability* of the event θ occurring given that i is true. This quantity is also called the *posterior* probability of θ given i.
- $p(i|\theta)$ is also a conditional probability: the probability of event i occurring given that θ is true. This can be interpreted as the *likelihood* function of θ for a given i: $p(i|\theta) = \mathcal{L}(i|\theta)$.
- the unconditioned probability $p(i)$ can be expressed as the weighted sum of all conditioned probabilities:

$$p(i) = \int \mathcal{L}(i|\theta)p(\theta)d\theta \equiv \mathcal{Z} \tag{10.45}$$

$\mathcal{Z}$ is called *the evidence*.

The Bayes' theorem (or Bayes' rule) states, in a deceivingly simple formulation, that

$$\boxed{p(\theta|i) = \frac{1}{p(i)}p(i|\theta)p(\theta)} \tag{10.46}$$

as long as $p(i) \neq 0$.

An example from medical field and a diagram may help gaining insight with this important tool. Consider a human population that may or may not have some sort of cancer and a medical test that returns positive or negative for detecting that disease. We know that 6% of the population is affected by this cancer.

So we set $\theta = Disease\ is\ True$ and $i = Test\ is\ Positive$: the parameters here are not only discrete variables but also are binary (true/false). The *prior* is $p(\theta) = 6\%$.

Diagnostic tests are not perfect: in 5 % of the tests a patient will not have cancer, but the test gives a positive result (*false detection, or false positive*); conversely, the test might not detect a cancer that is present (*false dismissal, or false negative*) in 18 % of the cases. The likelihood is thus $\mathcal{L}(i|\theta) = 0.82$. The situation is summarized in the tree diagram of Fig. 10.3. By working backward this tree, we can answer the question:

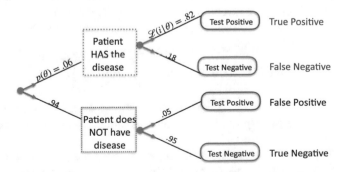

Fig. 10.3 A tree diagram summarizing the probabilities involved in the example given in the text

- what is the probability that a patient has the disease, given that he tested positive?

$$
\begin{aligned}
&p(\text{patient has disease| test is positive}) \\
&= \frac{\text{true positive}}{\text{true positive} + \text{false positive}} \\
&= \frac{.06 \cdot 0.82}{.06 \cdot 0.82 + 0.94 \cdot 0.05} = 51.1\%
\end{aligned}
\tag{10.47}
$$

A patient who tested positive has therefore a probability p = 51.1 % of actually having the disease.

Bayes and GW Detection

In the previous section, the symbol i has been associated to the data input, i.e. a metric strain *realization*. In this sense, it could be more appropriate to call $p(i)$ the *evidence* associated to the data input, while $p(\theta)$ is the *prior*, i.e. our level of a priori knowledge of the multiple parameters of the signal, before carrying on the experiment. Lacking any extra information on the parameters, a constant value is usually assumed for the prior, i.e. the probability value of a uniform distribution. Note that a uniform prior takes the value $p(\theta) = 1/interval$, and can only be assumed if the variable is defined on a bounded interval.

Using the definition of evidence and likelihood introduced above, the Bayes' theorem can be rewritten as

$$
p(\vec{\theta}|i) = \frac{1}{\mathcal{Z}}\mathcal{L}(i|\vec{\theta})p(\vec{\theta})
$$

$$
\text{Posterior} = \frac{\text{Likelihood} \cdot \text{Prior}}{\text{Evidence}}
\tag{10.48}
$$

where from now on $\vec{\theta} = (\theta_1 \ldots \theta_n)$ is a vector whose components are the waveform parameters of a model under scrutiny.

It sometimes happens that we are mostly interested in assessing the value of one of these parameters, giving up any information on the others: the conditional posterior probability of a parameter θ_k, is obtained by applying the marginalization procedure, that simply means integrating over all possible values, over all the other components of $\vec{\theta}$

$$
p(\theta_k|i) = \frac{1}{\mathcal{Z}}\mathcal{L}(i|\theta_k)p(\theta_k) = \int p(\vec{\theta}|i) \prod_{n \neq k} d\theta_n
\tag{10.49}
$$

It should be noted that if we marginalize over a parameter θ_1 in order to obtain the posterior for θ_2, the latter posterior has some uncertainty due to the assumptions (priors) made on θ_1.

In Fig. 10.4, we show a well-known, classic example: the posterior probabilities derived for the masses of the two black holes that were the sources of the GW150914 signal.

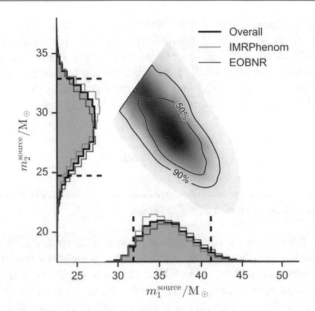

Fig. 10.4 Posterior probability density functions for the masses m_1 and m_2 (it is assumed by convention that $m_2^{\text{source}} < m_1^{\text{source}}$) of the two black holes of the coalescent binary system, which emitted the GW150914 signal. In the plot, we show the results obtained with two different family of waveforms: IMRPhenom in blue and EOBNR in red. In solid black, the *Overall* posterior is reported. The dashed vertical lines mark the 90% credible interval for the overall posterior. The two lines of the contour plot show the 50% and 90% credible regions of the posterior probability density function. Credits: Abbott et al. (2016b). Creative Commons License

The evidence Eq. 10.45 can be used to define to what extent this signal model is statistically supported by the data. Let us consider the case of a signal embedded in Gaussian noise:

$$\mathcal{L}(i|\vec{\theta}) = \frac{1}{2\pi\sigma^2} \exp\left(-\frac{|i - h(\vec{\theta})|^2}{2\sigma^2}\right)$$

where $h(\vec{\theta})$ is a given GW signal template and σ^2 the noise variance. We can assign evidences both in presence of a signal $\mathcal{Z}_h$ and in its absence, i.e. the presence of pure noise $\mathcal{Z}_n$:

$$\mathcal{Z}_h \equiv \int \mathcal{L}(i|\vec{\theta}) p(\vec{\theta}) d\vec{\theta} \qquad \mathcal{Z}_n \equiv \mathcal{L}(i|0)$$

The *Bayes ratio* BF_n^h is defined as

$$BF_n^h = \frac{\mathcal{Z}_h}{\mathcal{Z}_n}$$

the logarithmic value of this ratio is a handier quantity:

$$\ln(BF_n^h) = \ln(\mathcal{Z}_h) - \ln(\mathcal{Z}_n) \tag{10.50}$$

to indicate which model is preferred over the other: in this case for negative $\ln(BF_n^h)$ noise prevails on signal. A typical threshold for a strong evidence of signal presence is $\ln(BF_n^h) = 8$. A similar approach can be used to discriminate among different models: for example, to compare the case of waveforms with or without spins (case A and B). If the two priors $p(\vec{\theta}_A)$ and $p(\vec{\theta}_B)$ are significantly different, the Bayes ratio is modified by weighting the two evidences with the priors

$$BF_B^A = \frac{\mathcal{Z}_A}{\mathcal{Z}_B} \frac{p(\vec{\theta}_A)}{p(\vec{\theta}_B)} \tag{10.51}$$

10.10 Signals in Non-Gaussian Noise

The matched filter theory we have presented assumes the stochastic process $n(t)$ to follow a Gaussian statistic and to be stationary. It is a matter of fact, however, that the detectors are often affected by non-Gaussian, non-stationary noise sources. As a consequence, the signal searches suffer from reduced sensitivity, so it is crucial to understand how to deal with the non-Gaussian effects in order to maximize the yield of the available data. The non-Gaussian behaviour can be manifested as a change in the noise power spectral density (PSD): we refer to slow and continuous adiabatic drifts in the power spectrum occurring over minutes or hours; procedures to correct, in first approximation, the effect of the variation in the PSD of the background have been developed (Zackay et al. 2019; Abbott et al. 2020). The adiabatic drift of the power spectrum can be treated as a local stationary processes and a simpler approach is to divide the data into small chunks of time centred on time t_i, and compute a smoothed estimate for the power spectrum for each chunk $S_n(\omega, t_i)$. If the chunk is short the visualization of the signal in a time-frequency plot is poor and not efficient if based on the Fourier Transform (too few points for computing the discrete Fourier transform of the signal). A more efficient approach it is provided by a representation of the detector output using *wavelet* transform (Graps 1995), which extracts local spectral and temporal information simultaneously, while the Fourier transform captures global frequency information, i.e. frequencies that persist over the entire output stretch. In practice, the wavelet transform decomposes a function into a set of wavelets, wave-like oscillations that are localized in time. The formal mathematical definition of a continuous wavelet transform of the function $f(t)$ is

$$W(s, \tau) = \frac{1}{\sqrt{s}} \int_{-\infty}^{+\infty} f(t) \psi^* \left(\frac{t - \tau}{s} \right) dt \tag{10.52}$$

where the parameters $s \neq 0$ and τ are called the scale and translation parameters and ψ is a function of choice. Usually, s is associated to a sort of frequency content and τ to the location in the time domain. If we decrease s the wavelet becomes more squeezed, where τ shifts the wavelet along the signal $f(t)$. In practice, the basic idea is to compute how much of a wavelet is in a signal for a particular scale component and location. The calculation of the continuous wavelet transform is implemented

Fig. 10.5 An example of wavelet family: Morlet functions with fixed time duration and different frequencies

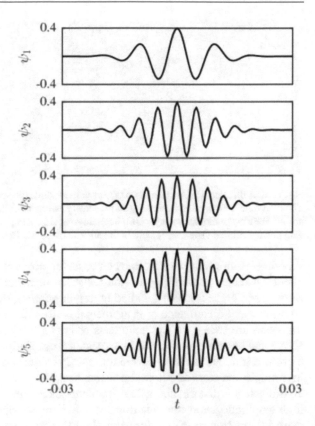

by taking discrete values for the scaling parameter s and translation parameter τ. The resulting wavelet coefficients are called wavelet series. The simplest and most efficient discretization method for practical purposes leads to the construction of an orthonormal wavelet basis

$$\psi_{m,n} = s_0^{-m/2}\psi(s^{-m}t - n\tau_0)$$

where m and n control the wavelet dilation and translation. Then, the wavelet series are calculated as

$$W_{m,n} = \int_{-\infty}^{+\infty} f(t)\psi^*_{m,n}(t)dt \qquad (10.53)$$

Any wavelet transform for which the wavelets are discretely sampled represents a valid discrete wavelet transform. It decomposes a signal into a set of mutually orthogonal wavelet basis functions. These functions differ from sinusoidal basis functions in that they are localized in time: they differ from zero only over part of the total signal length. Figure 10.5 shows the first five Morlet wavelets, a popular example of basis.

The quantitative assessments of non-stationarity is made by using discrete, orthogonal wavelet transforms. These can be visualized using a *scalogram*, see Fig. 10.6,

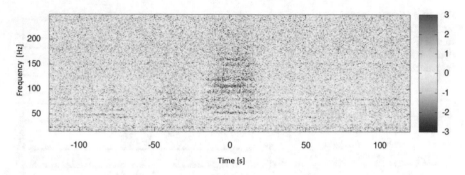

Fig. 10.6 A discrete wavelet transform was used to produce the scalogram of non-stationary data. The data were first whitened using a spectral density estimated from a stretch of 256 s data centred around the time marked here as $t = 0$. The scalogram shows that, for about 30 s around $t = 0$, the data had a significant spectral content in the 50–150 Hz band. Courtesy of the Virgo-Roma group

showing the amplitudes of the wavelet basis functions at each discrete time and frequency pixel. A second type of non-Gaussian behaviour is encountered when we have abrupt noise transients, called in jargon *glitches*, that can be caused by either environmental disturbance or instrumental malfunction. These non-Gaussian noise transients are tricky, as they manifests as bursts of excess power in the detectors above and beyond what would be expected from stationary Gaussian noise alone: therefore, glitches can be mistaken for real GW signals if they occur simultaneously in multiple detectors and limit the search sensitivity to many astrophysical signals. To mitigate this disturbance, a time-frequency excess power search is used to identify high-amplitude, short-duration transients that are not already vetoed on the basis of extra information provided, for example, by environmental monitors. To discriminate glitches from GW signals a χ^2 test is applied: the data and the GW waveform are sliced in several different frequency bands. Then, given a specific number of frequency bands n, a reduced χ^2 is computed:

$$\chi_r^2 = \frac{n}{2n-2} \sum_{k=1,n} \frac{1}{<h|h>} \left| <i|h_k> - \frac{<i|h>}{n} \right|^2$$

where, for reason of compactness, we have borrowed Dirac's bra-ket notation for the inner product of the functions

$$<a|b> = \int \frac{a(\omega)\, b^*(\omega)}{S_n(\omega)}\, d\omega$$

$|i>$ is the FFT of the input time series, $|h>$ is the template and $|h_k>$ is the subtemplate corresponding to the k-th frequency band. Values of χ_r^2 near unity indicate a very good match between data and model.

10.11   The Wiener-Kolmogorov Filter

Early gravitational wave data analysis was concerned with the detection of bursts originating from supernova explosions; at the epoch of the Weber-like resonant bars, the analysis was limited essentially to looking for impulsive events modelled by a Dirac-δ signal. This was also due to the relatively narrow detection band of the detector, that made it difficult to distinguish the shape of any transient signal other than a δ. For Earth-based interferometers, the detection band extends from a lower limit of few Hz up to 10 kHz. The broad band makes the detector more versatile and permits to study waveforms of transient signals beyond the δ function. To detect a delta-like signal in the resonant GW detector it was adopted the *Wiener-Kolmogorov* (WK) filter. This filter was independently introduced by two eminent mathematicians: N. Wiener, in the Western world and A.N. Kolmogorov in the USSR. The problem concerns how to perform a prediction in presence of stationary stochastic processes. In Wiener's particular case, the application studied was the prediction of the trajectory of an aircraft in order to direct the anti-aircraft shooting. In fact the filter theory had a strong development during World War II for the detection of radar signals, signals of known form: the received signal must have a shape similar to the emitted one and measuring the delay in reception we can establish the distance of the target.

Assume we have a signal $h(t)$, while $n(t)$ is noise, i.e. a random variable added to the output. We want to filter the stationary process $i(t) = h(t) + n(t)$ in order to extract information about the nature of the signal $h(t)$.

$\hat{h}$ is the signal estimation based on a linear combination of the process realizations $i_1, i_2, \dots i_n$ weighted by w_i $i = 1, 2 \dots n$ coefficients

$$\hat{h}(t) = w_1 i_1 + w_2 i_2 + \cdots + w_n i_n$$

Our goal is to find the optimum value of $\hat{h}$ at the time t using the information available about the process i.

The optimization is based on minimizing the estimation error, which is variance

$$E[\epsilon^2(t)] = E\Big[|h(t) - \sum_{n=-\infty}^{+\infty} w_k i_k|^2\Big] \tag{10.54}$$

It follows that the parameters w_i must be chosen to verify the condition

$$E\Big[\frac{\partial \epsilon}{\partial w_k}\epsilon\Big] = 0 \quad \forall k \tag{10.55}$$

This leads to the so-called *orthogonality principle*, which states that the minimum error is uncorrelated to each of the available data:

$$E\big[(h(t) - \hat{h}(t))i_k\big] = 0 \qquad \forall i_k$$

In other words, the variance of the error *is minimal* when ϵ is orthogonal to the sample of the data available at the filter's entry.

If the stochastic variable is continuous and so are its realizations (e.g. in the case of analog measurements), the estimation takes the form of an integral

$$\hat{h}(t) = \int_\alpha^\beta w(\xi)i(t-\xi)d\xi$$

and the application of the principle of orthogonality takes the form:

$$E\left[\left(h(t) - \int_\alpha^\beta w(\xi)i(t-\xi)d\xi\right)i(t)\right] = 0 \quad \forall \alpha < \xi < \beta$$

This can be written as

$$R_{hi}(t) = \int_\alpha^\beta R_{ii}(t-\xi)w(\xi)d\xi$$

In the frequency domain, this relationship leads to the conclusion that

$$W(\omega) = \frac{S_{hi}}{S_{ii}}$$

In this formulation, the filter is based on all the available data (the realizations), acquired both before and after the time t. With an abuse of language, this filter is also referred to as *non-causal or unworkable*. The Wiener filter is insensitive to the input signal phase, being the $w(t)$ function tied to auto-correlation functions. We report an example with historical flavour: the search for burst signals with resonant-mass GW antennas (see Chap. 8). The search consisted in detecting a sudden change of vibration amplitude against a background given by the sum of slowly varying (because of high Q) thermal noise and white amplifier noise. The best estimate of the pure noise output at a given time was given by a WK filter that used, with weights as shown in Fig. 10.7, both past and future samples: a *non-causal filter*.

10.12 Matched and WK Filter

The first attempts to use sophisticated statistical filters to improve the detection SNR were carried on in the search of GW signal with resonant detectors. The first algorithms applied by Joseph Weber were conceived to monitor the derivative of the detector output, to enhance the energy innovation in the detector. At the time of cryogenic resonant detectors, a few decades later, the data samples of the detector output were analysed by applying the Wiener-Kolmogorov filter (Bonifazi et al. 1978). As we have shown in the previous section, the filter aims to extract an estimate of a signal sequence from an observable data sequence by minimizing the mean square error (MSE). In addition the filter is defined under the assumptions of a noise, additive to the signal, which is treated as a stationary linear stochastic process with known spectral characteristics or known auto-correlation. The signal and the noise

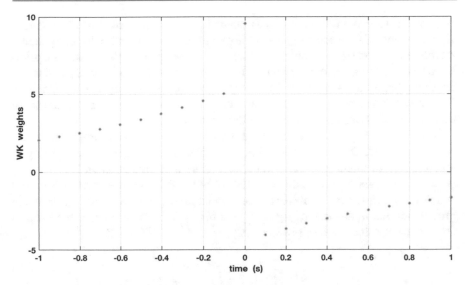

Fig. 10.7 The weights w_n of the Wiener Kolmogorov filter applied to the data of resonant-mass GW antennas in a search for short bursts. Applying these weights to the output data stream yields the best estimate of the noise at the sample $n = 0$. In abscissa, the sample number: numbers $n > 0$ refer to the future of the considered sample

are mutually independent and the stochastic process has zero-mean. At a first glance matched and Wiener-Kolmogorov filters seem very similar. Both are linear, but the matched filter is conceived to maximize the output SNR, while the Wiener filter minimizes the mean square error. MSE is a measure of the deviation between the desired and actual response, while SNR includes an additive noise term as well, which is missing in the MSE.

Generally, the two filters are applied having in mind different goals: the matched filter is used to the assess the *detection* of a known signal embedded in the noise, whereas Wiener filter applies to the *estimation* of an unknown signal with known power spectrum. In either case, the spectrum of the signal is assumed to be known. We also notice a couple of additional differences: when we consider the filter effect on the signal, we realize that the Wiener one partially affects it, while the matched filter doesn't. We can say that the matched filter is derived from the transient signal in the time domain, whereas the Wiener filter is derived from the signal and noise covariances.

10.13   Unbiased Methods to Detect Signals

When the GW impulsive signals are characterized by poor predictions of the emitted wave forms, the matched filtering, as used for the detection of inspiraling binaries, is clearly of little use, and robust methods for detecting this kind of signals are then required. In general, we are considering the case of signals with durations from milliseconds to seconds, frequencies 100 Hz–a few kHz, and a large range of waveforms.

A typical case is the GW emission associated to an asymmetric Supernova explosion. Filtering of such short signals in the output of interferometric detectors should therefore be as general or robust as possible, and designed with almost no a priori knowledge of the waveforms. Even the time duration of the signal, compared to the data sampling time, prevents us from adopting a Wiener filter tuned to a δ-function. It follows that we have to develop new filters, which are *sub-optimal*, in the sense that they are less efficient than the optimal linear filtering.

The simplest approach to search for a short burst consists of monitoring the noise power over time. The detector output is discretized by sampling at the frequency ν_s by an analog-to-digital converter. Then, in a interval of time duration T containing $N = T\nu_s$ data, we count the number of data (bins) n whose value exceeds a threshold equal to a multiple s of the noise standard deviation σ_n. If the noise is Gaussian, the probability that in absence of signal, a data bin x_i is by chance larger than $s\sigma_n$ (a event to be classified as a *false alarm*) is

$$P(|x_i| \geqslant s\sigma_n) = 2 \int_s^{+\infty} \frac{e^{-x^2/2}}{\sqrt{2\pi}} dx$$

The number of bins N_c which are by chance above threshold, follows a binomial distribution and the probability to have $N_c = n$ is

$$P(N_c = n) = \binom{N}{n} p^n \left[1 - p\right]^{N-n}$$

with $p = erfc(\frac{s}{\sqrt{2}})$.[10]

It can be shown that the reduced random variable $\hat{N}_c = (N_c - \mu_c)/\sigma_c$ with $\mu_c = Np$ and $\sigma_c^2 = Np(1 - p)$, follows a statistics, which is approximately normal when $Np > 5$ and $N(1 - p) > 5$. This method is reasonably simple and demands just two arbitrary choices, the window length T and the threshold s, and using the statistic we can asses the significance of a GW event by computing the probability to observe the event by chance.

Several, more sophisticated methods have been proposed, such as those based on *Time-frequency* strategies, a standard procedure in many areas of signal analysis. If we have hints on the duration and frequency band of the signal to be detected, we can adapt the excess power filter by comparing the power of the data, in the estimated frequency band Δf and for the estimated duration Δt, to the known statistical distribution of noise power. Under the usual hypothesis of stationary Gaussian noise of the detector output, the output power, sampled and whitened during the time window of observation, follows a χ^2 distribution.

---

[10] $erfc(x) = 1 - \frac{2}{\sqrt{\pi}} \int_0^x exp(-t^2) dt$ is the complementary error function.

If a signal of sufficiently large amplitude is also present in the detector output, the statistical distribution of the power output is distributed as a non-central χ^2 distribution.[11] The viability of the excess power method depends on the expected duration and bandwidth of the gravitational wave as well as on its intrinsic strength. We remark that, in order to implement this method, one needs to decide the range of frequency bands and duration to search over. Moreover, in the presence of *glitches*, it is essential to take advantage of the synchronized observation in various detectors in order to reject false alarms, since the method cannot distinguish between noise innovations and deterministic signals in any one detector.

A more general analysis of the detection process should start from few basic consideration: we consider the process by defining a decision rule for selecting one of two mutually exclusive hypotheses, H_0 (*null hypothesis*, the signal is absent) or H_1 (*alternative hypothesis*, the GW signal is present), about the available data that we express here as a n-dimensional vector $\vec{i}$.

Under either the H_0 or H_1 hypothesis, $\vec{i}$ is a realization of a stochastic process, described by the joint probability density $p(H_0|\vec{i})$ and $p(H_1|\vec{i})$, respectively. In the decision process, there is a probability to incur in two kind of errors

- select H_1 when H_0 is true, i.e. we decide that the signal is present and it is not true: *false alarm* with probability Q_0
- select H_0 when H_1 is true, i.e. we decide that the signal is absent when it is present: *false dismissal* with probability Q_1

According to the Neyman-Pearson criterion, we should construct our decision rule in such a way to have maximum probability of detection while keeping the false alarm probability below a set value. Thus, the decision rule can be based on the likelihood ratio Λ, defined as

$$\Lambda(\vec{i}) = \frac{p(H_1|\vec{i})}{p(H_0|\vec{i})}$$

We accept H_1 and reject H_0 when Λ is larger than a threshold fixed by a specified value of Q_0.

In the classical case of signal embedded in Gaussian white noise with zero mean, the probability densities are

---

[11] A χ^2 distribution is the sum of n independent terms, each one being the square of a gaussian random variable, with zero mean and unity variance. When the normally distributed variables have a mean other than zero, then the corresponding sum of squared terms yields a non-central χ^2 distribution of n degrees of freedom and the *non-centrality parameter* being the sum of the squared means of the normally distributed quantities.

$$p(H_0|\vec{i}) = \prod_{k=1,N} \frac{1}{\sqrt{2\pi\sigma^2}} \exp\left(-\frac{i_k^2}{2\sigma^2}\right)$$

$$p(H_1|\vec{i}) = \prod_{k=1,N} \frac{1}{\sqrt{2\pi\sigma^2}} \exp\left(-\frac{|i_k - \xi_k|^2}{2\sigma^2}\right) \tag{10.56}$$

where σ is the noise standard deviation, i_k are the generic component of the vector $\vec{i}$ and ξ_k the corresponding components of the signal. As usual, when dealing with products or ratios of exponential functions, it is convenient to define the logarithm of the likelihood ratio

$$\Lambda = log\left[\frac{p(H_1|\vec{i})}{p(H_0|\vec{i})}\right] = \sum_{k=1,N} \frac{1}{\sigma^2}\left(i_k\xi_k - \frac{1}{2}\xi_k^2\right) \tag{10.57}$$

We then look looking for the maximum (with respect to all possible templates ξ_i) of the functional Eq. 10.57

$$\Lambda_M(\vec{i}) = \max\left[\frac{p(H_1|\vec{i})}{p(H_0|\vec{i})}\right]$$

and comparing it to the chosen threshold. This test is called maximum likelihood ratio.

A relevant example: in order to assess the first detection of gravitational waves, GW150914, where we could not afford to be wrong, the false detection probability was set at $Q_0 < 2 \cdot 10^{-7}$, corresponding to 5.1 σ of the noise, or 1 spurious event every $2 \cdot 10^5$ years. The matched filter for binary coalescences yielded a SNR = 23.6.

10.14   Hunting Gravitational Wave Signals

An interferometer for the search of gravitational waves produces a continuous flow of data with a rate of 10 MBytes/s. The approach followed is to organize these data into structures, whose format is common to all interferometers throughout the world. The analysis is performed on the so-called *reconstructed h(t)*, an estimate of the interferometer input, computed by dividing the output $o(t)$ by the opto-mechanical transfer function of the whole interferometric apparatus.[12] The analysis *online* is intended to highlight any transient signal simultaneously detected in the various observing stations located throughout the world (*triggered search*). The information about the selected event is then transferred in the shortest possible time (*low latency*) to the astronomical observatories on Earth and in space, in an attempt to identify

---

[12] In coherence with our notation, we should denote the reconstructed input as $i(t)$. However, in this section we will abide by the use of the GW community that invariably refers to the *reconstructed h(t)*.

an electromagnetic signal (optical, X or gamma) in coincidence. In parallel to this process, all the data produced are sent to large computing centres where they are recorded on a permanent basis. This allows to refine the *offline* analysis both for the transient signals, for the search of continuous signals such as those emitted by a rotating neutron star and for the search of stochastic background of GW.

10.15   The Search for Continuous GW Signals

Rotating neutron stars, the main target of searches for continuous GWs, are extremely stable emitters, as we discuss in Chap. 12: we expect the emission of a pure, sinusoidal signal. Nevertheless, the task we face is trying to detect an *almost* monochromatic signal: almost, because, due to the relative motion of the detector with respect to the emitting source, the spectral purity of the signal is contaminated by the Doppler effect. As a third step in complexity, we will finally address the issue of signals with intrinsic frequency change (spin down). If the Doppler modulation were negligible, the analysis would be limited to the calculation of the Fourier transform of the data and the identification of a possible peak at a given frequency. The frequency resolution of a Fourier transform is set by the length of the observation time, $\Delta \nu_m = 1/T_{obs}$. By performing high-resolution spectra, we can assume the noise spectral density $S_n(\nu)$ not to vary in the resolution bandwidth, so that $\sigma_n^2 = S_n \cdot \Delta \nu_m$. The periodic signal, of amplitude h_0, will emerge above the noise with a power SNR:

$$SNR_{power} = \frac{|h_o|^2}{S_n} T_{obs} \qquad (10.58)$$

and the amplitude SNR grows as $SNR_h \propto \sqrt{T_{obs}}$.

Thus, we should consider stretches of data several months long to allow the signal to rise above the noise. However, we will see that data stretches so long pose a significant computational burden.

The Doppler effect complicates the analysis, modulating the signal in frequency and amplitude, thus spreading it in a frequency range. We need to determine the positions in the celestial sphere of the sources that are compatible with the Doppler correction to the signal frequency ν:

$$\nu(t) = \nu_o \left(1 + \frac{|\vec{v}|}{c} cos\phi \right) \qquad (10.59)$$

where $\vec{v}$ is the speed of the detector with respect to the source and ϕ is the angle between this vector and the line of sight of the source. $\vec{v}$ is the sum of two components: the orbital revolution around the centre of gravity of the solar system (SSB), with $|\vec{v}_{orb}| \sim 32$ km/s and the daily spinning around its axis, with $|\vec{v}_{spin}| \sim 0.45$ km/s at the equator. In the long term (of the order of a year), the significant effect is given by the orbital motion which determines a variation

$$\Delta \nu_{orbital} \sim 10^{-4} \nu_o$$

In the short term (from 1 hour to 1 day), the variation is dominated by the spin motion, even though the spin velocity magnitude is much smaller than that associated with the rotation motion around the terrestrial axis.

It is therefore convenient to subtract from the data the Doppler effect associated with the motion of the Earth, calculating it in the reference system with origin in the SSB.[13] In practice the Doppler effect is efficiently corrected in the time domain by changing the time stamp t of data samples according to the "Römer correction":

$$t' = t + \frac{r(t)}{c} cos\phi \qquad (10.60)$$

where r is the detector position in the SSB.[14]

Once these corrections have been applied, we proceed by calculating a Fourier transform on data collected over a year.

The data are obtained by sampling the signal at regular interval Δt_c, producing a series of N samples and on them we apply the discrete Fourier transform (DFT). An algorithm that speeds up this calculation is the *Fast Fourier Transform—FFT* that requires $O(Nlog(N))$ operations, vs $O(N^2)$ in a traditional DFT. In a modern computer this calculation requires a large dynamic memory (*cache* memory) to manipulate the N data at the same time. For example, to look for a signal at 1 kHz, we need to sample at least at twice that frequency (according to Nyquist's theorem on sampling), that is to $\Delta t_c = 5 \cdot 10^{-4}$ s. If we want to handle a year of data, we are faced with the problem of simultaneously manipulating $N \sim 6.3 \cdot 10^{10}$ data.

To simplify the calculation, if the value of the emission frequency is approximately known, we will be able to apply the heterodyne method, as discussed in the case of the *lock-in*: the data is multiplied by a sinusoidal signal at the carrier frequency ν_s, and a supplement filter is applied that eliminates the high-frequency signals. This action shifts the spectral band of interest around zero frequency, and the size $\delta\nu_s$ of this band is determined by the integration time of the applied filter. This corresponds to re-sampling the data at a much slower pace, and producing a drastic reduction of the number of data samples, on which the FFT can then be carried out. Applying this technique the requirements on CPU time and size of the *cache* are reduced to the point that the analysis can be performed even on a simple laptop.

The heterodyne technique is useful if the emission frequency and the position in the sky of the source are at least approximately known. In the absence of this information, we need to scan the whole sky assuming physically sensible values of frequencies. We are facing the case of a search across the full sky without a clue, called in jargon *full sky search or blind search*.

---

[13] The Doppler correction depends on the location in the sky of the source, because the angular momentum of the Earth associated with its orbital motion around the Sun is not parallel to the Earth's spin axis.

[14] We neglect here possible orbital motion of the source and all relativistic effects, that are instead considered in the refined analyses.

This search is performed dividing the sky into N_z zones, the extension of which is chosen taking into account the angular resolution of the detector, that increases with the SNR that, in turns, improves with the observation time. The resolution depends on the square of the signal frequency and increases with the fourth power of the observation time, $\left(\frac{\nu}{\Delta \nu_m^2}\right)^2$. On each patch, the *coherent* search requires to apply the Doppler correction, perform a huge FFT on the whole stretch of data and finally look for bins with excess power. Schutz calculated, for a search at 1 kHz and a relatively short observation time ($\sim$ 1 day), a number of zones $N_z \sim 10^{13}$ (Schutz 1991). We conclude that the coherent method for a blind search, optimal from the point of view of signal theory, is not viable due to the enormous computing power required.

It follows that it is necessary to develop sub-optimal statistical techniques, reducing as much as possible the losses in terms of SNR. All methods use the power from the Fourier transforms of short stretches of data, ignoring the phase information of the signal. In this sense, all these approaches are classified as *incoherent methods*. In the frequency bins where the signal is present, there should systematically be an excess of power. As Doppler effect imposes a frequency modulation on the signal, we expect the signal related excess power to migrate from bin to bin in successive Fourier transforms. The search algorithm must chase such frequency change.

In the so-called *stack-slide method*, one shifts the frequency bins of each Fourier transform to align the signal peaks and then adds the power.

We mention here another method based on the *Hough transform* (Palomba et al. 2005). The Hough transform is a method of digital image processing, useful to identify geometric shapes in images: it was invented in 1959 as an automatic tool to read out ionization traces left by charged particles in bubble chambers, and was later extended to identify arbitrary shapes such as circles or ellipses. It is a robust parameter estimator of multidimensional patterns in images and it finds many applications in astronomical data analysis. In essence, the Hough transform is an algorithm that converts from one set of variables, to another, more suited for the analysis of interest. In the original application, the conversion was performed from the coordinates $\{x_i, y_i\}$ of the points, in the bubble chamber image, darkened above a threshold, to the description {slopes intercepts} of the best lines[15] connecting those points.

To show how the Hough transform is used in this context of continuous GW searches, we consider a signal emitted at a given frequency ν and modulated, at reception, according to Eq. 10.59.

We project the position of the detector on the celestial sphere and use equatorial celestial coordinates: Declination δ and Right Ascension α;

- the coordinates δ_0, α_0 indicate the detector position, projected on the celestial sphere

---

[15] Note the plural in *lines*: if it were only one line, the problem would be trivially solved by a linear fit.

- α and δ are the right ascension and declination variables associated with the instantaneous value of the frequency that changes due to Doppler effect
- The angle ϕ, as defined in Eq. 10.59 is the angle between the detector velocity and the line of sight of the source and is computed using the relation:

$$cos(\phi) = cos(\delta)cos(\delta_o)cos(\alpha - \alpha_o) + sin(\delta)sin(\delta_o) \qquad (10.61)$$

We then look for the values of ϕ, i.e. of α and δ, that satisfy the relation Eq. 10.59 for the given value of v. As α and δ change, the locus of points drawn on the celestial sphere is a circle. Redefining

$$F(\alpha, \alpha_o) = cos(\alpha - \alpha_o) \qquad (10.62)$$

$$G(\delta, \delta_o, \phi) = \left[\frac{cos(\phi) - sin(\delta)sin(\delta_o)}{cos(\delta)cos(\delta_o)} \right] \qquad (10.63)$$

the Eq. 10.61 can be rewritten in the form:

$$F(\alpha, \alpha_o) = G(\delta, \delta_o, \phi) \qquad (10.64)$$

In this way, the points, or more appropriately the *pixels*, identified by the pair of values α, δ, for which the identity Eq. 10.64 is verified, will compose a circular annulus. In practice, using Eq. 10.64 we build a correspondence (a Hough transform) between the observational space, { frequency and time}, and the parameter space, the coordinates $\{\alpha, \delta\}$ of the source in the celestial sphere.

In the light of this observation, we proceed by dividing the data, related to a long period of observation, in many sub-periods. We choose the length of the sub-periods to be short enough.[16] that the discrete spectrum, obtained by calculating the spectral power of the interferometer signal $h(t)$, the *periodogram*, has a low-frequency resolution: a bin large enough that the signal does not suffer from the Doppler effect. Any signal will then be contained in a single frequency bin of the discrete spectrum; in the next periodogram the signal will move from one frequency bin to an adjacent one. Plotting together all the periodograms produced, we can build a time-frequency map. In that map, in the presence of a signal with large SNR, a signature like that shown in Fig. 10.8 will appear.

We now focus the attention on those particular frequency values of the discrete Fourier spectrum that exceed a predetermined threshold value. For each of these frequency values, the relative closed curve on the celestial sphere is traced. By performing repeated independent observations, at different times or with different detectors, several circles are produced and the various curves will intersect at one point of the celestial sphere, identifying the position of the source.

Finally, we address the additional complication that we face when the signal is not intrinsically monochromatic. The signal might exhibit an intrinsic frequency drift,

---

[16] The typical choice is on the order of an hour.

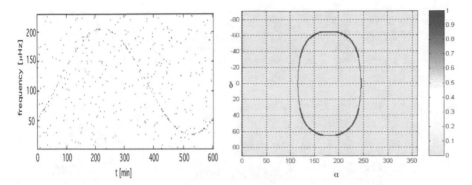

Fig. 10.8 Left: time-frequency map of simulated data in the presence of a signal with $SNR > 100$. Right: the data on the left, mapped into the $\{\alpha, \delta\}$ plane by the Hough transform, show a closed curve. Courtesy of the Virgo-Roma group

as expected from emitting neutron stars which spin down with time, or modulation due to its motion (nutation of the axis, or rotation around a companion). This is a very relevant complication: if there is no spin-down, the star is not losing energy, and does not emit GW ! These effects must also be included in the analysis. This is done by assuming as general model for the source frequency drifts, a power expansion of the GW signal frequency in terms of its time derivatives

$$\nu(t) = \nu(t_0) + \dot{\nu}\, t + \frac{1}{2}\ddot{\nu}\, t^2 + \dots$$

The unknown values of the source frequency and its time derivatives are additional dimensions of the parameter space, increasing the search complexity and the computation burden.

As title of example, we report here the upper limit of GW emission of isolated spinning neutron stars. In the plot Fig. 10.9, we show the upper limit of the amplitude of isolated spinning neutron stars of unknown position in the sky (Abbott et al. 2019). It is based on data collected by the two Advanced LIGO detectors; the data have been analysed by three different methods of search covering a frequency band from 20 to 1922 Hz. None of these searches has found clear evidence for a continuous GW signal. However, this result provides an interesting astrophysical limit on the neutron star eccentricity. We recall here Eq. 7.39, linking the amplitude h_o of the emitted wave to the eccentricity e of the compact spinning star:

$$e = \frac{c^4 r}{4G}\frac{h_o}{I_3 \Omega^2} \tag{10.65}$$

where r is the distance of the source from the Earth, I_3 the moment of inertia assumed to be 10^{38} kg m^2 and $\Omega = \pi \nu$ is the *rotation* angular frequency of the source. It follows that for sources at 1 kpc emitting 500 Hz, the ellipticity is below 10^{-6} and at 10 kpc the limit rises at 10^{-5}.

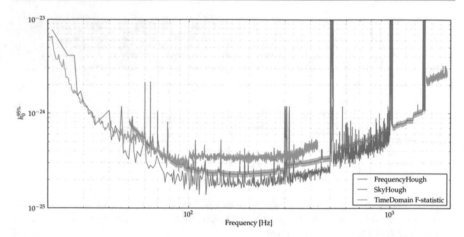

Fig. 10.9 Upper limits on the strain amplitude h_o at the 95 % of confidence level obtained from LIGO data, using three different incoherent methods. Note that in the vertical axis are reported not the spectral amplitude $\sqrt{S_h}$, measured in $Hz^{-1/2}$, but the dimensionless amplitude h_o. Credits: Reprinted figure with permission from Abbott et al. (2019). Copyright 2019 by the American Physical Society

10.16   Assessing the Detection in a Network of Antennas

It should now be clear to the reader that the detection of transient gravitational wave signals must be performed by exploiting information from multiple detectors.

To trace back from the detected signal to all the physical parameters that characterize the emission process, that is to solve the *inverse problem*, it is necessary to combine the data of five interferometers, taking into account their different position and orientation on the Earth. In terms of the SNR, the linear combination of the output signals must take into account the different noise level of the various antennas. This is done by combining the *whitened* data of each detector, or combining in the time domain using the Fourier anti-transform, the ratio of the collected data in the frequency domain divided by the relative spectral density of noise. The consistent combination of these data must take into account:

– the arrival time of the signal and therefore the phase in each detector,
– the amplitude of the signal response, which depends on the antenna orientation at that moment,

To apply a coherent (i.e. phase sensitive) combination method we need to construct the possible combinations of data that cover the space of parameters which characterize the response of the antennas to the signal. It is therefore a matter of processing the same data many times, each time assuming a different combination, choosing the one that sticks out in the statistical process constructed through all these combinations.

In addition, the same data, produced by the detector network, can also be combined in a *destructive* way, i.e. in order to cancel any signal, generating a *null data stream*. This allows us to apply a method in anti-coincidence: for each event of the constructive combination of data that exceed the detection threshold at a given moment, there must correspond the absence of a signal in the *null stream*. The most powerful algorithm to detect signals in an unbiased way is the *coherent Wave Burst* CWB method (Klimenko et al. 2008). In coherent methods, a statistic is built up as a coherent sum over detector responses. In general, it is expected to be "more optimal" (better sensitivity for the same false alarm rate) than the detection statistics of the individual detectors that make up the network. The method that we summarize here, was developed in the context of the GW search of burst signals: it combines all data streams of the independent detectors into one coherent statistic constructed in the framework of the constrained maximum likelihood analysis (Klimenko et al. 2005). To be be coherent, the data are combined taking into account the phase. The total likelihood ratio of the detector network to be maximized, similar to the definition of Eq. 10.57, is

$$log(\mathcal{L}) = \sum_{k=1}^{D} \sum_{i=1}^{N} \frac{1}{\sigma_k^2} \left(x_{ki} \xi_{ki} - \xi_{ki}^2 \right) \tag{10.66}$$

where k runs over the D detectors in the network and i runs over the N samples in the data sequence. We also assume that noises in different detectors are uncorrelated. The likelihood method introduced requires high computational load and use of memory. To reduce the computational burden, in the Coherent Wave Burst algorithm data are analysed applying a wavelet transform, which generates time-frequency pixels (s, τ), i.e. time intervals and frequency bands centred in a definite time and frequency. The square of coefficient $d(s, \tau)$ is the energy of the signal related to the time-frequency pixel which is associated to the pair (s, τ). CWB uses different decomposition levels for the same data stream, so to have different characterizations of the signal and to find the optimal one. After the application of the wavelet transform, the algorithm selects the most energetic pixels (*core*) and their neighbours. *Core* pixels are chosen if the corresponding energy is above a threshold which depends on the noise level.

The significant advantages of this coherent method are

- the sensitivity is not limited by the least sensitive detector in the network
- the maximum likelihood ratio statistic represents the total SNR of the GW signal detected in the network
- the null data stream allows to distinguish genuine signals from noise artefacts.

10.16.1 Historical Interlude: The Coincidences Method

The coherent analysis maximizes the SNR of the antenna network, but requires huge computational power. A much more manageable approach was the threshold coincidences method, used by Weber (1980) for the analysis of the data of its antennas and in all subsequent resonant detector experiments.

The analysis method applied by Weber is conceptually simple. The filtered output of each detector is examined to determine when the preset threshold is exceeded. This is how the each group produce the list of events that is then exchanged with the other groups, operating the various detectors. These lists are compared and searched for coincidences for temporal events falling into a coincident window Δt set a priori, i.e. before starting the analysis. This window is chosen taking into account the overall time resolution of the antenna network and all the possible directions of arrival of the gravitational signals.

This comparison is performed N_{shift} times, introducing at each step a time delay Δt_{shift} in one list with respect to the other. This procedure is needed to derive the probability of random coincidences, exploiting the same data with their possible non-stationary characteristics. In essence, we will find a given number of coincidences with zero delay n_c, which we compare with those obtained for $\Delta t_{shift} \neq 0$. In other words we tend to check if it happens that

$$n_c(\Delta t_{shift} = 0) \geqslant n_c(\Delta t_{shift} > 0) \tag{10.67}$$

This procedure allows us to estimate probabilities in an heuristic way, p_{exp}, by calculating the ratio between the number of times f for which the inequality Eq. 10.67 is verified on the total number of tests N_{shift}:

$$p_{exp} = \frac{f}{N_{shift}}$$

If the data selected by two antennas were realizations of a stationary process, the mean coincidence number $< n_c >_{random}$ should be deducible by applying the Poisson statistics:

$$< n_c >_{random} = N_1 N_2 \frac{\Delta t}{t_m} \tag{10.68}$$

where N_1 and N_2 are the numbers of selected events in the first and second antenna, t_m and the total coincidence data collection time (actual measurement time of the detector network). The possible discrepancy of the number observed at a delay with respect to what is foreseen by the Eq. 10.68, will then be subjected to a statistical test to quantitatively assess its consistency.

It is certainly faster and more robust, has a negligible computing weight and can be applied to data produced by detectors of a different nature such as resonant bars and interferometers (different in sampling time, bandwidth, filtering etc.). Some disadvantages of this method are:

- it cannot reveal events that are poorly reconstructed by filters and that are still immersed in noise;
- it works correctly if the filters and detection thresholds applied to the different detectors produce data and events of comparable sensitivity. This is not always possible, e.g. when the detectors are very different: one possible criterion is then to produce lists where the events are selected, rather than on their energy, by the choice of having a chosen number (same for all detectors) of events per unit time, with the intent to build an almost stationary data set of events. In any case, the standardization of the procedure on the selection of events among the various detectors is a crucial aspect for the application of the method.

Fig. 10.10 The EXPLORER resonant detector of the Universities of Rome, installed at CERN. The metal cube on the left contains the calibrator, an aluminium quadrupole of 14 kg rotating up 460 Hz

As an example, we report the case of the coincidence analysis carried on with EXPLORER, the resonant antenna of the Istituto Nazionale di Fisica Nucleare installed at CERN in Switzerland (Fig. 10.10), and the ALLEGRO detector of the Louisiana State University, in Baton Rouge (USA) (Astone et al. 1999).

Both antennas were sensitive to energy innovation of their filtered output of about $T_{eff} = 10$ mK[17] and the event detection threshold was set at 200 mK. The coincidence window was a choice of 41 s for a coinciding observation time of 121.96 days. At zero delay one coincidence was found for a value of $T_{eff} = 203$ mK. Assuming a Poisson statistic, the expected number of coincidences was 2.5, while applying the time shift method to the list of events the accidental coincidence number was found to be 1.55. The conclusion, statistically obvious, was that no impulsive event of gravitational waves with energy exceeding 200 mK had been observed during that period of observation time.

For the first undisputed detection of gravitational signals, we need to wait for the interferometers in their advanced configuration. The two LIGOs observed in 2015 the first event that appears in the data of the two detectors with a delay of 6.7 ms. This allows a rough identification of the sky area where the signal was generated, a strip in the sky of at least 10^3 square degrees.

---

[17] Antenna engineers like to express the energy associated to a stochastic signal in degrees Kelvin, $1 \, K = 1 \, J/k_B = 1.38 \cdot 10^{-23} J$. This is just like high-energy physicists do with the electron-Volt: $1 \, eV = 1 \, J /|e| = 1.6 \cdot 10^{-19} J$.

In 2017, with also Virgo in operation, the three interferometers identified the first coalescence event of neutron stars in a much narrower area, $\sim$30 square degrees, ushering in the era of multi-messenger astronomy.

10.17   Search for the Stochastic Background of GW

The Stochastic Background of Gravitational Waves (SGWB) is described in terms of the energy density ρ_{GW} of the GW background with respect to the closure density of the universe, as introduced in Eq. (7.58)

$$\Omega_{GW} = \frac{8\pi G}{3H_o^2 c^2} \nu \frac{d\rho_{GW}}{d\nu} \tag{10.69}$$

where H_o is the Hubble constant.[18]

The standard method of data analysis for the SGWB consists in cross-correlating the outputs of two GW detectors. If the noise in the two detectors is uncorrelated, the only non-zero contribution to the average cross-correlation will come from non-local disturbances, most likely the stochastic GW background. In practice, the typical approach is based on analysis methods in the frequency domain; instead of computing the cross-correlation of the GW strain of the two detectors, h_1 and h_2, we derive the corresponding bi-lateral spectrum S_{h_1,h_2}. Note that here the *signal* we are looking for is a spectrum S_{h_1,h_2} with dimensions Hz^{-1}.

S_{h_1,h_2} is related to Ω_{GW}; to obtain the explicit formula, we first have to compute the energy density of the gravitational waves ρ_{GW}, starting from the 00-component of the energy momentum tensor of the wave in GR. This computation is non-trivial and it implies an average over many wavelength. Here we skip the calculation and show the result (Maggiore 2018)

$$\rho_{GW} = \frac{4}{32\pi G} \int_{-\infty}^{\infty} (2\pi \nu)^2 S_{h_1,h_2}(\nu) d\nu \tag{10.70}$$

Then, using Eqs. 10.69 and 10.70 we conclude that

$$\Omega_{GW} = \frac{4\pi^2}{3H_o^2} \nu^3 S_{h_1,h_2}(\nu) \tag{10.71}$$

To search for a SGWB signal we consider the data stretches of two detectors $i_1(t)$, $i_2(t)$ and a matched filter q. In order to find the filter that maximizes the SNR, we introduce a figure of merit, call the *estimator* of the SBGW signal: the cross correlation of $i_1(t)$ and $i_2(t)$ weighted by the filtering function $q_\alpha(t, t')$. The latter is

---

[18] The typical assumption made when presenting the upper limits values of the LIGO-Virgo network, is $H_o = 67.9$ km s^{-1} Mpc^{-1}, the Hubble constant value from the observations of the Planck satellite).

built on the hypothesis that SGWB is expressed by the power law of spectral index α, given by Eq. 7.59 to $(-\infty, \infty)$. Thus, the SGWB estimator is

$$Y_{GW} = \int_{-T_{obs}/2}^{T_{obs}/2} dt \int_{-T_{obs}/2}^{T_{obs}/2} i_1(t) i_2(t') q_\alpha(t, t') dt' \tag{10.72}$$

Since T_{obs} is much higher than the wave travel time between the two detectors, we are justified in changing the limits on the second integral of Eq. 10.72 from $(-T_{obs}/2, T_{obs}/2)$ to $(-\infty, +\infty)$

Here too, it is simpler to compute this estimator in the frequency domain:

$$Y_{GW} = \int_{-\infty}^{\infty} d\nu \int_{-\infty}^{\infty} \delta_T(\nu - \nu') I_1^*(\nu) I_2(\nu') Q_\alpha(\nu') d\nu' \tag{10.73}$$

where

$$\delta_T(\nu - \nu') = \int_{-T_{obs}/2}^{T_{obs}/2} e^{-i2(\nu - \nu')t} dt = \frac{\sin[\pi(\nu - \nu')T_{obs}]}{\pi(\nu - \nu')} \tag{10.74}$$

is the finite-time approximation of the δ-function that converges to $\delta(\nu - \nu')$ for $T_{obs} \to \infty$. $I_1(\nu)$ and $I_2(\nu')$ are the Fourier transform of the two detectors data. Note that, because of the δ_T in Eq. 10.73, Y_{GW} will result linearly dependent on the observation time T_{obs}.

The filter in the frequency domain $Q_\alpha(\nu')$, computed following the dependence on α hypothesized by Eq. 7.59, is

$$Q_\alpha(\nu) = \Lambda_\alpha \frac{\gamma(\nu)}{\nu^3 S_{n_1}(\nu) S_{n_2}(\nu)} S_{h_1, h_2} \tag{10.75}$$

where Λ_α is a constant which depends on the spectral index α of the selected SGWB model. $S_{n_1}(\nu)$ and $S_{n_2}(\nu)$ are the noise spectral densities of the two detectors. $\gamma(\nu)$ is the *overlap reduction function*, which accounts for the reduction in sensitivity due to separation and relative misalignment between the two detectors. It is an adimensional function with maximum value $\gamma(\nu) = 1$, attained for two co-located and co-aligned detectors (Fig. 10.11). An explicit expression of $\gamma(\nu)$ requires a lengthy derivation (Flanagan 1993). Here we just report the result and discuss its physical meaning.

$$\gamma(\nu) = \frac{5}{8\pi} \sum_{k=+,\times} \int_{\Omega_s} F_1^k(\hat{n}) F_2^k(\hat{n}) e^{i2\pi\nu \frac{\hat{n} \cdot \Delta\vec{x}}{c}} d\Omega_s \tag{10.76}$$

$\hat{n}$ is a unit vector specifying a direction on a spherical surface surrounding the detectors. $\Delta\vec{x} = \vec{x}_1 - \vec{x}_2$ is the separation vector between the central stations of the two detector sites. k is the index used to specify the two GW polarization and the integration is performed over the solid angle Ω_s. The overall normalization factor $5/8\pi$ is chosen so that for a pair of coincident and co-aligned detectors $\gamma(\nu) = 1$ for

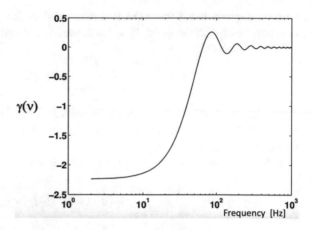

Fig. 10.11 Overlap reduction function for an unpolarized isotropic background for the two LIGO detector at Hanford and Livingston, at a distance $|\Delta\vec{x}| = c \cdot 10$ ms

all frequencies and the sum over polarization's k is appropriate for an unpolarized stochastic background. We note also that the integration over the whole two-sphere solid angle holds for an isotropic stochastic background, while the exponential factor accounts for the phase shift due to the propagation time of the wave between the two detectors. Finally, the quantity $F_1^k(\hat{n})F_2^k(\hat{n})$ is the product of the angular responses (antenna patterns) of the two detectors to both polarization waves (see Eq. 9.38).

The frequency dependence of the overlap function can be explained as follows: if the wavelength is comparable to or smaller than the separation between two detectors, the detectors will see different phases of the wave at the same time, and this phase difference will depend on the direction of propagation of the wave. Since the stochastic GW background is assumed to be isotropic, averaging over different propagation directions suppresses the sensitivity of a pair of detectors to high-frequency waves. For example, a wave whose wavelength is twice the distance between the two detectors will drive them π out of phase if it travels along the line separating them, but in phase if its direction of propagation is perpendicular to this line.

Just as in the case of periodic signals discussed in Sect. 10.15, a SGWB is always present in the detectors. Thus, it is not surprising that, to improve the SNR, we have to play with long observation time T_{obs}, i.e. long stretches of data. In order to evaluate the SNR, we assume optimal choice of the function $Q(v)$ and compute the variance of the estimator Eq. 10.73:

$$\sigma^2{}_{Y_{GW}} = T_{obs} \int_{-\infty}^{\infty} S_1(v)S_2(v)|Q_{(v)}|^2 dv \qquad (10.77)$$

It follows that the SNR is

$$SNR^2 = \frac{Y^2_{SGWB}}{\sigma^2{}_{Y_{GW}}} = T_{obs} \int_{-\infty}^{\infty} S_h^2 \frac{\gamma^2(v)}{S_1(v)S_2(v)} dv \qquad (10.78)$$

Note that SNR^2 depends linearly on T_{obs}. It follows that our accuracy in determining the spectral properties of S_h grows with $\sqrt{T_{obs}}$ and the strain amplitude of the SGWB with $\sqrt[4]{T_{obs}}$.

At present, no SGWB signal has been detected by the network of interferometers on Earth. The international collaboration has set various upper limits, depending on the assumptions made on the spectral index α. We quote here the most recent bounds, at the moment of writing, (Abbott 2021), on the energy density of the stochastic background, obtained with a 95% confidence in the frequency band 20–86 Hz.

- Assuming a flat frequency dependence, namely $\alpha = 0$, the constraint is $\Omega_{GW}(\nu) <$ 1.7×10^{-7}
- For spectral index $\alpha = 2/3$ we have $\Omega_{GW}(\nu) < 1.3 \times 10^{-7}$
- Finally, for $\alpha = 3$ the bound is $\Omega_{GW}(\nu) < 1.7 \times 10^{-8}$. This was computed using a bandwidth extending up 300 Hz, and setting $\nu_{ref} = 25$ Hz the parameter appearing in Eq. 7.59.

10.18 Conclusive Notes

Today, the special and general-purpose software systems for different research projects and with different features have been established in various laboratories around the world, to make a solid base for the gravitational experiments. Some of these tools are available to everybody, also outside the collaborations, so that anyone can try his/her hand at finding, in the public-domain data, undetected merging, or periodic signals or analysing the background.
See for instance:
https://www.gw-openscience.org/about/
https://asd.gsfc.nasa.gov/archive/astrogravs/docs/mldc/lisa_data_analysis.html

References

Abbott, B.: Upper limits on the isotropic gravitational-wave background from Advanced LIGO and Advanced Virgo's third observing run. Phys. Rev. D **104**, 022004 (2021)

Abbott, B., et al.: GW151226: observation of gravitational waves from a 22-solar-mass binary black hole coalescence. Phys. Rev. Lett. **116**, 241103 (2016a)

Abbott, B., et al.: (the LIGO and Virgo collaborations): properties of the binary black hole merger GW150914. Phys. Rev. Lett. **116**, 241102 (2016b)

Abbott, B., et al.: All-sky search for continuous gravitational waves from isolated neutron stars using Advanced LIGO O2 data. Phys. Rev. D **100**, 024004 (2019)

Abbott, B.P., et al.: A guide to LIGO Virgo detector noise and extraction of transient gravitational-wave signals. Class. Quantum Grav. **37**, 055002 (2020)

Astone, P., et al.: Search for gravitational radiation with the Allegro and Explorer detectors. Phys. Rev. D **59**, 122001 (1999)

Bonifazi, P., Ferrari, V., Frasca, S., Pallottino, G.V., Pizzella, G.: Data analysis algorithms for gravitational-wave experiments. Nuovo Cimento **1C**, 465 (1978)

Dhurandhar, S.V., Schutz, B.F.: Filtering coalescing binary signals: issues concerning narrow banding, thresholds, and optimal sampling. Phys. Rev. D **50**, 2390 (1994)

Flanagan, È.È.: Sensitivity of the laser interferometer gravitational wave observatory to a stochastic background, and its dependence on the detector orientations. Phys. Rev. D **48**, 2389 (1993)

Graps, A.: An introduction to wavelets. IEEE Comput. Sci. Eng. **54**, 50 (1995)

Klimenko, S., Mohanty, S., Rakhmanov, M., Mitselmakher, G.: Constraint likelihood analysis for a network of gravitational wave detectors. Phys. Rev. **D 72**, 122002 (2005)

Klimenko, S., Yakushin, I., Mercer, A., Mitselmakher, G.: A coherent method for detection of gravitational wave bursts. Class. Quantum Grav. **25**, 114029 (2008)

Maggiore, M.: Gravitational Waves, vol.1: Theory and Experiments. Oxford University Press, Oxford (2018)

Palomba, C., Pia, Astone P., Frasca, S.: Adaptive Hough transform for the search of periodic sources. Class. Quantum Grav. **22**, S1255 (2005)

Papoulis, A.: Probability, Random variables and Stochastic Processes. Mac Graw Hill, New York (1965)

Ross, S.W.: Introduction to Probability and Statistics for Engineers and Scientists, 6th edn. Academic Press (2021)

Schutz, B.F. (ed.): Gravitational Wave Data Analysis. NATO ASI Series, Springer Science+Business (1989)

Schutz, B.F.: Data processing, analysis, and storage for interferometric antennas. In: Blair, D.G. (ed.) The Detection of Gravitational Waves. Cambridge Univ. Press, Cambridge (1991)

Thrane, E., Talbot, C.: An introduction to Bayesian inference in gravitational-wave astronomy: parameter estimation, model selection, and hierarchical models. Publ. Astron. Soc. Aust. **36**, e010 (2019)

Weber, J.: The search for gravitational radiation. In: Held, A. (ed.) General Relativity and Gravitation, vol. 2. Plenum Pub., New York NY USA (1980). Edited by A. Held. NY: Plenum Press, p. 435, 1980

Zackay, B., et al.: Detecting Gravitational Waves in Data with Non-Gaussian Noise (2019). arXiv:1908.05644

Zubakov, V.D., Wainstein, L.A.: Extraction of Signals from Noise. Dover Publication (1962)

Space Detectors of GW

11

In this chapter we review the space-based detectors, dedicated to observing gravitational waves (GW) in the low frequency band 10^{-4} to 1 Hz. A few projects exist, either proposed or actively pursued. Among these, LISA (Laser Interferometer Space Antenna) a mission of the European Space Agency (ESA) presently under active development, is by far the most advanced and robust: we shall thus focus on LISA and describe its main features, but most of the technological developments apply as well to the other detectors. Indeed, LISA has become a name to indicate a class of space-based GW detectors which share the use of multiple spacecrafts in a triangular formation with arm lengths in the order of Gm.

We first review the scientific motivation for a space-based detector, i.e. the interesting sources available in the low frequency band. LISA is a complex instrument, and relies on some key technologies: we describe the main experimental challenges that will make the mission feasible (the *enabling* technologies) and the brilliant results reported by LISA-Pathfinder, the technological demonstrator mission flown in 2015–2017. Finally, we mention the other space-GW projects presently under consideration.

11.1 Why a Space-Based Detector?

Gravitational-wave (GW) detection in space will probe the GW spectrum that is inaccessible from the Earth. Indeed, as shown in Chap. 9, seismic and Newtonian noise prevents Earth-based interferometers from operating at frequencies below a few Hz: 10 Hz for present instruments, 1–3 Hz for future, planned detectors. While seismic and other low frequency noises can be handled with ever-improving isolation and shielding, the Earth gravity-gradient noise is fundamental, and couples directly to the mirrors. Investigation of the sub-Hz region of the GW spectrum can only be

carried out from space: a space-based detector is immune from these noise sources
and can be made very long, thus exploring the frequency range down to 20 μHz. We
recall that an antenna is best matched to waves with $\lambda \sim 2L$; to observe GW in the
sub-Hz band we then need $L \sim c/2f \sim 10^{10}$ m: such length can only be achieved
in open space. Beside, a space interferometer will not require an arm-long vacuum
enclosure!

LISA will complement ground-based observatories such as LIGO-Virgo-KAGRA,
which detect gravitational waves in the Hz to kHz range.

11.2   The Gravitational Universe—Scientific Motivations

The region of the spectrum below 1 Hz is very rich in interesting GW sources, and
crucial advances in the understanding of strong-field phenomena are expected from
their observation. The survey of the mHz sky promises to detect tens of thousands
of individual astrophysical sources ranging from White Dwarf binaries in the Milky
Way to mergers of massive black holes (BHs) at red-shifts extending to the very early
universe, beyond the epoch of reionization. Before we detail the scientific goals of a
LISA mission, we need to recall Eq. 7.56

$$\nu \approx \frac{c^3}{5\pi G M} \approx 1.3 \cdot 10^4 \frac{M_\odot}{M} \quad \text{Hz} \tag{11.1}$$

that shows how binary systems of mass larger than $10^4 \, M_\odot$ emit GW at frequencies
below 1 Hz. A space-based, low frequency detector is thus suited to observe massive
and supermassive ($M > 10^6 \, M_\odot$) black holes.

The GW astrophysics achievable in the mHz band can be described in terms
of the eight Scientific Objectives that were spelled out in the proposal (Danzmann
et al. 2017), submitted to ESA in 2017.

The physics, astrophysics and cosmology that can be explored by LISA are subject
of a vast and ever-evolving literature. A thorough review is Amaro-Seoane (2012). For

Fig. 11.1 A sketch of the
LISA sensitivity curve, with
a pictorial representation of
the main classes of sources it
can detect. The y-axis is
expressed in *characteristic
strain* $= \sqrt{\nu \cdot S_h(\nu)}$.
Courtesy of A. Sesana

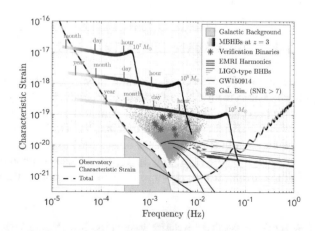

the purpose of discussing the sources described in this section and their detectability, it helps to put forward the expected sensitivity curve for LISA, Fig. 11.1, that will be derived and discussed in following sections.

- **[Science Objective 1]: Study the formation and evolution of compact binary stars in the Milky Way**.
  Numerous (millions) compact binaries in the Milky Way galaxy emit continuous GW signals, that are nearly monochromatic in the source reference frame. This population of Galactic Binaries consists mostly of white dwarfs, but also of combinations of neutron stars and stellar-mass black holes The orbital motion of the detector imparts a characteristic frequency and amplitude modulation (as described in Chap. 12) on the GW signal, that allows us to constrain the extrinsic properties of some of the systems. Based on current estimates for the population of compact stars and assuming LISA sensitivity as shown in Fig. 11.1, tens of thousands Galactic Binaries will be individually resolved with measured masses, orbital parameters, and 3D locations
  Several binaries are currently known, and many more will be discovered in the near future by GAIA[1] and other astronomy space missions. These systems are completely characterized by e.m. observations, and their masses and orbital periods $P = 2/f_{GW}$ are measured to high accuracy, so that the GW signal from these sources can be accurately predicted. They are guaranteed multi-messenger sources, called *verification binaries*, that can be used to calibrate the detector. At low frequencies, Galactic Binaries are thought to be so numerous that individual detections will be limited by confusion with other binaries yielding a stochastic foreground or confusion signal: LISA will have a GW noise! As shown in Sect. 7.6, a chirp GW signal passes through the sensitive band of Earth-based interferometers in the last fraction of a second before merging[2] and can be observed for up to a few hundred oscillations. A low frequency detector like LISA instead can observe sources, orbiting with periods of hours to a few seconds, for a very long time, accumulating up to 10^5 oscillations, harvesting a wealth of information about the "early" inspiral phase.
- **[Science Objective 2]: Trace the origin, growth and merger history of Massive BHs.**
  The origin of Massive BHs powering active nuclei and lurking at the centres of today's galaxies is unknown. LISA observations will probe massive black holes over a wide, almost unexplored, range of red-shift and mass, covering essentially all important epochs of their evolutionary history. LISA sources are coalescing massive binary black holes, among the loudest sources of GW in the Universe. They are expected to appear at the *cosmic dawn*, at red-shift $z \gtrsim 11$, when the first galaxies started to form. LISA will also explore black holes at red-shifts

---

[1] GAIA: Global Astrometric Interferometer for Astrophysics.
[2] A notable exception is the BNS event GW 170817, that had a duration of approximately 100 s.

$3 \lesssim z \lesssim 5$, when the star formation rate in the Universe and the activity of Quasi Stellar Objects (QSOs) and Active Galactic Nuclei (AGN) was highest.

Current studies predict masses for their seeds in the interval $10^3 - 10^5 \ M_\odot$ and formation red-shifts between $10 \lesssim z \lesssim 15$. They then grow up to $10^9 M_\odot$ and more through two mechanisms: accretion and repeated merging, thus participating in the clustering of cosmic structures. These events show up in the LISA frequency spectrum, from a few 10^{-5} Hz to a few 10^{-1} Hz. Mergers and accretion influence the spin of Massive BHs in different ways, thus informing us about their way of growing. The GW signal is transient, lasting from months to days down to hours. The signal encodes information on the inspiral and merger of the two spinning BHs and the ring-down of the new Massive BH that is formed. LISA will provide opportunities to probe the birth and growth of massive black holes and their host galaxies at red-shift ranges and for halo mass ranges that are not accessible with other techniques.

- **[Science Objective 3]: Probe the dynamics of dense nuclear clusters using EMRIs.**

  Extreme Mass Ratio Inspirals (EMRIs) describe the long-lasting (months to years) inspiral and plunge of stellar-mass black holes or neutron stars, into a Massive BH of $10^5 - 10^6 \ M_\odot$ in the centre of galaxies. The large difference in mass between the two objects results in a highly complex, highly relativistic orbit with multiple, time dependent frequency components. The stellar object spends 10^3 to 10^5 orbits in close vicinity of the Massive BH, and the orbit displays extreme forms of relativistic precessions, as described in Chap. 5. The large number of orbital cycles allows ultra precise measurements of the parameters of the binary system as the GW signal encodes information about the space-time of the central massive object. LISA will observe tens to hundreds of these EMRI events, yielding very precise tests of General Relativity in the strong-field regime (Babak et al. 2017) and also providing unique insight into the population and dynamics of these objects in galaxies of the local Universe ($z < 2$). LISA may also be able to detect GWs from the capture, and eventual disruption, of individual White Dwarfs by Massive BHs in the nearby Universe, leading to an exciting new multi-messenger source.

- **[Science Objective 4] Understand the astrophysics of stellar-mass black holes.**

  The LIGO-Virgo discovery of stellar-mass black holes in the mass range from 10 to 30 $M_\odot$, merging in binary systems in the nearby Universe, has stimulated a new science objective for a space detector: based on the inferred rates from the LIGO-Virgo detections, LISA will be able to observe hundreds of such Stellar-mass BH binaries, but with larger orbital separations, i.e. days or months before the merging. These rotating sources lose energy by emitting GW, and that leads to a slow decrease of their orbital separation and to an increase of the orbital frequency. Stellar-mass BHs will therefore be studied by LISA for years before their orbital frequency increases to the point that they exit the LISA band and enter the measurement band of ground-based detectors, where their final orbits and merger will be observed. It will be therefore possible to observe the same coalescence with both space and ground instruments, offering the intriguing possibility of multi-band GW observations.

- **[Science Objective 5]: Explore the fundamental nature of gravity and black holes.**
As mentioned before, LISA measurements of Massive BH Binaries and EMRIs will enable us to perform tests of GR in the strong field and dynamical regime. Measurements that can be performed include:
  – Observation of the post-coalescence ring-down of Massive BH binaries will verify if the resulting objects are the black holes of General Relativity, testing the *no-hair theorem*.
  – Test of the presence of emission channels not contemplated by GR, like dipole radiation, to unprecedented accuracy. It will be done by detecting Stellar-mass BH binaries, which appear in both the LISA and LIGO-Virgo frequency bands.
  – Verify the propagation properties of GW: speed and dispersion
  – Dark Matter search: by precise observation of the orbital motion of the "light" element in an EMRI, we can investigate the presence of non-spherical halos, e.g. of axions, around the central black hole. These precision tests require loud signals, that is, Massive BH Binaries with SNR> 100 in the post-merger phase or EMRIS with SNR > 50.
  Observations with LISA will offer opportunities to explore topics of fundamental physics in the extreme gravity regime, where the gravitational interaction is enormous and curvatures are large and dynamically changing: nature of black holes, speed of gravity, Equivalence Principle, nature of Dark Matter and Dark Energy and more are discussed in a thorough review (Barausse 2020)

- **[Science Objective 6]: Probe the rate of expansion of the Universe.**
Let us introduce the concept of *standard sirens*: some GW sources, like the inspiral of super massive black hole binaries, can yield a measurement of the distance, as pointed out by Schutz (1986) and discussed in Sect. 7.6. In analogy with the *standard candles* of e.m. astronomy, these sources are called *standard sirens*, with reference to the fact that measuring GW on Earth resembles more sound recording than image picturing. Simultaneous observation in the e.m. spectrum can also provide the red-shift z and a measure of the Hubble Constant, as it was done for the event GW170817 (Abbott et al. 2017b). Observation of a number of binaries with a total mass $M > 50 M_\odot$ in the near universe will allow determination of the Hubble constant to better than 0.02% without e.m. counterparts. Different sources can provide data at different scales: Stellar-mass BH binaries for $z < 0.2$, EMRIs at intermediate scales, $z < 1.5$ and Massive BH Binaries for the largest distances: $z < 6$.

- **[Science Objective 7]: Understand stochastic GW backgrounds and their implications for the early Universe and TeV-scale particle physics.**
One of the LISA goals is the direct detection of a stochastic GW background of cosmological origin in the presence of stochastic foregrounds. Probing a stochastic GW background of cosmological origin provides information on new physics in the early Universe. The signal spectrum gives an indication of its origin, while an upper limit allows to constrain models of the early Universe and particle physics

beyond the standard model. For these investigations it is crucial the availability of a particular combination of the LISA signals (the Sagnac combination) that is, at least partially, insensitive to the GW signal: this allows to separate the GW background from instrument noise. The sensitivity should be enough to constrain the spectral shape of the GW background in the LISA band, as well as to probe the gaussianity, polarization and level of anisotropy of the measured background.

- **[Science Objective 8]: Search for GW bursts and unforeseen sources.**
  The potential for discovery is probably the strongest motivation for exploring this spectral region, a yet-to-be observed window on the Universe. Possibilities include both astrophysical sources, like intermediate-mass black holes, and cosmological sources, like possible GW backgrounds from inflation or early universe phase transitions, and cosmic string bursts, etc. Distinguishing unforeseen, unmodeled signals from possible instrumental artifacts will be crucial in exploring new astrophysical systems or unexpected cosmological sources: this is one of the main challenges of the mission.

  The science addressed by LISA is extremely rich and covers many different domains of astrophysics. As shown in Fig. 11.1, many of the LISA sources have extremely high SNR, so that their detection will happen even with a minor changes in sensitivity or in the astrophysical estimates of the signals.

The LISA science case is robust, rich and bound to investigate new science.

11.3   50+ Years from Conception to Launch: A Little History

The idea of a space-based GW detector was first proposed in 1982 by P. Bender and J. Faller A six-page paper, with hand-drawn figures (Faller et al. 1985) describes the project, named LAGOS,[3] and works out most of the key ideas there are still today at the heart of a space interferometer, including arm length of 10^6 km (i.e. 1 Gm), heliocentric orbit at 1 AU, drag-free operation (see below), down to the details of data transmission to Earth. However, they proposed a LIGO-like interferometer, with a central station trailing the Earth on the ecliptic, and the two end-mirror stations rotating around it. Several missions were proposed to both ESA and NASA in the 1990s: they were based on six-spacecraft design, orbiting around the Earth, like SAGITTARIUS[4] or OMEGA[5] or around the Sun, like LISAG.[6] The latter was proposed as a medium-size mission to the European Space Agency in 1993. The design was then refined to a triangular configuration of three identical spacecrafts (S/C) with three 5 Gm arms, orbiting the Sun. In 1997 the two space agencies joined forces on LISA. The design of the mission progressed, in the frame of ESA's Cosmic

---

[3] Laser Antena for Gravitational radiation Observation in Space.
[4] Spaceborne Astronomical Gravitational-wave Interferometer To Test Aspects of Relativity and Investigate Unknown Sources.
[5] Orbiting Medium Explorer for Gravitational-wave Astrophysics.
[6] Laser Interferometer Space Antenna for Gravity.

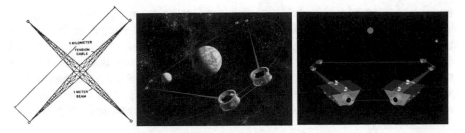

Fig. 11.2 Sketches of some pre-LISA proposals: from the left: the Gravitational Wave Interferometer (GWI), with an Aluminium structure to be extruded in space, 1 km armlength; OMEGA, orbiting the Earth and LISAG, close to the final design, but with 6 spacecrafts. Credits: NASA

Vision 2015–2025 programme and, in 2002, a technological demonstrator mission, called ELITE, then SMART-2 and finally LISA-Pathfinder (LPF), was approved (Fig. 11.2).

In 2011 NASA, due to budget cuts and increasing costs of other programs, withdrew from all astrophysics missions, including LISA. ESA tried to redesign the program on a smaller budget, with a "descoped" mission called eLISA/NGO[7] : this acronyms are still found in some of the literature. But the story has a happy end. On Feb 2016 the discovery of GW by ground detectors was announced and LPF flew in 2016–17 fulfilling its scope. The LISA technology passed the test with full marks as we will show in Sect. 11.7. On the wake of these two successes, the LISA Consortium submitted the proposal (Danzmann et al. 2017) in response to ESA's call on the theme *The Gravitational Universe*. ESA approved LISA as the third large mission (L3) of its program and NASA joined back in as a junior partner. Launch is scheduled for 2035, astrophysics data will flow in after roughly 18 months of transfer.

11.4   The LISA Constellation

An interferometer, as seen in Chap. 9, requires 3 distant stations. A space interferometer thus requires at least 3 spacecrafts, located as far as possible from each other and from the Earth, to reduce tidal perturbations; it is a natural evolution to arrange them in an equilateral triangle. In this way, each spacecraft becomes the central station of one interferometer and the end mirror for other two (see Fig. 11.5). Thus, with the same number of stations needed for a simple, Michelson-like interferometer, we can have three detectors, two of them independent with respect to the GW signal. It yields a clear advantage in terms of redundancy, sensitivity (despite having a $\pi/3$ angle between the arms, instead of the optimum $\pi/2$) and, in particular, reliability: should one-arm fail, we would still have one interferometer. Moreover, two interferometers make it possible the simultaneous measurement of both polarizations, allowing stronger tests of general relativity. Finally, by summing the phase differ-

---

[7] Evolved LISA/Next Gravitational Observatory.

ences around the three sides of the triangle, we obtain a Sagnac signal (see Chap. 13) that is insensitive to GW. Unlike terrestrial detectors, where the output contains (at present) GW information only with rare, sporadic short burst, LISA will be flooded with continuous GW signals; for this reason, the Sagnac combination offers the opportunity to better characterize the instrumental noise and calibrate the detector *without the signal*. This triple interferometer configuration is so promising that it is also being considered for the *Einstein Telescope*, the next generation of Earth-based interferometers.

A key ingredient of LISA design is the so-called *smart orbit*: the triangular configuration is maintained, in an almost rigid way, by placing the spacecrafts on three independent orbits with the following characteristics (Fig. 11.3)

(a) It is convenient to define, as an expansion parameter, the ratio
$\alpha \equiv L/2R$, where $R = 1\ AU \simeq 1.5 \cdot 10^{11}$ m and L is the interferometer arm, i.e. the distance between two spacecrafts. As of today, $L = 2.5$ Gm, having oscillated between 1 and 5 Gm in the past proposals. This length choice is based on considerations that will be discussed in the following sections. So $\alpha = 8.3 \cdot 10^{-3}$

(b) Each spacecraft orbits the Sun at an average distance $R = 1\ AU$, on a slightly elliptic orbit.

(c) The orbit, in first approximation (neglecting perturbations) is purely Keplerian.

(d) The orbital plane has a small inclination $\tan i \simeq \alpha \simeq 0.5°$.

(e) The inclination and the 1 year period determine the ellipticity[8]

$$e \sim \frac{\alpha}{\sqrt{3}} + \frac{2\alpha^2}{3}$$

(f) The three orbits are identical, but shifted by $2\pi/3$ with respect to each other.

(g) The distance from the earth, about 20° or 50 Gm, is chosen as a compromise between long-term stability of the constellation (minimization of Earth perturbations) and communications requirements.

We have reviewed in Sect. 1.3 how the equations of motion of a generic body orbiting around the Sun, like anyone of our three spacecrafts, can be expressed in terms of the eccentric anomaly ψ, defined by the equation

$$\psi - e \sin \psi = \Omega t;$$

where $\Omega = 2\pi/T = \sqrt{\frac{GM_\odot}{R^3}}$ is the mean orbital velocity.

---

[8] The inclination and ellipticity reported here are approximations, for small values of α, of exact relations, for the particular case considered here: $\tan i = \frac{\alpha}{1+\alpha/\sqrt{3}}$ and $e = \sqrt{1 + \frac{2\alpha}{\sqrt{3}} + \frac{4\alpha^2}{3}} - 1$.

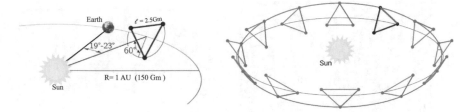

Fig. 11.3 Left: a sketch of the LISA constellation. On the right, a sketch of its motion of rotation and heliocentric revolution. The constellation size and the orbit inclination are not to scale

Each spacecraft of the LISA constellation has its own eccentric anomaly, phase shifted (see previous item f) by $\pm 2\pi/3$ with respect to the others:

$$\psi_k - e \sin \psi_k = \Omega t - \frac{2\pi(k-1)}{3} \qquad (k = 1, 2, 3) \qquad (11.2)$$

So, we define the phases $\phi_k \equiv \frac{2\pi(k-1)}{3}$

We can now reconsider the equations of motion (1.6) and derive the motion of each spacecraft by simply applying a rotation of $\phi_k \equiv \frac{2\pi(k-1)}{3}$ around the Z axis, to those equations (Pucacco et al. 2010):

$$
\begin{aligned}
X_k &= [R(\cos \psi_k - e) \cos i] \cos \phi_k - [R\sqrt{1 - e^2} \sin \psi_k] \sin \phi_k \\
Y_k &= [R(\cos \psi_k - e) \cos i] \sin \phi_k + [R\sqrt{1 - e^2} \sin \psi_k] \cos \phi_k \\
Z_k &= R(\cos \psi_k - e) \sin i
\end{aligned}
\qquad (11.3)
$$

Many corrections must be applied to these first-order, Keplerian solution, to account for all sorts of perturbations: Earth tidal field, relativistic effects, Sun quadrupole, just to mention the time-independent ones. Nevertheless, Equations 11.3 contain all the relevant features of the constellation motion. The ensemble of these three independent orbits describes the motion of the spacecrafts that move in such a way to form an *almost rigid equilateral triangle*, with the following characteristics[9] :

- The plane of the triangle maintains a constant inclination of $\pi/3$ over the ecliptic plane,[10] with constant illumination from the Sun, thus avoiding periodic thermal effects that could affect the sensitive components like optics and sensors.
- The triangle centre moves on the ecliptic, trailing the Earth by an angle chosen at $\sim 20°$, or 50 Gm. The distance to Earth is rather stable, thereby reducing the

---

[9] The motion of the LISA constellation can be visualized through the animated gif: https://it. wikipedia.org/wiki/File:LISA_motion.gif or through the video by the Max Planck Institute for Grav. Physics: https://www.youtube.com/watch?v=x-k112InxfY. Both of them are largely not-to-scale to improve the view of the orbits.

[10] The optimum value is actually $\frac{\pi}{3} + \frac{5}{8}\alpha$, given by a slightly different orbit inclination i, but we shall neglect these subtleties.

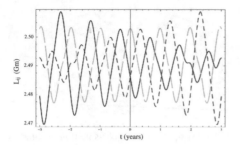 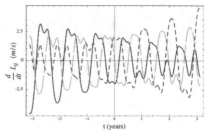

Fig. 11.4 Right: Flexing of LISA arms over a 6-year period: the three arms lengths change in different ways. The oscillations are not purely periodic due to the *drift away* effect. Left: velocities $\dot{L}_{ij}$ of the spacecrafts along the line of sight. Adapted from Pucacco et al. (2010)

problems for radio communication, as antenna design and transmission power. This distance is however not exactly stable, due to the Earth tidal perturbation: the centre drifts closer and then away from the Earth at the rate of few 10^8 m/year, depending on initial conditions. This sets a hard limit of some 10–20 years on LISA lifetime.

- The spacecraft constellation also rotates around its centre, with the same one-year period. These rotations will induce a double modulation on the signal, allowing identification of the source to better than $1°$.
- Basic orbital mechanics cause the constellation to maintain its size and shape *almost* constant for the nominal duration of the mission, allowing us to operate LISA with no need, in principle, for periodic station keeping manoeuvres, thus avoiding the use of thrusters that can spoil the free fall of the spacecraft.

However, the word *almost* in the previous sentence must be seriously considered: the tidal forces on the constellation (the spacecrafts have different instantaneous distances from Earth and Sun) cause the distances between the satellites to change by roughly 2%, resulting in a differential velocity, in the direction towards the opposite spacecraft, of up to 10 m/s and in slight changes of the constellation shape (Fig. 11.4).

This change of arm length, called *flexing* has important consequences in the design of the detector:

- We no longer have an equal arm interferometer.
- There is a Doppler shift of the signals due to the rate of change of arm length.
- The angles between the spacecrafts, nominally $60°$, change (*breathing* $\pm 1°$) forcing a realignment of the telescopes pointing the distant spacecrafts.

In the next section we shall see how to deal with these issues.

A refined analysis must also consider perturbations: the Earth and other planets, the Sun quadrupole, tides, relativistic effects. Modern *integrator* codes solve the orbits numerically to the required accuracy, but the methods of celestial mechanics also allow an analytical evaluation of these effects.

11.5   LISA: The Instrument

Mission Concept: Each spacecraft houses two test masses (TM), kept as closely as possible in free fall, that form the reference points for an interferometric measurement of the inter-spacecraft distance. The TMs take the place of the mirror in ground interferometers: a change in the distance between two distant TM is the signature of the passage of GW. The hardware on the spacecraft must provide shielding to external, non-GW disturbance and measure the TM separations.

An interferometer in space is not a simple task and each LISA spacecraft is a complex instrument: a detailed description of the numerous sub-units can be found in Jennrich (2009). We shall follow, in this section, the instrument description as outlined in the proposal (Danzmann et al. 2017), but the reader should be aware that the technologies are continuously evolving.

Each of the three spacecrafts contains two identical scientific units, serving the two arms pointing at the other two spacecrafts. As shown in the simplified drawing of Fig. 11.5, each unit contains:

- a free falling test mass (TM) inside a gravitational reference sensor (GRS)
- an infrared laser system
- an optical bench to condition the light and form optical beat notes between the local laser and lasers from other units
- a phase metre to measure the phase evolution of each of these beat signals
- a telescope of 30 cm diameter to send and receive the laser beams to and from the other spacecrafts.

The two assemblies are mounted in a common frame that allows rotation of each assembly about the vertical axis by about 2° in order to track the breathing of the constellation vertex angles.

The Observable: Phase Shift Versus Frequency Shift

Ground-based interferometers are operated in the Long Wavelength Limit, $L \ll \lambda$, and the response to GW is usually defined, by the geodesic deviation equation, in the frame associated with the beam splitter. We then chose, in Sect. 9.6.1, an inertial frame covering the whole instrument, which simplifies the calculations.

For LISA, the Long Wavelength Limit does not hold over much of its frequency band: 10^9 m $\leq \lambda \leq 10^{13}$ m. The computed detector response must then account for time delays in the response of the instrument to the waves, and travel times along beams in the instrument. It is convenient to describe the interaction in the "TT" (transverse-traceless) frame which can be seen as "co-moving" with the GW: in this frame the coordinate distance between the Test Masses does not change (but the metric does!).

In standard interferometers the light from the two arms is combined optically and the phase of each beam impinging on the recombining beamsplitter is not known. In LISA, instead, each incoming beam is combined optically with a reference beam

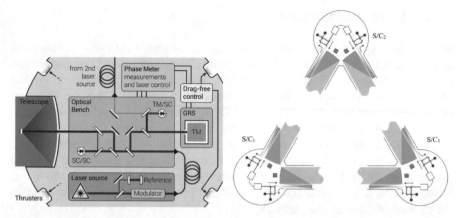

Fig. 11.5 Left: schematic block diagram of one instrumentation unit of a LISA mission: each spacecraft contains two such units, pointing to and receiving light from the two other spacecrafts. Credit: The LISA Consortium (2013). On the right, a sketch of the spacecraft constellation: each spacecraft has two units, connected by a fibre optics back link, a simplified science (or long-arm) interferometer and the telescopes pointing the other, distant spacecrafts. The grey squares represent the Test Masses. Credit: Jennrich (2009)

individually, so that the phase of the incoming light is individually measured and recorded. Recombination of the phases of the two beams is performed off-line. Thus, we will regard LISA not as constituting one or more conventional Michelson interferometers, but rather as an array of six one-arm delay lines between the test masses.

The detector response is best described in terms of observed differential frequency shifts (Doppler shifts) of the laser frequency, rather than in terms of phase shifts as used in ground-based interferometry; these data streams can obviously be derived one from the other.

$$\frac{f_{rec} - f_{emit}}{f} = \frac{1}{2}\left[h_{rec}(t) - h_{emit}(t - L/c)\right] \tag{11.4}$$

The interferometric measurements in LISA are based on heterodyne, where two lasers with respective frequencies f_L and $f_L + \delta f$ are interfered to yield a beat note[11] at the frequency δf, in the MHz region: the phase is then detected on this lower frequency signal. Unlike ground interferometers, beam recombination does not take place on-site. LISA produces six *single link* time series describing one-arm, one-way phase evolution. These are separately recorded, and the full three arms, six test masses, arrangement of LISA is off-line reconstructed by a composition of such basic elements.

---

[11] Appendix A might be useful in refreshing some concepts about modulation.

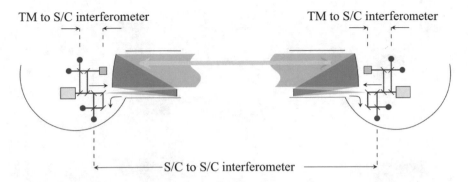

TM to S/C interferometer   TM to S/C interferometer

S/C to S/C interferometer

Fig. 11.6 Schematics of the distance measurement between two test masses, split in three different interferometers. Adapted from Jennrich (2009)

11.5.1 Inter Spacecraft Communication—The Phasemeter

For practical reasons, the distance between TMs is measured by summing the signals from three interferometers: the position of each TM is measured with respect to its own optical bench, and a third interferometer measures the inter-satellite distance, from one optical bench to the other. Such a partition of the measurement would be normally avoided, as it increases the detection noise. However, the noise budget for LISA is dominated by the contribution of the shot noise in the measurement between the spacecraft, and the amount of noise added by this technique is negligible (Fig. 11.6).

Reflection interferometry, as used in LIGO-Virgo, cannot work for a Gm path length, for the following reason: to point the light beam towards the *far spacecraft*, LISA will use infrared ($\lambda = 1064$ nm) laser light, with an emitted power $P_E = 2$ W, and a telescope with diameter $D = 30$ cm. The beam divergence, for a diffraction limited telescope, is

$$\theta_d = \frac{\lambda}{D} = \frac{1.064\ \mu m}{30\ cm} = 3.5\ \mu rad$$

When the beam emitted by spacecraft 1 (S/C_1) reaches S/C_2 at the far end of the arm, the beam diameter is $2\theta_d L \sim 9$ km, and the telescope collects a fraction P_{in} of the emitted power

$$P_{in} = \left(\frac{D}{2\theta_d L}\right)^2 P_E = \left(\frac{D^4}{4\lambda^2 L^2}\right) P_E \sim 0.7\ nW \qquad (11.5)$$

making it unthinkable to reflect the light back. The adopted alternative approach is the transponder scheme with offset phase locking (Fig. 11.7).

The received light on spacecraft S/C_2 is combined on the optical bench with light derived from the local, transmitting laser (2), to produce a heterodyne beat note in the MHz region. The phase of the beat note can be measured with high precision ($\mu rad/\sqrt{Hz}$) by an electronic *phasemeter*: it contains the information about the

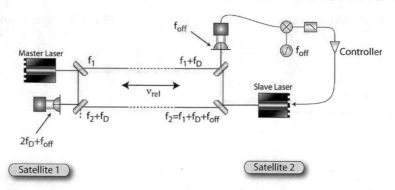

Fig. 11.7 A simplified scheme of the phase-lock system aboard the two spacecrafts of GRACE-Follow On. The light beam emitted by laser 1 (master) has frequency f_1; it is received on S/C_2 with a Doppler frequency shift $f_D = v_{rel}/\lambda$; here, a second shift f_{off} is added to it. The laser 2 (transponder) is locked to this frequency, and sends it back to S/C_1; here it is received with an additional shift f_D (identical, if the light travel time is short); interfering the received beam with the local oscillator f_1 leaves a low frequency (MHz) signal at frequency $2f_D + f_{off}$. As f_{off} is known, the measurement yields the value of $2f_D$, hence the relative velocity v_{rel}. Courtesy of G. Heinzel (Albert Einstein Institute)

Doppler shift from the relative motion of the two spacecrafts. By controlling the frequency of the laser (2), we can make the phase of the transmitted beam (2) a true copy of the phase of the received light. When received back on spacecraft S/C_1 and recombined with the local oscillator, the resulting beat note contains f_{off}, the Doppler shift f_D due to arm flexing and the signal f_{GW} due to the GW, as well as some noise due to local accelerations.

With the same procedure, we can phase-lock the two lasers on the same spacecraft and, in the end, all six lasers can be locked to a single, designated master laser. To operate such a scheme, we need two phase measurements for each arm and one between the two lasers of each spacecraft. In addition, the position of each test mass with respect to the optical bench must be measured interferometrically, requiring six more phase measurements. A total of fifteen phase measurements thus carry the complete information on the relative motion of the spacecrafts.

This operating mode makes it convenient to consider LISA not as one or more conventional Michelson interferometers, but rather a closed array of six one-arm delay lines between the test masses.

Communication among LISA spacecrafts is achieved by imprinting, with an Electro-Optic Modulator, several weak auxiliary modulations on the transmitted laser light: clock noise, bi-directional timestamp synchronization, absolute range distances and other data are exchanged among the spacecraft.

The transponder technique has been tested in the lab and, more recently, in space: as mentioned in Sect. 1.7 the geodesic mission GRACE-Follow On (Abich et al. 2019), launched in 2018, has an inter-satellite optical link that works just like that designed for LISA. Although the distance between the satellites is only 250 km, the instrument successfully passed the crucial tests of beam acquisition, phase locking and interferometer alignment. It routinely operates as an alternative to the microwave link (heritage of GRACE), with improved sensitivity.

11.5.2   The Optical System

The Interferometric Measuring System uses optical benches (OBs) which are constructed, in the lab of University of Glasgow, using the *hydroxide-catalysis bonding* technique (van Veggel and Killow 2014): all components (mirrors, beam splitters, plates...), made of fused silica, are aligned and bonded to a fused silica baseplate in the lab via a procedure that uses an aqueous hydroxide solution to form a chemical bond between oxide or oxidisable materials (e.g. SiO_2, sapphire, silicon and SiC). This method forms strong, thin bonds without using epoxy nor glue: the bench results in a quasi monolithic object, with components aligned to $\sim 10\,\mu$m precision. The technique has been used in space for GP-B (Sect. 5.8.1) and LISA-Pathfinder (Sect. 11.7) as well as in LIGO, GEO and Virgo for the mirror suspensions.

Each optical bench (Fig. 11.8) hosts one *science interferometer* for the received light from the distant spacecraft, one *local, or test mass interferometer* which monitors the position and orientation of the test mass, and a reference interferometer. Construction techniques for the optical bench with the required alignment accuracy ($\sim 10\,\mu$m) and pathlength stability in orbit ($pm/\sqrt{Hz}$) have been demonstrated with LISA-Pathfinder (LPF). The main laser field is injected via a single mode optical fibre and distributed via several beam splitters and mirrors to the different interferometers and additional sensors such as a power monitors. A signal of few mW is also exchanged between the two optical benches on each spacecraft via a bi-directional

Fig. 11.8 The optical bench of LISA-Pathfinder: all optical elements are bonded with the hydroxide-catalysis technique. In the background the Test Mass can be seen, past the optical window of the vacuum enclosure and past the coupling hole of the Electrode Housing. The picture is taken from the position that eventually hosted the second GRS. Integration of the sub-systems was carried out by Airbus Defence and Space GmbH (Friedrichshafen), that has also provided this photo

backlink: either an optical fibre, or a free beam path between both OBs. The OB has optical interfaces with the test mass (local interferometer) on one side and the telescope (science interferometer) on the other side. Each telescope has an aperture of about 30 cm diameter and serves simultaneously both to transmit and receive the beams along the respective arm. The amount of backscattered transmitted light into the receiving path must be minimized, and this is achieved through an off-axis design with up to 6 curved reflectors, some aspherical, requiring a surface figure accuracy of ~ 30 nm. The test mass interferometer measures the distance between the test mass and the optical bench by reflecting light off the TM and combining this measurement beam with a local oscillator on the optical bench.

Beside the breathing of the constellation mentioned above, another issue need to be addressed when doing interferometry in space: the light travel time between spacecrafts is $L/c \sim 8$ s: light should therefore be sent not towards the present position of the far spacecraft, identified by the wavefront of the incoming beam, but in the direction where the far spacecraft will be in a time L/c from now. This task is performed by the so-called Point Ahead Angle Mechanism, an actuator based on a rotatable mirror that must steer the beam without introducing any extra optical path or angular jitter.

Differential Wavefront Sensing

The laser beams in the science and local interferometers are monitored by InGaAs quadrant photodiodes: each device has a sensitive area (2 mm in diameter) divided in 4 quadrants, separately read. The phasemeter processes the signals from each segment using the Differential Wavefront Sensing (DWS) technique (Pierce et al. 2008): proper combinations of the four outputs ϕ_i provide both the optical path difference, as a sum of all quadrants, and the angle, horizontal and vertical components, between the interfering wavefronts, as shown in Fig. 11.9

$$
\begin{aligned}
\phi_{avg} &\propto (\phi_A + \phi_B + \phi_C + \phi_D) & \text{path difference} \\
\alpha_y &\propto (\phi_A + \phi_B) - (\phi_C + \phi_D) & \text{top-bottom angle} \\
\alpha_x &\propto (\phi_A + \phi_C) - (\phi_B + \phi_D) & \text{left-right angle}
\end{aligned}
\tag{11.6}
$$

The application of Differential Wavefront Sensing for long inter-spacecraft links has been tested on GRACE-Follow On. This scheme provides pitch and yaw angular readouts of the TM with respect to the OB (local interferometer), and of the spacecraft with respect to the incoming beam, respectively. These signals will be used in the feedback loops described further on.

11.5.3 The Gravitational Reference Sensor

The Test Masses (TMs) are the nominal reference points of the experiment, the elements in free fall along geodesics lines, whose distance is modified by the GW. Geodesic motion is made possible by the technique called *drag-free motion* of satellites:

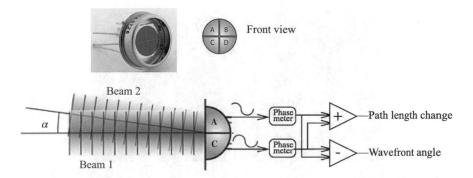

Fig. 11.9 Top: a quadrant photodiode (Otron Sensor) and a schematic indicating the four outputs. Bottom: schematic of the Differential Wavefront Sensing. Two beams (flat profile) interfere on the photodetector at an angle α: the phase difference is constant in average but varies along the photodiode surface: one quadrant photodiode reads the path difference and two inclination angles between beam 1 and beam 2. If the two beams have slightly different frequencies ($\delta f \sim 10$ MHz), all signals oscillate at δf, and a demodulation scheme can be applied

- *If the TM moves with respect to the spacecraft, it is the spacecraft that must be recentered around the Test Mass.*

Drag-free operation was proposed by Lange (1964), when he was a graduate student; it was first tested on the spacecraft TRIAD I (De Bra et al. 1974). 30 years later it was used, and improved, by GP-B (Sect. 5.8.1) and GOCE (Sect. 1.7) and, as we shall see in Sect. 11.7, by LPF. We shall return to the technicalities of drag-free operation at the end of this section, but the concept will be needed earlier on.

The sub-system composed of the Test Mass and the device that measures and control its position takes the name of *Gravitational Reference Sensor.* They must be immune from any possible disturbance: to keep the test mass in free fall, we must measure the position and orientation of the test mass with respect to the spacecraft and command micronewton thrusters so that the TM remains in its nominal position with respect to the spacecraft.

As two TMs are hosted on the same spacecraft, we must give up the goal of having the spacecraft follow both TMs on all their rotational and translational degrees of freedom (DoFs). Each TM will be thus left in free fall and followed by the spacecraft only in the direction facing the far spacecraft. This is called the x-direction of a local reference system, the one that carries all the science information. All other DoFs, both translational and rotational, are acted upon, via capacitive actuators, to keep their position and alignment. The spacecraft following two TMs along two different directions would be intuitive if the directions were orthogonal, but is quite feasible also with a $60°$ angle between the two x axes. We thus need a set of capacitors around each TM, hosted in a cubic hollow box, to monitor the motion on five DoFs and to actuate on them: the *Electrode Housing.*

The LISA Test Masses Each TM is a cube, 46 mm in side, made of an Au-Pt alloy, chosen for its high density and low magnetic susceptibility; it is gold plated

Fig. 11.10 The Test Mass (left) and the Electrode Housing (right) of LISA-Pathfinder. On the TM the recesses hosting the grabbing mechanism (on the edges) and for the soft release (centre of top face) are visible. On the electrode housing we can see some of the x and y electrodes and the entry holes for the plunger and the laser beam at the face centres, for the caging fingers at the edges. Photos courtesy of OHB-Italia

for high reflectivity, as it constitutes one end mirror of the local interferometer, that measures its position with respect to the OB, i.e. to the spacecraft. It has a mass $m_{TM} = 1.96$ kg and is machined with a few indentation that mate with the caging device that keeps it in position during launch. The TM and the Electrode Housing used in LISA-Pathfinder are shown in Fig. 11.10 and no relevant modifications are expected for LISA.

The Gravitational Reference System is composed of the TM, the Electrode Housing surrounding it, and some additional devices to serve numerous purposes:

- It shields the TM and limit stray forces, to allow the free fall requirement on the x axis: ~ 3 fm/s^2/Hz$^{1/2}$ at 1 mHz.
- It is padded with capacitor plates: two facing each TM side, form six capacitance bridges (two bridges for each pair of cube sides) with the TM as the central electrode. Others six plates are needed to inject a 100 kHz "pump" voltage used to polarize the bridges. These are simultaneously used for both *readout and actuation*: different combinations of the twelve bridge signals provide information and exert electrostatic force or torque on five DoFs. The x DoF is read by the local interferometer and no force is actuated on it, to preserve its free fall, but the electrostatic readout is available as a backup.
- The capacitor gaps are wide: 3–4 mm depending on the axis, to reduce force noise from stray electrostatics and residual gas effects;
- The actuation *authority*,[12] when operating in *science mode* (with maximum sensitivity and reduced range, see below) must be enough to compensate linear accelerations of nm/s^2 and angular accelerations of 10 nrad/s^2.
- The GRS comprises the Grab-Release-Position Mechanism mentioned above: a first set of "large fingers" to hold the TM in position with a force of $\sim 1kN$ at launch, and a second "soft" set (the plunger) to release it into free fall. This must

---

[12] *Authority* is the maximum force or torque that an actuator can provide.

be done with a velocity as little as 5 μm/s, to prevent the TM to go banging into the GRS side, before it is captured by the electrostatic actuation.

- No mechanical contact between TM and the housing is allowed, as it would provide a path for external disturbances and spoil the free fall. However, cosmic-rays charge the TM at the rate of 5 to 50 positive elementary charges per second and a charged TM is prone to electrostatic stray forces $q\vec{E}$. These are generated both by fluctuating charge interacting with the native stray fields (patch effect), and by net charge interacting with fluctuating actuation fields, on both TM and electrodes. Discharging is achieved by shining UV photons that, via photoelectric effect, extract electrons from the surfaces of TM or Electrode Housing, as needed. The work function of a clean Gold sample is about 5 eV and as consequence photons with wavelength $\lambda \leq 255$ nm are required to discharge the mass. In LPF photons were provided by Hg vapours lamps, while for LISA the use of UV-LED is foreseen, due to lighter weight, lower power consumption, longer lifetime and fast ON-OFF time. The short response time permits to turn on the UV for a short fraction of the injection period ($T = 10\,\mu s$), catching the pump field at its maximum, thus improving the yield of the electron extraction.
- A clear path must be provided for the interferometer light beam that measures the x position of the TM.
- The GRS will be positioned in a high vacuum chamber. Once in orbit, the chamber will be vented to space vacuum to further reduce the residual gas pressure, down to $1\,\mu Pa$. Residual gas is a source of Brownian noise from molecular impact on the TM surface.
- The input of external disturbances, like thruster noise, from the spacecraft to the TM must be avoided, but cannot be reduced to zero: typical paths for this non-contact leakage are the residual gas, the actuation electric fields and especially the local gravity gradient. These forces can be cancelled at the equilibrium position of the TM, but the jitter of the "noisy" spacecraft is coupled to the TM when displaced from equilibrium: $F_x/m = -\omega_s^2 x$; the parameter ω_s^2 takes the name *parasitic stiffness* and account for this coupling. This relation should be generalized to all 6 DoFs, generating a stiffness matrix that accounts for *cross-talk*, i.e. couplings with other displacements and rotations. Recalling that a Doppler shift $\delta f \sim v/c$ is caused if the TM moves, we take the time derivative of Eq. 11.4 adding the spurious accelerations:

$$\frac{\dot{f}_{rec} - \dot{f}_{emit}}{f} = \frac{1}{2}\left[\dot{h}_{rec}(t) - \dot{h}_{emit}(t - L/c)\right] + \frac{1}{c}\left[a_{rec}(t) - a_{emit}(t - L/c)\right]$$
(11.7)

where each noise acceleration is resulting both from forces acting on the TM and from leakage of the spacecraft noise motion X_{tm}:

$$a = \frac{F_{tm}}{m} - \omega_s^2 X_{tm}$$
(11.8)

- The GRS is complemented with a number of auxiliary sensors that monitor temperature, magnetic fields, cosmic ray flux.

The Front End Electronics. A specially designed electronics measures the capacitance change in all the bridges and provides, to the same capacitors, audio band voltages to generate the electrostatic force or torque actuations, as required by the feedback loop. When operated in its high sensitivity "Science Mode", with the range limited to $\pm 100\,\mu m$, its noise level is roughly $1\,aF/\sqrt{Hz}$ ($10^{-18}F/\sqrt{Hz}$!). This corresponds to a displacement sensitivity as low as $1.2\,nm/\sqrt{Hz}$ and a rotation sensitivity as low as $83\,nrad\sqrt{Hz}$.

Drag-Free operation and Micro-Newton Thrusters. Geodesic motion of the TM (or free fall) is crucial for a GW detector, and it must be achieved by shielding the TM from any external disturbances, like radiation pressure, micro-meteorites or solar wind fluctuations, that are present even in deep space. The spacecraft acts as a shield for the Test Mass, protecting it from any non-gravitational drag. This *drag-free* motion is achieved with the use of micro-thrusters: when the distance of the TM changes with respect to the shield, the thrusters activate to move the spacecraft in such a way to keep the test mass always at the centre of the Electrode Housing. The thrusters must be able to exert a force as small as $10\,\mu N$, with noise of $0.1\,\mu N/\sqrt{Hz}$. Several technologies have been developed:

- Colloid thrusters: small droplets of a ionic liquid are produced and ionized by an electro-spray process, accelerated in an electrical field and ejected from the thruster. Developed at NASA JPL, and tested on LPF, are now commercially available. However, they provide a small amount of thrust, no more than $15\,\mu N$.
- Field Emission Electric Propulsion: a method based on field ionization of a liquid metal (Caesium, Indium or Mercury) and subsequent acceleration of the ions by a strong electric field. Emission from a tip or a slit is achieved with fields of $\sim 10^9$ V/m. They were considered for LPF but, due to reliability concerns, were eventually replaced by
- Cold Gas thrusters: they use the expansion of an inert, pressurized gas through a nozzle to generate thrust. It is a simple and reliable propulsion system, with the drawback of needing a larger mass of propellant than the other considered technologies. They were space-tested by GAIA and later adopted by Microscope and LPF.
- Radio-frequency Ion Thrusters: an r.f. field ionizes xenon atoms to form a plasma. The heavy positive ions are then accelerated by an electrostatic field and ejected to cause thrust. The plasma is then neutralized by adding electrons a from a neutralizer, which prevents the satellite from becoming charged. Still under development, they would require substantially less propellant mass than cold gas.

The spacecraft and the two TMs constitute a 18 DoFs system, that must be properly kept centred and oriented, i.e. towards the incoming beam. The thrusters and the control law (see Appendix B) commanding them constitute the Drag-Free and Attitude Control System (DFACS), a key component of LISA.

11.6   Noise and Sensitivity

The sensitivity of LISA is limited by three main noise sources in different frequency regions, as summarized by the spectral sensitivity of Fig. 11.11.

11.6.1 Acceleration Noise

The limitation to the sensitivity at frequencies below approximately 3 mHz is given by residual acceleration noise acting on the test masses, that is expected to be limited to $2.4 \cdot 10^{-15}$ m/s^2/$\sqrt{\text{Hz}}$. Converting to displacement, and then to strain, produces a typical $1/\omega^2$ behaviour that becomes dominant at frequencies below approximately 3 mHz. The detailed requirement for the acceleration noise spectrum is: (Babak et al. 2021)

$$S_a(\nu) = (2.4 \cdot fm/s^2/\sqrt{Hz})^2 \cdot \left[1 + \left(\frac{0.4 mHz}{\nu}\right)^2\right] \cdot \left[1 + \left(\frac{\nu}{8mHz}\right)^4\right]$$

allowing the noise to grow below 400 μHz, where the $1/f$ behaviour occurs, and above 8 mHz, where cross-talk from other DoFs can leak into the x sensing.

Detailed noise budget analyses count dozens of sources of acceleration noise. We mention here the most relevant:

– Brownian motion, due to impact of residual gas molecules on the TM. Due to small gaps (mm) between the TM and its housing, multiple bounces on the walls enhance this effect.
– Electrostatic forces: the TM and the facing electrodes are overall equipotential, but inevitably exhibit *patch effect*, i.e. the presence of charge domains on the surfaces, due to contamination and/or varying work function. As charge accumulates and fluctuates on the TM for effect of cosmic rays, fluctuating stray forces appear.

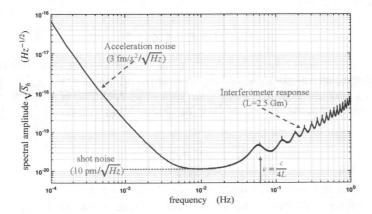

Fig. 11.11 The expected sensitivity curve $\sqrt{S_h(\nu)}$ for LISA

Hence the need of periodically discharging the Test Mass and of compensation of d.c. stray fields.

- Thermal gradients can appear across the TM, due to thermal radiation pressure or radiometric effects. A temperature difference turns into a net force. A fluctuating temperature difference across the TM of $10\,\mu\mathrm{K}/\sqrt{\mathrm{Hz}}$ would be a concern.
- Magnetic forces, arising from fluctuating fields, of either interplanetary o local (spacecraft) origin. These fields produce a fluctuating magnetic moment in the TM, that interacts with any magnetic gradient present in the spacecraft.
- Gravity fluctuations: any change in local gravity (caused, e.g. by the consumption of thruster fuel, or by settling of the structures) will produce a pull on the TM.
- DC force and torques: continuous forces and torques are applied to the TMs to compensate for the different local gravity and, in general, for all the DC terms affecting the TM. A small fraction of these feedbacks, applied to any DoF can leak into the "GW sensitive" x axis. This is a particularly serious issue for the actuation on the ϕ rotation, because it uses the same electrodes than x, and for the y translation, due to the fact that the y directions in the two TMs are not parallel, as shown in Fig. 11.5. Moreover, non-diagonal stiffness elements can couple any DoF of a noisy spacecraft motion into x motion.

Many of these noise sources have been tested, measured and analysed in the LISA-Pathfinder mission, described in Sect. 11.7.

11.6.2 The Shot Noise

- Several noise sources, related to the optical measuring system, are summarized under the label *Shot Noise*: the three interferometers (long-arm, TM readout, reference), thermal noise in the optical components, etc. The shot noise, see Sect. 9.6.3, has a white spectrum with a $1/f^2$ tail below ~ 2 mHz, and the most relevant term is that of the inter-spacecraft interferometer. When expressed in GW h units, it takes the form of Eq. 9.46:

$$S_h^{(shot)}(\nu) = \frac{1}{L^2} \frac{\hbar c \lambda}{2\pi\eta P_{in}}$$

where P_{in} is the detected light power and η the photodetector quantum efficiency. It is worth noticing that, for LISA and any Gm long interferometer with a given emitted power P_E and telescope diameter D, the shot noise contribution is independent of the arm length L: indeed, by combining this last relation with Eq. 11.5 we obtain

$$S_h^{(shot)}(\nu) = \frac{2\hbar c \lambda^3}{\pi\eta P_E D^4} \tag{11.9}$$

showing that the shot noise spectrum only depends on the emitted laser power and the telescope diameter. The interferometer displacement sensitivity is around $10\,\mathrm{pm}/\sqrt{\mathrm{Hz}}$ for each arm, dominated by shot noise in the long-arm interferometer.

This is about nine orders of magnitude worse than what achieved by ground-based detectors in the "sweet spot" above 100 Hz, due to the different environments of the two detectors: in LISA the test masses are in motion, and the received light is measured in nW rather than tens of W. The shot noise from the local interferometers and from the back link is negligible with respect to this. Equation 11.9 and current technology set the shot noise limited GW sensitivity at $10^{-20}\sqrt{\text{Hz}}$. This noise is dominant for intermediate frequencies, between approximately 3 and 30 mHz.

11.6.3 Sensitivity at Higher Frequencies

The sensitivity is reduced for frequencies above 30 mHz: not due to the insurgence of other noise sources, but to the transfer function of the detector. Computing the antenna pattern of a space detector is a more complex task than what done in Sect. 9.6.2 for terrestrial interferometers, for two main reasons: we cannot apply the long-wavelength approximation and the detector moves (revolution and rotation) during the measurement, that can last months. One has to be satisfied with computing quantities averaged over time, directions and polarizations, and even so a part of the calculation must be carried out numerically (Larson et al. 2000). The essential feature of this computation is that the effect of gravitational waves starts to cancel out as soon as the optical path in the detector (2L) approaches a multiple of the half wavelength of the GW. As mentioned above, for $\lambda < 4L$ multiple wavelengths of the GW fit into the arms, causing partial cancellation of the signal. Actually, the detector response should be zero at all frequencies that satisfy the relation $\nu = n\, c/4L$ with n integer. This is not quite the case, as can be seen in Fig. 11.11: the peaks raise well above the noise level, but remain finite.[13] This is the only benefit of arm flexing: due to the unequal length of the 3 arms in LISA, signal cancellation does not take place on all the links.

We are now ready to discuss the value of the inter-satellite distance, or arm length L. Intuitively, it should pay off to extend the distance as much as it is practical (e.g., 5 Gm of the original 1998 LISA proposal); however, we have just shown that the high frequency rising and the peaks of S_h start at a frequency $\nu = c/4L$ inversely proportional to the length. On the other hand, the acceleration noise δL_{acc} at low frequency is due to local disturbances, and is independent of the distance L from the other spacecrafts: hence, the strain noise $\delta L_{acc}/L$ scales with $1/L$. A longer arm improves sensitivity at the low end of the band but reduces the bandwidth on the high end. The shot noise has no role in this optimization, being independent of L, as shown in Eq. 11.9. The choice $L = 2.5$ Gm balances these opposite requirements, and yields an expected noise curve for LISA of Fig. 11.11.

---

[13] We remind the reader that the noise spectral density $S_h(f)$ is closely related to the inverse of the SNR.

11.6.4 Frequency Noise and Time Delay Interferometry

The noise breakdown described so far has the overlooked frequency fluctuations of the laser sources; to include them, we rewrite Eq. 11.7, adding a term that depends on frequency noise f_n:

$$\frac{\dot{f}_{rec} - \dot{f}_{emit}}{f}(t) = \frac{1}{2}\left[\dot{h}_{rec}(t) - \dot{h}_{emit}(t - L/c)\right] +$$

$$+ \frac{1}{c}\left[a_{rec}(t) - a_{emit}(t - L/c)\right] + \frac{1}{f}\left[\dot{f}_{n,rec}(t) + \dot{f}_{n,emit}(t - 2L/c)\right] \quad (11.10)$$

where $L/c = 8.3$ s, is the one-way travel time between two spacecrafts.

Let us briefly recap the influence of the laser frequency noise in equal arm interferometers: the relevant quantity measured by an interferometer is the phase difference between the two recombined beams: $\Delta\phi = \frac{4\pi f}{c}\Delta L$, where f is, as usual, the laser frequency and ΔL the static arm length difference. Any time-varying change of frequency δf mimics a displacement δL as: $\delta f/f = \delta L/\Delta L$.

If we now move from time domain fluctuations to their spectral densities: $\delta f \rightarrow \sqrt{S_f}$ and $\delta L/L \rightarrow \sqrt{S_h}$, we can see the contribution of frequency fluctuations to the overall detector noise:

$$\sqrt{S_h^{freq}} = \frac{\sqrt{S_f}}{f}\frac{\Delta L}{L} \quad (11.11)$$

For ground-based detectors, ΔL is controlled at a fixed value of the order of 1 m (the Schnupp asymmetry, defined in Appendix B.5). The laser in Virgo has reached a frequency fluctuation spectrum as low as $\sqrt{S_f} \sim 10^{-6}\ Hz/\sqrt{Hz}$, thanks to a complex system that includes a mode cleaner cavity 144 m long and even the lock to a 3 km long Fabry-Perot arm cavity. This produces an equivalent strain noise $\sqrt{S_h^{freq}} \sim 10^{-24}/\sqrt{Hz}$.

Very different are the operating conditions in LISA: the laser frequency is stabilized by an on-board cavity to about 30 $Hz/\sqrt{Hz}$ at the operating frequencies (mHz). Moreover, arm flexing forces path differences as large as $\Delta L \sim 10^7$ m, yielding an equivalent strain $\sqrt{S_h^{freq}} \sim 10^{-15}/\sqrt{Hz}$, orders of magnitude larger than LISA target sensitivity $\sqrt{S_h} = 10^{-20}/\sqrt{Hz}$

Locking to the long-arm interferometer, in analogy to what is done in LIGO and Virgo, has been considered, but it would not give all the needed improvement. This apparently hopeless situation has been saved by an ingenious method of data processing (Tinto and Dhurandhar 2014): the Time Delay Interferometry (TDI). In essence, TDI requires, in the off-line data post-processing, to appropriately time-shift the single link Doppler signals coming from different arms, and recombine them in such a way to reconstruct n equal arm Michelson interferometer.

To catch the flavour of the TDI, let us consider the round trip of two light beams collected in spacecraft n.3 (S/C_3); in the LISA arm 1, connecting $S/C_1 - S/C_3$, and arm 2, reaching to S/C_2[14]: their lengths are $L_1 = cT_1$ and $L_2 = cT_2$ respectively. The two beams are collected, on S/C_3, at time t: each carries the Doppler information about frequency noise, that we indicate with F_n, imprinted on the beam at the time of emission and at the time of detection. Moreover, there will be a Doppler signature h_i from the GW, and n_i from any other noise source. We then have two time series:

$$y_1(t) = F_n(t - 2T_1) - F_n(t) + h_1(t) + n_1(t)$$
$$y_2(t) = F_n(t - 2T_2) - F_n(t) + h_2(t) + n_2(t) \qquad (11.12)$$

The two beams, emitted at different times, carry a different frequency noise, that is not cancelled as it happens in standard interferometry. Indeed, by taking the difference of the two signals, we see that only the frequency noise at detection time t cancels out:

$$y_1(t) - y_2(t) = F_n(t - 2T_1) - F_n(t - 2T_2) + h_1(t) - h_2(t) + n_1(t) - n_2(t) \qquad (11.13)$$

We now consider the same data, but shifted in time: take the time series y_1 at time $t - 2T_2$ and the series y_2 at time $t - 2T_1$: we shift the data of each beam by the round-trip light time in the opposite arm. We obtain

$$y_1(t - 2T_2) - y_2(t - 2T_1) =$$
$$= F_n(t - 2T_1) - F_n(t - 2T_2) +$$
$$+ h_1(t - 2T_2) - h_2(t - 2T_1) + n_1(t - 2T_2) - n_2(t - 2T_1) \qquad (11.14)$$

This combination has the exactly the same frequency noise than the previous one, Eq. 11.13. If we now subtract Eq. 11.14 from Eq. 11.13, we generate a new data set, that is immune from laser frequency fluctuations:

$$X \equiv [y_1(t) - y_2(t)] - [y_1(t - 2T_2) - y_2(t - 2T_1)] \qquad (11.15)$$

This defines the TDI *first-generation* variable X, the first and simplest of an ample, and ever growing, family of TDI data sets. Analogous definitions can be given for the beams "recombined" at the other spacecrafts, called Y and Z. In order to gain some physical insight, we rearrange the terms of this definition of the data set X

$$X(t) \equiv [y_1(t) + y_2(t - 2T_1)] - [y_2(t) + y_1(t - 2T_2)] \qquad (11.16)$$

We can interpret X as the difference of the Doppler signal arising from two fabricated light paths, not as simple as the usual round trip along one arm, as shown in Fig. 11.12b: the first term in square bracket can be read as the path of a light beam

---

[14] Here we derogate from the usual TDI convention of naming the arm with the number of the opposite spacecraft; we are just presenting an example, and privilege clarity and immediacy over completeness.

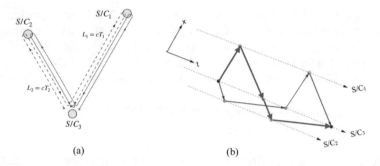

Fig. 11.12 a A pictorial image of first-generation TDI, balancing the light paths for stationary end stations. Adapted from Tinto and Dhurandhar (2014) under Creative Commons (2021). **b** A different sketch of first-generation TDI, with a space-time visualization. Adapted from Muratore et al. (2020). Both diagrams have been modified to adhere to the spacecraft naming convention used here

generated in S/C_3, reflected[15] off S/C_2, returning at S/C_3, reflected off S/C_1 and finally detected at S/C_3. The second term represents a beam reflected first off S/C_1 and then off S/C_2. The travel time of these two synthetic beams is the same, and when recombined, the frequency noise cancels exactly.

We remark that the lines of Fig. 11.12 do not represent real beam paths: in each arm we only have two beams (one in each direction). The lines can rather be intended as the paths of relative frequency changes. The algebra of TDI is usually expressed by means of the *time-shift operators* :

$$D_i y(t) \equiv y(t - T_i) \tag{11.17}$$

so that, for example, Eq. 11.16 is rewritten as

$$X(t) = (1 - D_2 D_2) y_1(t) - (1 - D_1 D_1) y_2(t)$$

A point needs to be addressed: TDI requires knowledge of the inter-spacecraft distances cT_i, to better than 1 m, to effectively cancel frequency noise.[16] The absolute inter-spacecraft distances are determined to the required ~ 10 cm accuracy using an auxiliary modulation on the laser beams.

The above derivation holds for arm lengths that are different, but stationary. However, another consequence of arm flexing is the relative motion of the spacecrafts along their mutual line of sight. This velocity, that can be as large as 5 m/s, imprints an additional Doppler shift on the exchanged light beams, that must be accounted for and cancelled. For this purpose, *second generation TDI* combinations have been developed, with synthetized beams much more complex than that in Eq. 11.16, using

---

[15] "Reflected" is to be intended as "transponded".
[16] The arm flexing occurs on time scales of one revolution, that is 1 yr, as shown in Fig. 11.4; during a light travel time, $2T_i \simeq 16.6$ s, the constellation is virtually still.

up to 24 links: in this case, the delay operators no longer commute. A detailed analysis of TDI is beyond the purpose of this book, a comprehensive description of the algebraic approach is given (Tinto and Dhurandhar 2014).

TDI has been experimentally verified in the lab, with delays in fibre of 3 km (Mitryk et al. 2012), but the exhaustive test, over the Gm distances, will have to wait for the flight of LISA.

11.7   Paving the Way: LISA-Pathfinder, the Technology Demonstrator

The experimental challenges outlined in the previous section are huge, and great would be the risk of launching LISA without proper testing the technological solutions. Here is a first list:

- Stability of the optical set-up: laser, photodetectors, phasemeter.
- Alignment of the interferometer.
- Release of the TMs from lock with low enough momentum to allow electrostatic position control.
- Quality of free fall: residual forces on the TMs of electrical, magnetic, thermal, cosmic-rays origin.
- Electric charge control on the TM.
- Thruster noise, stability of the feedback that re-centres the spacecraft on the TM.
- Self-gravity balance, gravity gradient on the TM.

All these issues require free fall for a proper validation. Extensive test were performed on torsion pendulums, that allow *quasi free fall* on one DoF to evaluate residual forces (Cavalleri et al. 2009). A special pendulum, with two *soft* DoFs (one rotation and one translation) was developed (Bassan et al. 2016) to investigate possible cross-talk between x and other DoFs. Although relevant upper limits were set, these were partial tests compared to real free fall on 6 DoFs. Fortunately, all of these noise contributions are independent of the long-arm length of LISA as they occur within each single spacecraft. That is why ESA promoted LISA-Pathfinder, a space mission with the purpose of paving the way for LISA: testing all of the above technologies and verifying the noise levels that can be achieved in space.

The driving concept of LISA-Pathfinder (LPF) was to squeeze a LISA arm from 2.5 Gm down to 38 cm: hosting two TMs (call them TM1 and TM2) on the same spacecraft, it was possible to monitor their distance and validate both the optical measuring system and the free fall inside the GRS, by measuring the residual force (actually, acceleration) on the TMs. An interferometer measured the position x_1 of TM1 with respect to the optical bench and another the distance Δx between TM1 and TM2, both along the x direction connecting the two centres of mass:

$$o_1(t) \equiv x_1(t) - x_{sc}(t) + n_1(t)$$
$$o_{12}(t) \equiv x_2(t) - x_1(t) + n_2(t) \tag{11.18}$$

Fig. 11.13 A schematic of
LISA-Pathfinder: the optical
bench is positioned between
TM1 and TM2; The
measurement of Δx drives
the electrostatic feedback
forces on TM2, applied by
the GRS electrodes. The
measurement of x_1 is instead
the error signal driving the
drag-free control loop that
uses the μN thrusters to exert
forces on the spacecraft.
Credits: reprinted figure with
permission from Armano et
al. (2016). Copyright 2016
by the American Physical
Society

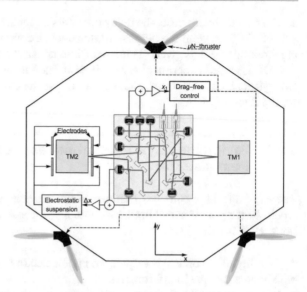

where x_{sc} is the spacecraft displacement, with respect to an inertial frame and n_i are
readout, additive noise contributions, mostly laser shot noise.

LPF was launched on Dec. 3, 2015 and reached, in a 50 day trip, a wide orbit around
the L1 Lagrange point of the Earth-Sun System.[17] Then, the two Test Masses were
successfully released into free fall and locked into a complex feedback system: as
described above, both TMs are electrostatically controlled on 5 DoFs, but free (in
principle) along the x direction. However, the spacecraft cannot follow both TMs in
their motion along x, thus only one TM can be kept in geodesic motion:
- TM1 was left in free fall, with the micro-thrusters controlling the spacecraft position. The *drag-free control loop*, with 1 Hz bandwidth, nulls the output o_1 of the first
interferometer.
- TM2 was instead actuated by the GRS electrodes in order to keep constant the
distance TM1-TM2. The *electrostatic suspension*, a control loop with 1 mHz bandwidth, nulls o_{12}.
We have here a dynamical system with 18 DoFs, controlled by the Drag-Free Attitude Control System, a sophisticated control law loaded on the on-board computer.
The diagram of Fig. 11.13 may help clarifying the set-up.

The main scientific goal of LPF is assessing the differential acceleration of the
two TMs:

$$\Delta g(t) \equiv \frac{F_2(t)}{m} - \frac{F_1(t)}{m} \tag{11.19}$$

---

[17] The Lagrange points are defined in a footnote of Sect. 6.6.1.

We write the equations of motion for the two TMs:

$$\ddot{x}_1 = \frac{F_1}{m} - \omega_{s1}^2(x_1 - x_{sc})$$

$$\ddot{x}_2 = \frac{F_2}{m} - \omega_{s2}^2(x_2 - x_{sc}) + \frac{F_{es}}{m} \qquad (11.20)$$

where ω_{si}^2 ($i = 1, 2$) are the stiffnesses, mentioned in Sect. 11.5.3, that feed residual spacecraft motion to the TM.

Combining Eqs. 11.19, 11.20 and 11.18 leads us to:

$$\Delta g = \ddot{o}_{12} - \frac{F_{es}}{m} + (\omega_{s2}^2 - \omega_{s1}^2)o_1 + \omega_{s2}^2 o_{12} + n_{IFO} \qquad (11.21)$$

where n_{IFO} is the combined noise of the two interferometers, that contains, among other terms, the second time derivative of n_2. Equation 11.21 shows that, in order to have a fair estimate of the differential acceleration, one must subtract all known and measurable contributions, like those introduced by the stiffnesses (that are evaluated with a calibration procedure) and by the applied feedback. Other effects, not reported in this equation but accounted for in the analysis are the centrifugal forces due to spacecraft rotation (about $2°/$ day, due to orbiting around L1) and pick-up of the noisy spacecraft motion on other DoFs, due to residual misalignment of Electrode Housings and optical bench.

The LPF mission requirement relaxed the LISA goal, on the measured residual acceleration, by a factor 10 both in frequency and in noise spectral density: $30\,\mathrm{fm/s^2}/\sqrt{Hz}$ at 1 mHz.

One reason of the relaxed sensitivity goal is the presence of the electrostatic suspension; indeed, it will not be present in LISA, where the spacecraft will be free to follow each TM along its x axis, the direction of the long arm; the feedback controls were expected to be the limiting source of noise at low frequency, proportional to the amount of DC fields, mainly gravitational, they balance. In LPF these fields turned out to be small as a result of accurate design and positioning of the masses in the spacecraft; the reduced electrostatic suspension controls allowed for the excellent performances of the LPF mission.

Figure 11.14 shows the acceleration noise measured by LPF. Since its first day of operation (blue curve), the mission requirement, marked in dark grey, was largely overcome, and the sensitivity approached the level required for LISA.

The mission lasted 15 months, and numerous experiments were carried out to identify and characterize the various sources of noise. The apparatus was fine tuned, e.g. by lowering the pressure via continuous venting to outer space, and by reducing the feedback authority on TM2.[18] The ultimate performance of LPF (Armano et al. 2018) is shown by the red curve in Fig. 11.14: it was achieved also by lowering the

---

[18] This was possible because the local gravity turned out to be extremely well balanced: no serious gradient was measured on TM2.

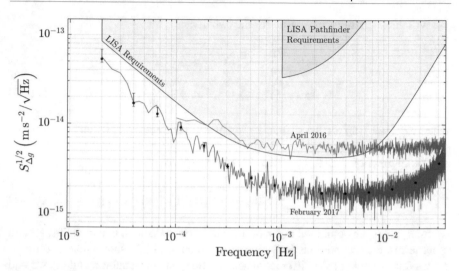

Fig. 11.14 Initial (blue) and best (red) performance of LISA-Pathfinder. Unlike sensitivities usually shown in the GW community, here the plotted variable is the differential acceleration Δg, i.e. the deviation from perfect free fall. The shaded areas show the acceleration noise requirements for LISA and, relaxed by a factor 10, for LPF. Credits: reprinted figure with permission from Armano et al. (2018). Copyright 2018 by the American Physical Society

temperature by 12 degrees, to about 284 K. The interferometer noise was measured to be $\sqrt{S_0} = 35$ fm/$\sqrt{\text{Hz}}$, well below the LISA requirement: when converted to acceleration (our variable of interest here) it grows with frequency as $\sqrt{S_0}\,\omega^2$ and dominates the spectrum for $f > 20$ mHz, barely visible at the right edge of Fig. 11.14.

LISA-Pathfinder was an extremely successful mission, that cleared the field from many potential LISA stoppers. Some technological issues could not be tackled in a single-spacecraft mission, e.g.:

- High stability telescopes
- Constellation acquisition and keeping over a Gm baseline
- High precision spacecraft attitude control
- High accuracy phase metres: the technique has been successfully tested on GRACE-FO, but operation on a 2.5 Gm distance is more demanding
- Frequency stabilization of lasers, exchange of frequency standards among the spacecrafts

All these will be tested with the full three-spacecraft, long-arm constellation, but the road towards LISA has been definitely smoothed by LPF and GRACE-FO.

11.8   Other Space GW Missions

The Europe-USA project LISA is, at the moment, the most advanced space project for observing GWs, but by no means the only one. The Chinese space-GW community is simultaneously pursuing two different missions (Gong et al. 2021), TianQuin and Taiji, while the Japanese community has proposed the ambitious DECIGO observatory; we briefly describe these projects.

The amount of effort devoted into these studies might close the gap with LISA, so that the exciting perspective of having two or more observatories active at the same time (mid 2030s) is not ruled out.

11.8.1 DECIGO

The most futuristic proposal is probably the Deci-hertz Interferometer Gravitational Wave Observatory (DECIGO) (Kawamura et al. 2021), a project of the Japan Aerospace Exploration Agency (JAXA) that aims at filling the frequency gap between LISA (sensitive up to $\sim 1\,Hz$) and the ground-based interferometers, AdVirgo and aLIGO, that have sensitivity starting at $10\,Hz$. In this frequency band there are several sources of interest: signals from coalescences of galactic binaries should emerge from the confusion noise (the GW noise mentioned in Sect. 11.2), but the detector is mainly designed to investigate the GW background, both cosmic and astrophysical, thanks to its ambitious design.

The main element of DECIGO closely resembles LISA: a cluster of three spacecrafts connected by laser beams and free fall of the TMs. To achieve sensitivity in the $0.1–10\,Hz$ band, the arms of the interferometers will have length $L = 10^6\,m$, about a thousand times shorter than LISA: this should allow the use of Fabry-Perot cavities (see Sect. 9.5) in the arms, with a finesse $\mathcal{F} = 10$. As a consequence, the TM must be mirrors, $1\,m$ in diameter and $100\,kg$ of mass.

The most striking feature of DECIGO is its redundance: there will be four identical clusters, for a total of twelve spacecrafts: three in heliocentric orbit, 120° apart, to maximize the angular resolution for source localization. The fourth cluster is colocated (same centre of mass) with one of the previous, to perform correlation measurements for the detection of primordial GW. Indeed, the noise of each cluster is independent, while GW signals are common to the two clusters: correlation of the two outputs would improve sensitivity. It should be noted that the same four-cluster configuration was proposed to NASA around 2004, with the name Big Bang Explorer.

JAXA envisions a pathfinder mission, B-DECIGO, composed of only one cluster, with shorter arms ($100\,km$) and smaller mirrors ($30\,cm$ diameter, $30\,kg$), to be launched in the 2030s.

11.8.2 TianQin

TianQin is a space-based GW detector developed by two Chinese teams at Sun Yat-Sen University (in Guangzhou) and Huazhong University of Science and Technology (in Wuhan). It resumes a concept put forward for OMEGA and other proposed missions, like GADFLI[19] and gLISA[20] : a triangle of spacecrafts on a geocentric orbit, which is easily reachable for implementation and operation. The orbit radius is 10^5 km, yielding an arm of the equilateral triangle $L = \sqrt{3} \cdot 10^5$ km; the frequency sensitivity band of the detector overlaps with that of LISA near 10^{-4} Hz and with that of DECIGO near 0.1 Hz. TianQin aims at bridging the frequency gap between LISA and DECIGO and might complement LISA in the high frequency regime, that is relevant to the search for intermediate-mass BHs. The launch of TianQin is foreseen around 2035. The constellation plane is inclined in such a way as to maximize sensitivity towards a verification signal: the white dwarf binary RX J0806.3+1527, also called HM Cancri69–73. TianQin was proposed before the discovery of GW, and the goal was to aim at a guaranteed detection; it is still a valid choice, as this source assures a proper calibration of the detector.

The choice of geocentric orbits has relevant advantages: reducing the launch cost, shortening the signal transfer duration, simplifying telecommunications and more easily guiding satellites through global navigation, laser ranging and so on. However, there are also important drawbacks, that LISA's heliocentric orbits are immune to:

- the thermal stability of the spacecraft, affected by the variation in the sunlight direction relative to the orbital plane or eclipses due to the Earth or the Moon temporarily blocking the sunlight
- the constellation stability: distortions in the equilateral triangle are induced by the gravitational disturbances, especially from the nearby Earth-Moon system
- the problem of a steady power supply because of the eclipses of the Earth and the Moon when the constellation passes through their shadows.

All these issues are object of careful, ongoing studies by the TianQin team and will probably impact on the duty-cycle of the observatory operations.

TianQin project started in 2016 with an aggressive roadmap in four steps: Step 0: laser ranging technology, has been fulfilled with the launch in 2018 of a laser ranged satellite to the L2 Lagrange point and the construction of a Earth ranging station with single-photon sensitivity. Step 1: TianQin-1 is a satellite with inertial sensing, micronewton propulsion, drag-free control and laser interferometry technologies with in-orbit experiments; beside, it aims to test the temperature control technologies and centre-of-mass measurements of the satellite. It was successfully launched in 2019 and achieved all its goals, measuring a residual acceleration of $5 \cdot 10^{-11}$ m/s^2/$\sqrt{\text{Hz}}$ at 50 mHz. Step 2 is a two-satellite mission similar to GRACE-FO, to test inter-satellite interferometry and map the gravity field. It is scheduled for launch in 2025. The final Step 3 is the full constellation and is planned for deployment in 2035, with an exciting perspective of simultaneous observations with LISA.

---

[19] GADFLI: Geostationary Antenna for Disturbance-Free Laser Interferometry.
[20] gLISA: Geosynchronous Laser Interferometer Space Antenna.

11.8.3 Taiji

Similar to LISA, Taiji is a project of the Chinese Academy of Sciences, very similar to LISA: a three-satellite constellation in a heliocentric orbit ahead of the Earth by about 18–20° (unlike LISA, that trails the Earth), as a compromise between orbit injection costs and Earth-lunar system disturbance. The arm length is $L = 3 \cdot 10^9$ m, longer than LISA by 20%, allows a better sensitivity at frequencies around 0.01–0.1 Hz. Goals for displacement noise and acceleration noise are roughly the same than LISA's, and so are the scientific objectives.

A pre-pathfinder mission, Taiji-1, was launched in 2019. Taiji-2, a pair of technology demonstration satellites designed to cover almost all of the Taiji technologies except TDI, is scheduled to launch in 2024. They will fly at a distance $L = 5 \cdot 10^8$ m on the same orbit, and carry the same payload, than the final Taiji observatory. Taiji is expected to launch around 2033: if both Taiji and LISA launch according to plan, they will have a few years of overlapping observing time, leading to amazing capabilities in source localization: while LISA or Taiji alone have an angular resolution of about 1 deg^2, the combination of 2 or 3 networks could lead to resolutions of the order of 10^{-3} deg^2 because they will operate as a VLBI interferometer (see Sect. 6.6.1), with a baseline of 10^{11} m.

11.9 Final Remarks

We must remark that, while the noise spectrum for LISA rests on solid experimental background, sensitivities and timelines for other detectors appear, at the present state of technology, quite optimistic. Observation of GW from space is an exciting perspective for the mid 2030s. The quantity and variety of sources has been outlined in Sect. 11.2. The long signal duration (days to months for many sources) will allow detailed determination of the source parameters, and precise estimate of the coalescence time, months or even years afterwards, with uncertainties better than 10 s and 1 deg^2 (Sesana 2016). This prediction capability will give advance warning to ground GW detectors and will allow the prepointing of telescopes to realize coincident GW and multiwavelength electromagnetic observations of the final seconds before merging. Time coincidence is critical, as it was shown by the first multi-messenger detection (Abbott et al. 2017a) of a Binary Neutron Star Merger. Figure 11.16 pictorially shows

Fig. 11.15 A synoptic view of all future space mission for the observation of GW. Credits Gong et al. (2021)

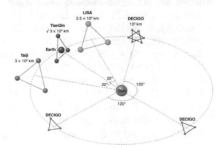

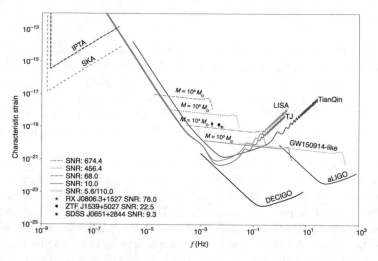

Fig. 11.16 A sketch of combined sensitivities for the GW observatories of the 2030s. The green lines mark the signal evolution of black hole binaries, that will first cross the LISA sensitivity band, to end up, weeks or years later, in the VIRGO-LIGO band. The SNR for some TianQin sources is also shown. Note the y-axis units, the "characteristic amplitude" $\sqrt{\nu \cdot S_h}$. Credits Gong et al. (2021)

this concept: the green lines show the frequency evolution of signals from BH binary coalescences.

Multimessenger astronomy will be even more fruitful when signals from neutrino detectors will complement GW and electromagnetic observations.

References

Abbott, B.P., et al.: Multi-messenger observations of a binary neutron star merger. ApJL **848**(L12), 1–59 (2017a)

Abbott, B., et al.: GW170817: observation of gravitational waves from a binary neutron star inspiral. Phys. Rev. Lett. **119**, 161101 (2017b)

Abich, K., et al.: In-orbit performance of the GRACE follow-on laser ranging interferometer. Phys. Rev. Lett. **123**, 031101 (2019)

Amaro-Seoane, P., et al.: Low-frequency gravitational-wave science with eLISA/NGO class. Quantum Grav. **29**, 124016 (2012)

Armano, M., et al.: Sub-Femto-g free fall for space-based gravitational wave observatories: LISA pathfinder results. Phys. Rev. Lett. **116**, 231101 (7pp) (2016)

Armano, M., et al.: Beyond the required LISA free-fall performance: new LISA pathfinder results down to 20 μHz. Phys. Rev. Lett. **120**, 061101 (10 pp) (2018)

Babak, S., et al.: Science with the space-based interferometer LISA. V: extreme mass-ratio inspirals. Phys. Rev. D **95**, 103012(21) (2017)

Babak, S., Hewitson, M., Petiteau, A.: LISA sensitivity and SNR calculations. Technical Note LISA-LCST-SGS-TN-001. arXiv:2108.01167 (2021)

Barausse, E., et al.: Prospects for fundamental physics with LISA. Gen Relativ Grav. **52**, 81 (2020)

Bassan, M., et al.: Approaching free fall on two degrees of freedom: simultaneous measurement of residual force and torque on a double torsion pendulum. Phys. Rev. Lett. **116**, 051104 (2016)

Cavalleri, A., et al.: Direct force measurements for testing the LISA Pathfinder gravitational reference sensor. Class. Quantum Grav. **26**, 094012 (10pp) (2009)

Danzmann, K., et al.: LISA Laser Interferometer Space Antenna - A proposal in response to the ESA call for L3 mission concepts (2017). arXiv:1702.00786

De Bra, D.B., Dassoulas, J., Kershner, R.B.: A satellite freed of all but gravitational forces: "TRIAD I". J. Spacecr. **11**, 637 (1974)

Faller, J.E., et al.: Space antenna for gravitational wave astronomy. In: Proc. Colloquium "Kilometric Optical Arrays in Space" ESA-SP226 (1985)

Gong, Y., Luo, J., Wang, B.: Concepts and status of Chinese space gravitational wave detection projects. Nat. Astron. **5**, 881–889 (2021)

Jennrich, O.: LISA technology and instrumentation. Class. Quantum Grav. **26**, 153001 (2009)

Kawamura, S., et al.: Current status of space gravitational wave antenna DECIGO and B- DECIGO. Theor. Exp. Phys. **2021**, 05A105 (2021)

Lange, B.: The drag-free satellite. AIAA J. **2**, 1590 (1964)

Larson, S.L., Hiscock, W.A., Hellings, R.W.: Sensitivity curves for spaceborne gravitational wave interferometers. Phys. Rev. D **62**, 062001 (2000)

Mitryk, S.J., Mueller, G., Sanjuan, J.: Hardware-based demonstration of time-delay interferometry and TDI-ranging with spacecraft motion effects. Phys. Rev. D **86**, 122006 (2012)

Muratore, M., Vetrugno, D., Vitale, S.: Revisitation of time delay interferometry combinations that suppress laser noise in LISA. Class. Quantum Grav. **37**, 185019 (18pp) (2020)

Pierce, R., et al.: Intersatellite range monitoring using optical interferometry. Appl. Opt. **47**, 5007–5019 (2008)

Pucacco, G., Bassan, M., Visco, M.: Autonomous perturbations of LISA orbits. Class. Quantum Grav. **27**, 235001 (2010)

Schutz, B.F.: Determining the Hubble constant form gravitational wave observations. Nature **323**, 310 (1986)

Sesana, A.: Prospects for multiband gravitational-wave astronomy after GW150914. Phys. Rev. Lett. **116**, 231102 (2016)

The LISA Consortium: The Gravitational Universe (2013). - arXiv:1305.5720

Tinto, M., Dhurandhar, S.V.: Time-delay interferometry. Living Rev. Relativ. **24**, 1 (2021)

van Veggel, A.M.A., Killow, C.J.: Hydroxide catalysis bonding for astronomical instruments. Advan. Opt. Technol. **3**, 293–307 (2014)

Pulsar as Gravitational Laboratory

12

Pulsars are the most advanced laboratory for gravitational physics. While tests in the Solar System probe the "weak field" regime, with a *compactness* parameter $U = 2GM/rc^2 \lesssim 10^{-5}$, and observations of gravitational waves provide information about strong regime only in the very few instants preceding the merging, pulsars offer long term observations of gravity in the "intermediate" or "not weak" regime, with $U \sim 0.3$ ("strong" is reserved to black hole physics, where compactness reaches unity). Pulsars have proven, in recent years, to be fantastic tools to probe relativistic gravity, offering measurements of unprecedented accuracy.

We list here the topics addressed in this chapter, together with some in-depth review papers to expand and integrate our brief overview:
– we first recap our still vague knowledge of pulsar structure: see the reviews by Lorimer (2008), Lattimer (2015)
– we describe the method of pulsar timing and introduce the Post-Keplerian orbital elements: see Becker et al. (2018), the review paper (Taylor 1992) and the book by Lorimer and Kramer (2005).
– we show how accurate pulsar timing has allowed to collect a wealth of significant experimental results, setting stringent limits to PPN parameters and reinforcing our faith in GR: see Wex (2014) and Manchester (2015)
– finally, we describe the Pulsar Timing Array (PTA) projects to detect gravitational waves at nHz frequencies: see Lommen (2015), Perrodin and Sesana (2018), Tiburzi (2018) or Dahal (2020).

The original version of this chapter was revised: Chapter have been updated with the correction. The correction to this chapter can be found at https://doi.org/10.1007/978-3-030-95596-0_15

F. Ricci and M. Bassan, *Experimental Gravitation*, Lecture Notes in Physics 998, https://doi.org/10.1007/978-3-030-95596-0_12

12.1   Introduction to Pulsars

Pacini (1967) predicted that a rotating neutron star with a magnetic field would emit
a beam of electromagnetic radiation that can be detected when the beam propagates
along the Earth line of sight, a *Pulsar*. Pacini's paper was published 30 years after the
speculations of Walter Baade and Fritz Zwicky about the existence of stars super-
compressed by gravity, the neutron stars (Baade and Zwicky 1934). At the same
time, in the summer of 1967, Jocelyn Bell Burnell, graduate student at Cambridge
University, made the first detection of a radio pulsar. Bell and her adviser Antony
Hewish published their findings in February 1968 (Hewish et al. 1968). The discovery
set off a race to find more pulsars and, in 1974 Hewish (but not Bell), was awarded
a share of the Nobel Prize in Physics for the discovery.

A pulsar is a compact star spinning around an axis different from that of its
magnetic field, so that the emitted beam of electromagnetic radiation sweeps the sky
like a lighthouse beam (Fig. 12.1). When the beam crosses the line of sight of the
observatory, a pulse is detected. The pulse profile encodes information about neutron
star physics, its magnetic field, the interstellar medium and much else.

Pulsars are born, following the gravitational collapse of a giant ($10 - 25 \; M_\odot$)
stellar progenitor, with rotation periods of the order of fractions of a second, with magnetic fields of $B \sim 10^8 - 10^{10}$ T (Shapiro et al. 1983). The current model describes
pulsar emission assuming a rotating dipole magnetic field, due to a misalignment
of the rotation and dipole axis. Particles, mostly $e^\pm$, accelerated by the field, emit
radiation, according to the Larmor equation, and this radiated energy produces a loss
of rotational kinetic energy.

It is widely accepted that the rapid rotation of the neutron star, combined with its
super-strong magnetic field, leads to a large electric field that accelerates particles
to extremely high energies. These *primary* particles emit gamma rays which decay
into *secondary* electron-positron pairs, and the emission in the radio bandwidth

Fig. 12.1 Schematic figure
(not drawn in scale) of the
magnetic dipole model of a
pulsar. Electrons and
positrons from particle
cascades are accelerated in
the magnetosphere. They
stream along the open
magnetic field lines and emit
radio pulses. Credits:
Lorimer and Kramer (2005)

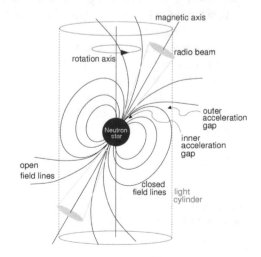

is attributed to these outflowing secondary pairs. But the scientific debate on the emission mechanism of a pulsar is still open.

However, the radiation pulses are too bright to be explained in terms of incoherent emission (plasma particles radiating independently of each other). The brightness values sustain the hypothesis of a *coherent* emission. In the early 1970s, two emission models have been proposed: the *coherent curvature emission* and the *relativistic plasma emission*. Curvature emission occurs because particles are accelerated as they follow the curved magnetic field lines, and this accelerated motion causes the emission of radiation: the coherence is due to particle bunching. However, no realistic hypothesis on the mechanism to bunch the particles has been proposed. The other model assumes a relative streaming motion in the outflowing pair plasma, leading to plasma instability in which peculiar electromagnetic waves grow (the Languimir waves[1]). The radio emission is then attributed to the partial conversion of the energy of these waves into escaping radiation.

A significant fraction of the observed pulsars, about 10%, are components of a binary system. According to the evolutionary scenario known as neutron star recycling, an old pulsar can acquire mass and angular momentum from a low-mass ($M \leqslant 1\,M_\odot$) companion star: at the end of a long accretion phase, some of these neutron star can have rotation periods of few milliseconds: for this, they are called millisecond pulsars (MSP). Recycled MSPs also have slower spin-down rates than classic pulsars: their rotation is significantly more stable than the normal pulsars, so their pulse arrival times can both be measured and also predicted with high accuracy (Hobbs and Shi 2017). It can be proven that the surface magnetic field grows with the rotation period and its derivative: $B_s \propto (P_s\,\dot{P}_s)^{1/2}$. Therefore, these recycled pulsars have lower magnetic fields than classic pulsars, $B_s \sim 10^4$ T. MSPs are very old systems, often located in globular clusters, an environment that favours the formation of binaries and thus the recycling mechanism (Fig. 12.2).

Thanks to the high stability of their rotation period, pulsars are used as clocks to perform experiments in the relativistic regime of gravity. Moreover, pulsars in binary systems are ideal system to test the effects of General Relativity and to look for deviations from the predictions of Einstein's theory. Foster and Backer (1990) showed how a comparison of timing observations from multiple millisecond pulsars (a spatial array of pulsars) could be used to search for gravitational waves (GWs), building on previous intuitions by Sazhin (1978) and Detweiler (1979).

---

[1] The Langmuir waves are rapid oscillations of electron density in the plasma, whose frequency depends only weakly on the wavelength of the oscillation. The *plasmon* is the quasi-particle resulting from the quantization of these oscillations.

Fig. 12.2 Known pulsars shown in a $P - \dot{P}$ diagram, red circles indicate a binary pulsar. Courtesy of Norbert Wex

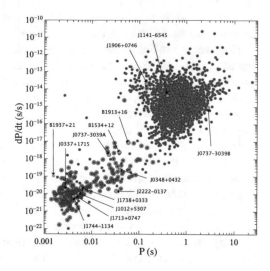

12.2   Pulsar Timing

Pulsars are often referred to as *cosmic clocks*, as it is possible to predict the pulse times-of-arrival (TOA) with high accuracy: this is the key quantity of interest in the powerful technique known as *pulsar timing*. The best pulsars to be exploited are those with a small repetition rate, to make the averaging operation easier, and those with high electromagnetic flux S, i.e. higher signal to noise ratio.

Pulsar timing consists of regularly monitoring the TOAs of a known pulsar for several years, on a weekly or monthly basis (Perrodin and Sesana 2018). For high-precision experiments, pulsars should have predictable spin evolution. There are three main requirements that need to be met, to enable accurate predictions of the arrival times

- The pulsar must be a very stable clock but, as mentioned in Sect. 12.1, it loses rotational energy via magnetic dipole radiation, and therefore it spin down. Millisecond pulsars have slower spin-down rates than classic pulsars, so that their rotation is significantly more stable. In addition, the spin-down might suffer from irregularities: for example, there are phenomena of sudden accelerations, called glitches, that can be explained by a slight decoupling of the crust from the core and therefore by differential rotation, at the end of which the star resumes its slowdown.
- The shape of the integrated pulse profile must be stable in time: pulsars are intrinsically weak radio sources and, typically, the individual pulse signal is below the radiometer noise of the telescope. Therefore, it is usually necessary to add up thousands of individual pulses, with a technique called *folding*, that increases the signal-to-noise ratio. Remarkably, for a given pulsar, even though the individual pulses have different shapes, the integrated profile is usually very stable, on time scale of years, for any observation at the same radio frequency.

- The timing model of the pulsar, the *ephemeris*, must be well known. It is a set of parameters that describe the pulsar spin and spin-down, its orbital parameters, its astrometry, and the dispersive influence of the ionized InterStellar Medium (ISM) along the line of sight to the pulsar. Pulsars emit mainly in the radio band: at these frequencies, ISM influences the TOA of the signals, through two main effects:
  - Dispersion: pulses observed at higher radio frequencies arrive earlier at the telescope than their lower frequency counterparts. This is due to the frequency-dependent dispersive effect of the ionized ISM on radio pulses. This is quantified by the dispersion measure (DM), defined as the integrated column density of free electrons along the line of sight:

$$DM = \int_0^d n_e \, dl, \tag{12.1}$$

  where d is the distance to the pulsar and n_e is the free electron number density[2]
  - Scintillation: in addition to being magnetized and ionized, the ISM is highly turbulent and inhomogeneous. These irregularities cause variations in the signal intensity. This effect is known as interstellar scintillation. Since the effect decreases with the signal frequency as $1/f^4$, the problem is tackled by dividing the wide observation radio band in many narrower bands, where the shape remains stable (Lommen 2015);

Generating ephemeris for a given pulsar is an ongoing, iterative effort: the first version is typically obtained at the time of discovery and provides approximate estimate of the pulsar spin, position, and DM. A precise knowledge of the timing model can be achieved through an iterative procedure of data correction and pulsar parameters estimations. When a stable set of parameters is obtained, we can compute expected TOAs based on our best-known models for the pulsar motion. Taking the difference between the observed TOAs and the expected ones, we obtain the *timing residuals* that, if the model is correct, should be scattered around a zero mean. Model parameters are continuously adjusted under this constraint to minimize the timing residuals.

We now try and clarify this method through an example (Perrodin and Sesana 2018): assume that an observing campaign is performed, with a radio telescope, on a given pulsar. For each observation, we first obtain an integrated pulse profile PP that is statistically stable and has a suitably high signal-to-noise ratio (Fig. 12.3). In pulsar timing, average TOAs for each observation are computed via a cross-correlation of each integrated pulse profile with a high SNR reference template S, which is typically obtained from the superposition of many earlier observation at that particular observing frequency (Eq. 12.4). $PP(t)$ is a scaled and shifted version of the template $S(t)$, with added noise, $n(t)$

$$PP(t) = a + b\, S(t - \tau) + n(t), \tag{12.2}$$

---

[2] In the odd units adopted by pulsar scientists, d is expressed in pc and n_e in cm^{-3}.

Fig. 12.3 A sequence of 100 pulses from the 253-ms pulsar B0950+08. Consecutive pulses are plotted vertically to show the variability in individual pulse shape and arrival time. The lowest curve is a cumulative profile: the average of 1188 pulses (5 min) and is quite stable and periodic. Credits: Stairs (2003)

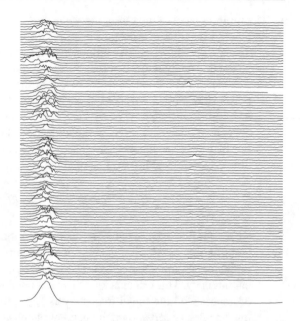

where a is an arbitrary offset and b is a scaling factor. The time shift τ, that must be added to the observation time to optimize the overlap between the profile and the template, yields the TOA. However, our observations are performed from a non-inertial frame: we use telescopes located on the Earth, spinning and orbiting the Sun. Before any analysis, we need to transfer the TOAs measured with the observatory clock (*topocentric arrival times*), to an inertial reference frame: conventionally, we use the Solar System Barycentre (SSB), where the *Temps coordoneè barycentrique* is defined. We must, therefore, carry out the following *barycentric correction*, the transformation between the TOA measured at the telescope, TOA_{topo}, and the TOA_{SSB}:

$$TOA_{SSB} = TOA_{topo} - k\,\frac{DM}{f^2} + \Delta_{R_\odot} + \Delta_{E_\odot} + \Delta_{S_\odot} \qquad (12.3)$$

The first correction is the de-dispersion discussed above: k is a constant, DM is the dispersion measure (Eq. (12.1)) and f is the observing frequency. Unfortunately, the interstellar medium is not static, so DM must be measured at different epochs and updated.

The term $\Delta_{R_\odot}$, the dominant correction, is the *Roemer delay* and accounts for the varying distance of the telescope from the SSB

$$\Delta_{R_\odot} = \frac{\vec{r}\cdot\hat{n}}{c} + \frac{(\vec{r}\cdot\hat{n})^2 - |\vec{r}|^2}{2cd} \qquad (12.4)$$

where $\vec{r}$ is the vector pointing from SSB to Earth and $\hat{n}$ is the unit vector pointing from SSB to the pulsar. The second term in Eq. 12.4 is due to the propagation of the spherical radio waves from the source to the telescope and contains the pulsar distance d: so far, it has been accurately measured for only five pulsars.

The term $\Delta_{E_\odot}$ accounts for the relativistic delay due to Earth motion and the gravitational red-shift caused by other celestial bodies in the Solar System. Its time derivative is (Burgay et al. 2021)

$$\frac{d\Delta_{E_\odot}}{dt} = \sum_k \frac{Gm_k}{c^2 d_{k,\oplus}} + \frac{v_\oplus^2}{2c^2} - const \tag{12.5}$$

where the Earth velocity $v_\oplus$ is relative to the SSB.

The term $\Delta_{S_\odot}$ accounts for the *Shapiro delay*, (see Sect. 5.4.1) acquired by photons travelling in the gravitational field of the Sun and planets (mostly Jupiter).

All these corrections produce our best estimate of the Solar barycentric TOA: due to the variability of DM, clock corrections must regularly be updated (Lynch 2015).

Once the barycentric TOAs t has been derived, we can compute the pulse number N that represents a "counter" for the number of pulsar rotations, i.e. of received radio pulses

$$N(t) = N_0 + \nu_0 (t - t_0) + \frac{1}{2} \dot{\nu}_0 (t - t_0)^2 + \frac{1}{6} \ddot{\nu}_0 (t - t_0)^3 + ... \tag{12.6}$$

where N_0 is the pulse number at the reference time t_0, while ν_0 and $\dot{\nu}_0$, $\ddot{\nu}_0$ are respectively the pulsar frequency and its time derivatives at t_0. The right-hand side of Eq. (12.6) is the Taylor expansion of the pulsar spin.

The timing procedure requires several parameters (position, proper motion, rotation frequency and higher derivative, etc.) that are not known a priori (or they are initially known with limited precision). Given a minimal set of starting parameters, a least square fit is needed to match the measured arrival times to pulse numbers, according to Eq. (12.6). Therefore, we minimize the expression

$$\chi^2 = \sum_i \left(\frac{N(t_i) - n_i}{\sigma_i}\right)^2, \tag{12.7}$$

where n_i is the nearest integer to $N(t_i)$ and σ_i is the TOA uncertainty in units of pulse period. The aim is to obtain a phase-coherent solution that accounts for every single rotation of the pulsar between two observations. Starting off with a small set of TOAs, the data set is gradually expanded (Fig. 12.4).

Ideally the residuals $r_i = N(t_i) - n_i$ should be randomly distributed with zero average. A different trend can suggest that one or more parameters are incorrectly evaluated: for example a linear trend (the residuals increase linearly over time), implies a wrong assumption about the frequency ν_0. A periodic modulation of the residuals with a period ~ 1 year implies that the position of the pulsar, which appears in the Roemer delay, is not well identified (Fig. 12.5).

Moreover, since we are considering the signals emitted and received in two inertial frames, we have also to evaluate and subtract the contribution due to the galactic acceleration acting both on SSB and the pulsar (Burgay et al. 2021). Summarizing, the goal of pulsar timing is to develop a model of the pulse phase as a function of

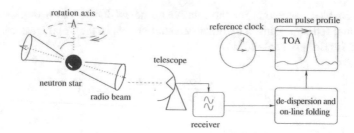

Fig. 12.4 Diagram showing the basic concept of pulsar timing observation. Credits: Lorimer (2008)

Fig. 12.5 Timing residuals as produced by errors in some parameters. The model pulsar has a pulse frequency of 300 Hz (3.3 ms period) and is in an eccentric (e = 0.4) binary orbit of period 190 days. The reference epoch, for period, position and binary phase, is at the middle of the plotted range. Courtesy of Marta Burgay

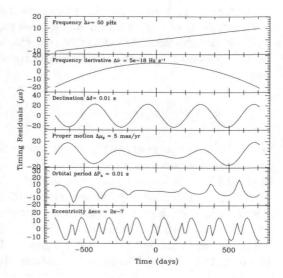

time, so that all future pulse arrival times can be predicted with good accuracy. All these continuous estimates and corrections allow for extremely precise measurement of the model parameters, illustrating the power of pulsar timing.

12.3   Post Keplerian Parameters

If the pulsar is part of a binary system, and orbits with its companion star around their centre of mass, we have to extend the timing model to take into account the modulation caused by the pulsar motion (Fig. 12.6). Therefore, TOAs must be converted from the pulsar proper time to the Binary System Barycenter (BSB) time. Other terms must be added to Eq. 12.3

$$t_{SSB,BSB} = t_{SSB} + \Delta_{R_{Bin}} + \Delta_{E_{Bin}} + \Delta_{S_{Bin}} \tag{12.8}$$

where the terms on the right represent corrections similar to those seen for the Solar System (Earth motion, red-shift, Shapiro delay) but are now referred to the pulsar-

Fig. 12.6 A sketch of the timing problem when dealing with a binary pulsar: When computing the TOA, we must apply corrections for both the motion of the Earth about the SBB and that of the pulsar about the Centre of Mass (CoM) of the binary system. Horizontal lines indicate planes of constant coordinate time. Courtesy of Norbert Wex

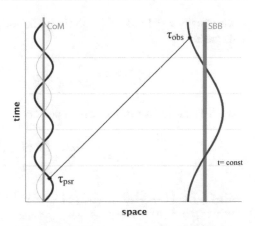

companion system. In the following, we label with the subscript p all quantities referred to the pulsar, and with c those referred to the companion star, while b indicates the binary system.

Modelling the TOAs of a binary pulsar allows us to derive the value of five Keplerian parameters (see Chap. 1): the pulsar orbital period $P_b = 1/\nu_0$; the semi-major axis (projected on a plane Π normal to the line of sight) x; the eccentricity e; the longitude of periastron ω and the epoch of passage of periastron T_0. These parameters are the same as those obtained with normal spectroscopic binaries, but the use of pulsar timing allows a much higher precision (Burgay et al. 2021). Kepler's third law allows us to derive a relation between these parameters and the masses of the stars, m_p and m_c, called the *Mass Function*

$$f(M) = \frac{(m_c \sin i)^3}{M^2} = \frac{4\pi^2 x^3}{G P_b^2} \cdot (1 - e^2)^{3/2} \qquad (12.9)$$

where $M = m_p + m_c$ and i is the angle between the plane of the orbit and the plane Π defined above. Moreover, pulsar binary systems are often in a regime of strong gravity, and relativistic phenomena take place that cannot be described by the Newtonian-Keplerian dynamics, like the precessions. Therefore, additional parameters have to be included in the timing model: they are called **Post Keplerian** (PK) Parameters. We list them here with their expression derived in the framework of General Relativity, expressing the *star masses in units of solar mass* and defining $T_\odot \equiv \frac{GM_\odot}{c^3} = 4.92549\ \mu s$. The five PK parameters are:
– the rate of advance of periastron

$$\dot{\omega} = 3 \left(\frac{2\pi}{P_b}\right)^{5/3} (T_\odot M)^{2/3} \frac{1}{(1 - e^2)} \qquad (12.10)$$

We repeat here the warning given in Chap. 1: in this context, ω is an angle, <u>not</u> an angular frequency !

– the gravitational red-shift and time dilation parameter (No relation with the PPN parameter γ !)

$$\gamma^* = e \left(\frac{P_b}{2\pi}\right)^{1/3} T_\odot^{2/3} M^{-4/3} m_c (m_p + 2m_c) \qquad (12.11)$$

– the time derivative of the orbital period

$$\dot{P}_b = -\frac{192\pi}{5} \left(\frac{2\pi T_\odot}{P_b}\right)^{5/3} f(e) \frac{m_p m_c}{M^{1/3}} \qquad (12.12)$$

with $f(e) = \left(1 + \frac{74}{24}e^2 + \frac{37}{96}e^4\right)(1 - e^2)^{-7/2}$

Two parameters are related to the Shapiro delay, that can be expressed, in terms of the pulsar anomaly θ, by:

$$\Delta_{S_{bin}} = 2r \, log\left(\frac{1 + e \, cos\theta}{1 - s \, sin(\omega + \theta)}\right)$$

– the range r, in GR, is:

$$r = T_\odot m_c \qquad (12.13)$$

– and the shape s

$$s = \frac{a}{c} \left(\frac{P_b}{2\pi}\right)^{-2/3} T_\odot^{-1/3} M^{2/3} m_c^{-1} = sin \, i \qquad (12.14)$$

All these PK parameters are obviously null in Newtonian dynamics. In particular, the period of the orbit and the longitude of the periastron are constant. The exact dependence between the PK parameters and the orbital ones depends on the adopted theory of gravity. The general PPN expressions for the five PK parameters can be found in Will (2018), box 12.1. The PK parameters are indeed functions of the orbital parameters, which are determined with great accuracy by the timing procedure, and of the unknown masses of the two stars; this suggests a method to evaluate the two independent variables m_c and m_p. For a given gravity theory, the equation defining one PK parameter versus m_c, m_p (for GR, any of the Eqs. 12.10–12.14) will describe a curve in a mass-mass cartesian plane. For example, the precession of periastron gives the linear constraint $M \equiv m_c + m_p = const$, with the constant precisely set by Eq. 12.10 in terms of measured parameters. If another PK parameter is plotted on the same graph, the two curves will intersect in a point of coordinates (m_p, m_c) which determines the star masses. If another PK parameter is measured, it also can be plotted on the same diagram: the gravity theory successfully passes the falsification test only if the third line intercepts the other two in the same point. In real life, the uncertainties associated with every PK parameter will broaden each curve into a band, and the intersection defines the allowed region for the mass values. If $N > 2$

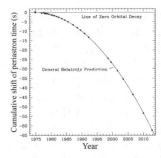
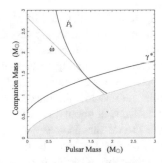

Fig. 12.7 Left: the shift in the periastron passage of the double neutron star system *PSR 1913+16*, plotted over four decades. The shift results from orbital energy loss due to the emission of gravitational radiation. The plot shows the perfect agreement between the observed orbital decay (black dots) and the prediction by GR (solid line). The lack of data in the period 1992–1998 was due to an overhaul of the Arecibo observatory. Credits: Weisberg and Huang (2016). Right: Mass-Mass diagram for the *PSR 1913+16*. The grey zone is forbidden by the Mass Function and the constraint $\sin i \leqslant 1$. The three curves, describing the values of $\dot{\omega}$, γ^*, $\dot{P}_b$ as computed within GR, meet in one point, yielding precise values for the two star masses. Adapted from Weisberg and Huang (2016)

parameters are known, $N - 2$ independent test of a theory can be performed. In the case of the first binary pulsar system discovered, *PSR 1913+16*, three PK parameters have been measured: $\dot{\omega}$, γ^*, $\dot{P}_b$. These parameters, evaluated as functions of the two masses according to GR, have been plotted, with ever increasing accuracy, since the 1980s, as in Fig. 12.7, where error bands have been omitted for clarity. The three curves intersect, General Relativity passed the test and the masses were precisely evaluated:

$$m_1 = 1.4398 \pm 0.0002 \qquad\qquad m_2 = 1.3886 \pm 0.0002$$

12.4 Two Famous Binary Systems

12.4.1 PSR1913+16 and the Existence of Gravitational Waves

The first binary pulsar, PSR 1913+16, was discovered in 1974 by R. Hulse and J. Taylor using the Arecibo Telescope (Hulse and Taylor 1974). It is the most famous and most extensively studied binary system, also because it brought the first indirect confirmation of the existence of GWs.

PSR 1913+16 is a pulsar located about 5 kpc away from the Earth, in the Aquila constellation. The radio pulse has a repetition time $P \simeq 59$ ms, but this varies by about one part in 1,000 with a periodicity of 7.75 hours. These variations were interpreted as due to a Doppler shift effect associated with the orbital motion about a companion with velocities of the order of 10^{-3} c. At first, Hulse and Taylor derived a velocity curve versus time, which deviated from a a sinusoidal dependence. The curve distortion was the crucial element permitting to compute the celestial parameters of the system. By a detailed fit of the velocity curve under the assumption of a Keplerian two-bodies orbit, they derived the parameters reported in Table 12.1.

The orbit is an ellipse of eccentricity 0.62 and the binary system is composed of two neutron stars, each having a mass of about 1.4 solar masses, with a semi-major axis separation of only 2.8 solar radii.[3] The minimum separation, at periastron, is about 1.1 solar radii, while the maximum separation, at apoastron, is 4.8 solar radii. The orbit is inclined at about 45° with respect to the line of sight.

Soon after the pulsar discovery, it was realized that the system is a natural laboratory for providing GR tests to high precisions levels. The first example is the relativistic precession of periastron: the orientation of periastron changes by about 4.2° per year in direction of the orbital motion (in January 1975, it was oriented so that periastron occurred perpendicular to the line of sight from Earth).

This amount can be compared with the GR perihelion advance of the Mercury planet, 43 arcsec per century: the precession this system achieves in 1 day is as much as Mercury does in a century.

Table 12.1 reports recent determinations of Keplerian and post-Keplerian parameters for the binary system PSR 1913+16, referred to the *epoch* MJD 52984.0.[4] As the authors wrote: "It is interesting to note that since the value of Newton's constant G is known to a fractional accuracy of only $\sim 10^{-4}$, masses can be expressed more accurately in solar masses than in kg". Indeed, the accuracy on $T_\odot$ is $\sim 10^{-6}$.

A binary star system is a rotating mass quadrupole: therefore, GR predicts it to emit gravitational radiation; the loss of energy due to the GW emission will result in a shrinking of the orbit, and in the decrease of the orbital period $\dot{P}_b$. The measurements of the orbital decay span from 1974 to nowadays: since the early 1980s, it was clear that the observed orbital decay of this binary system was in agreement with the GR predictions, with a precision that, today, exceeds 99.5%. The total power of the GWs emitted by this system is calculated to be 7.35×10^{24} W. With this energy loss, the orbital period decreases by 76.5 μs/year, the semimajor axis decreases by 3.5 meters per year and the calculated lifetime up to the final merger of the two neutron stars is some 300 million years.

This observation put an end, 30 years before the detection in 2015, to the scientific debate on the existence of gravitational waves. The results proved that GWs are not just a mathematical feature of the Einstein field equation, terminating the debate on the physical nature of this phenomenon. It played a central role toward the approval of the construction of the Earth-based interferometers. Hulse and Taylor received the 1993 Nobel Prize for Physics in recognition of their achievement.

Inserting these values in Eq. 12.12, we derive the GR prediction $\dot{P}_b^{GR}$. Comparing with the observed values $\dot{P}_b^{obs}$, we get (Weisberg 2016)

---

[3] Another odd unit of astronomers: measuring star dimensions in terms of the solar radius, $R_\odot = 6.957 \cdot 10^8$ $m = 1/215AU$, as defined in 2015 by the IAU. It is indeed suggestive to think of a two-star system rotating in an orbit that contains just 3 Suns.

[4] The Modified Julian Date (MJD) counts the number of days since midnight on November 17, 1858; MJD 52984 is December 11, 2003.

Table 12.1 Coordinates, Keplerian orbital elements and post-Keplerian parameters of PSR 1913+16. In parenthesis, the uncertainty on the last digit. Adapted from Weisberg et al. (2010), Weisberg and Huang (2016)

Observed parameters	Value
Epoch T_0	MJD 52984.0
Right ascension α	$19^h 15^m 27^s.99928(9)$
Declination δ	$16°00'27".3871(13)$
Signal frequency f	$16.940537785677(3)$ Hz
Signal frequency derivative $\dot{f}$	$-2.4733(1) \cdot 10^{-15}$ Hz/s
Fitted Keplerian parameters	
Orbiting period P_b	$0.322997448918(3)$ days
Eccentricity e	$0.6171340(4)$
Projected semi-major axis $x \equiv a_p \sin i /c$	$2.341776(2)$ s
Argument of periastron ω_0	$292.54450(8)$ deg
Orbital velocity of stars at periastron v_p (with respect to the centre of mass)	450 km/s
Orbital velocity of stars at apoastron v_a (w.r.t centre of mass)	10 km/s
Post-Keplerian parameters	
Rate of advance of periastron $\dot{\omega}$	$4.226585(4)$ deg/yr
Red-shift parameter γ^*	$4.307(4)$ ms
Orbiting period derivative $\dot{P}_b$	$-2.423(1) \cdot 10^{-12}$
Mass of companion m_c	$1.3867(2) M_\odot$
Mass of pulsar m_p	$1.4414(2) M_\odot$

$$\frac{\dot{P}_b^{obs}}{\dot{P}_b^{GR}} = 0.9983 \pm 0.0016 \tag{12.15}$$

12.4.2 The Double Pulsar PSR J0737-3039

The best binary pulsar system to test gravity theories is, to this day, *PSR J0737-3039 A and B*, for four diverse reasons: the two stars are very close to each other, $x_A/c = 1.42$ s, $x_B/c = 1.51$ s, leading to motion at almost relativistic velocities in a very strong gravity field; their orbital plane is almost parallel to the line of sight, $i \approx 88°$, so that the system is seen edge-on, and the effect of the Shapiro delay is enhanced; last and foremost, both neutron stars have been seen as pulsars and therefore we had two "clocks" to perform timing on. To give one example of how strong relativistic effects are in this binary system, we mention the precession of periastron: $\dot{\omega} = 16.9°/yr$. This is more than 10^5 times larger than the $43\ arcsec/century$ of Mercury!

For *PSR J0737-3039*, all five PK parameters are measured, and Fig. 12.8 shows how well their determination agrees with GR. In this system, one additional rela-

Fig. 12.8 Mass-Mass diagram for the double pulsar *PSR J0737-3039*. The constraints of the 5 PK parameters are shown, as well as the mass ratio R (derived from the projected semi-major axes, separately determined) and the spin precession rate Ω_B. Two grey areas are excluded by the constraint $\sin i \leqslant 1$ for each mass. The insert is a zoom on the intersection region. Adapted from Lyne et al. (2004)

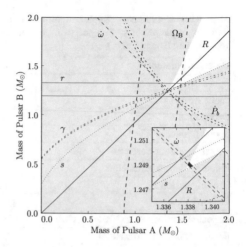

tivistic effect has been measured, the rate of the geodetic precession Ω_B for the spin of star B.

PSR J0737-3039 was discovered in May 2003. Just 3 months later, many parameters of both pulsars had been determined to a great accuracy, as shown in Table 12.2. In binary systems with only one visible pulsar, it takes years of accumulated data to achieve these measurements. The precision of these values has been largely improved since this first determination.

Since March 2008, the pulses of star *B* are no longer observable: due to the spin precession, the radio beam is no longer shining on the Earth. But don't despair: it should reappear in 2035.

12.4.3 Other Pulsars

More than 2800 pulsars have been discovered since 1967,[5] and about 10 % of these belong to a binary system. At least one pulsar, J0337+1715 is known to belong to a triple system (Voisin et al. 2020). Each one has its peculiarity, and timing observations have played a role in improving our knowledge of astrophysics and gravitational theory. Just to mention a few:

- PSR J1738+0333 has been used to set limits on PPN parameters, as discussed in the next section;
- the triple system PSR J0337+1715 has constrained SEP violations and improved the constraint on the Brans-Dicke parameter $\omega > 140\ 000$;
- PSR J0348+0432 and its white-dwarf companion have set limits on scalar-tensor theories;

---

[5] An ever-updated catalogue can be found at: https://www.atnf.csiro.au/research/pulsar/psrcat/.

Table 12.2 Keplerian and post-Keplerian parameters for the Double Pulsar, evaluated within 3 months (Lyne et al. 2004) from the discovery of PSR J707B. The distance is estimated from the dispersion measure

Observed parameters	PSR J0737−3039A	PSR J0737 3039B
Right ascension (J2000)	$07^h 37^m 51^s.247(2)$	−
Declination (J2000)	$-30°39'40''.74(3)$	−
Pulse period P (ms)	22.69937855615(6)	2773.4607474(4)
Pulse period derivative $\dot{P}$	$1.74(5) \cdot 10^{-18}$	$0.88(13) \cdot 10^{-15}$
Dispersion measure DM $(cm^{-3}pc)$	48.914(2)	48.7(2)
Fitted Keplerian parameters		
Orbital period P_b (day)	0.102251563(1)	−
Eccentricity e	0.087779(5)	−
Epoch of periastron T_0 (MJD)	52870.0120589(6)	−
Longitude of periastron ω (deg)	73.805(3)	73.805 + 180.0
Projected semi-major axis $x = \frac{a \sin i}{c}$ (s)	1.41504(2)	1.513(4)
Post-Keplerian parameters		
Advance of periastron $\dot{\omega}$ (deg/yr)	16.90(1)	−
Gravitational red-shift parameter γ^* (ms)	0.38(5)	
Shapiro delay parameter s	$0.9995(-32, +4)$	
Shapiro delay parameter r (μs)	$5.6(-12, +18)$	
Orbital inclination from Shapiro s (deg)	87(3)	
Distance from SSB (kpc)	∼0.6	
RMS timing residual (μs)	27	2660
Stellar mass $(M_\odot)$	1.337(5)	1.250(5)

- PSR J0348+0432, with a massive (2 $M_\odot$) pulsar in a relativistic orbit has constrained the emission of dipolar GW;
- PSR J0437-4715 give interesting bounds on $\dot{G}$.

12.5   Limits on PPN Parameters from Binary Pulsars

In Chap. 6, we have described the PPN formalism to test and compare different theories of gravity in the weak-field regime of the Solar System. Pulsar observations do not directly constrain the parameters γ or β but can provide a wealth of information useful in constraining many of the others PPN parameters, that GR sets to zero: indeed, these observations currently place the strongest constraints on several parameters. However, the PPN formalism cannot be applied, as is, to binary systems of compact objects, because the hypotheses of weak field and slow motion break down in these strongly gravitating systems. However, we can extend this formalism to the binary stars, by applying some straightforward changes (Will 2018), if the following conditions are verified: (a) the two stars are sufficiently far away from each other (distance much larger than their diameters), so that tidal interactions can be neglected and (b) their velocities are well below c.

It is thus customary to label the strong-field parameters with a *hat*: so, e.g. the parameters α_i are modified as $\hat{\alpha}_i$. This is the standard choice in the literature and could cause a notation conflict, as we use the *hat* to indicate unit vectors. We rely on the common sense of the reader to distinguish in the very few ambiguous cases.

We can assume that the strong-field corrections manifest themselves as additional terms in a Taylor series of the star compactness $U = 2GM/rc^2$ of both the pulsar p and the companion c:

$$\hat{\alpha}_i = \alpha_i + \alpha_i'(U_c + U_p) + \alpha_i''(U_c \cdot U_p) + \dots$$

In the weak field limit, we then recover the usual PPN values α_i.

12.5.1 Limits on the Parameters $\hat{\alpha}_i$

The parameters α_i, α_2, α_3 account for possible violation of the Local Lorentz Invariance (LLI). This means that, in the case of any $\hat{\alpha}_i \neq 0$, there should be an observable effect due to the relative motion with velocity $\vec{w}$ between the reference frame and the universal preferred rest frame. The review by Shao and Wex (2012) has more details on this topic.

In this game, we first need to determine the preferred reference system to evaluate $\vec{w}$. The most obvious choice is the frame in which the Cosmic Microwave Background (CMB) shows no dipole anisotropy: we assume the CMB to be at rest in this frame. The pulsar barycentre velocity with respect to the CMB is measured in two steps: first, the velocity of the pulsar with respect to the SSB and then, the velocity of the SSB with respect to the CMB.

$$\vec{w} = \vec{v}_{BSB-SSB} - \vec{v}_{SSB-CMB} \qquad (12.16)$$

The timing technique, described above, does not provide information about the radial velocity of the system; therefore, experiments to test the LLI, performed over a binary system with a pulsar and a White Dwarf (WD) are to be preferred, so that the radial velocity can be obtained from the spectroscopic analysis of the visible companion.

We need to recall here the definition of the *eccentricity vector* $\vec{e}$: a dimensionless vector with magnitude equal to the orbit's eccentricity e and direction pointing from apoapsis to periapsis. In a tightly interacting binary system, $\vec{e}$ rotates due to the precession of periastron, but its modulus is constant.

Fig. 12.9 If LLR is violated, the eccentricity vector is the sum of a rotating vector and a fixed vector pointing normal to the velocity $\vec{w}$ with respect to the preferred frame

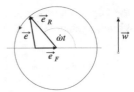

Limits on PPN parameter $\hat{\alpha}_1$

If LLI is violated, it has been shown (Shao and Wex 2012) that, for small eccentricity orbits ($e < 10^{-6}$), a non-vanishing α_1 would induce a forced eccentricity component $\vec{e}_F$ that adds to the rotating component $\vec{e}_R$ (see Fig. 12.9):

$$\vec{e}_F = \frac{\hat{\alpha}_1}{4c} \frac{m_p - m_c}{m_p + m_c} \left(\frac{2\pi}{P_b}\right)^{2/3} \frac{GM}{\dot{\omega}} \hat{J} \times \vec{w} \qquad (12.17)$$

We call Ψ the angle between $\vec{w}$ and the unit vector $\hat{J}$ of the total angular momentum. So, the total eccentricity $e = |\vec{e}_R + \vec{e}_F|$ changes its magnitude in time. Therefore, the binary orbit changes from a less to a more eccentric configuration and back on the time scale of the periastron advance:

$$T_e \approx \frac{2\pi}{\dot{\omega}} \simeq 10^4 years \cdot \left(\frac{P_b}{1\,day}\right)^{5/3} \left(\frac{2M_\odot}{M}\right)^{2/3}$$

This perturbation of the binary orbit is actually an extension of the Nordtvet effect discussed in Chap. 6.

Unfortunately, the angle Ψ depends on the longitude of the ascending node Ω which is not, usually, a directly measurable variable. Moreover, the angle between $\vec{e}_F$ and $\vec{e}_R$ is unknown therefore it is not possible to predict how the value of e will change. However, if the system is old and $\dot{\omega}$ is large enough so that the periastron has completed several turns, this angle can be assumed as a uniformly distributed variable between 0 and 2π.

The best binary pulsar system to test the value of $\hat{\alpha}_1$ is J1738+0333 (PSR J1012+5307 is pretty good too): it has a short orbital period, $P_b \sim 0.35$ days, extremely low orbital eccentricity $e < 10^{-7}$ and a precession rate $\dot{\omega} \approx 1.6°/yr$: this means that during the observation time, now over a decade, the rotating eccentricity has swept an angle wide enough to make a change in eccentricity measurable. The outcome of the analysis of *PSR J1738-0333* data permits to set the limit

$$\hat{\alpha}_1 = -0.4^{+3.7}_{-3.1} \times 10^{-5} \qquad (12.18)$$

This result is five times better than that obtained with observations in the Solar System with LLR and, even more important, also constrains strong-field effects.

Limits on PPN parameter $\hat{\alpha}_2$

Analogously to what shown for $\hat{\alpha}_1$, it was proven that if $\hat{\alpha}_2 \neq 0$, a precession of the angular momentum vector $\vec{J}$ would occur around the *absolute* velocity $\vec{w}$ of the binary system. The angular precession rate is given by

$$\dot{\Omega}(\hat{\alpha}_2) = -\frac{\hat{\alpha}_2}{2} \left(\frac{2\pi}{P_b}\right) \left(\frac{w}{c}\right)^2 \cos \Psi$$

$$\approx -\hat{\alpha}_2 \, (0.066°/yr) \left(\frac{P_b}{1 \, day}\right)^{-1} \left(\frac{w}{300 \, \text{Km/s}}\right)^2 \cos \Psi \qquad (12.19)$$

This precession would cause the plane of the orbit, normal to $\vec{J}$, to change its orientation with respect to the line of sight, leading to a change of the value of the projected semi-major axis $x = a \sin i$.

$$\frac{\dot{x}}{x} = -\frac{\hat{\alpha}_2}{4} \left(\frac{2\pi}{P_b}\right) \left(\frac{w}{c}\right)^2 \frac{\cos i}{\sin i} \sin 2\Psi \, \cos \theta \qquad (12.20)$$

where θ is the angle between the direction of the ascending node and the projected component of $\vec{w}$ on the orbital plane. x and $\dot{x}$ in Eq. 12.20 are directly measurable by the timing procedure, the inclination angle i can be obtained rewriting the mass function, Eq. 12.9, as a function of the mass ratio m_p/m_c. This latter ratio can be derived from the star velocities of the binary system since it is composed by a pulsar and a WD. This precession could be masked by other effects that cause a variation in the observed x

$$\left(\frac{\dot{x}}{x}\right)_{TOT} = \left(\frac{\dot{x}}{x}\right)_{P_b} + \left(\frac{\dot{x}}{x}\right)_{PM} + \left(\frac{\dot{x}}{x}\right)_{GR} + \left(\frac{\dot{x}}{x}\right)_{\hat{\alpha}_2} \qquad (12.21)$$

The first term accounts for the orbit shrinking due to the loss of energy via GWs emission. The contribution of this term can be estimated from Kepler's third law: being proportional to $\dot{P}_b/P_b$ it is usually of order $10^{-19}s^{-1}$, about 4 order of magnitude smaller than the relevant time scale. The second term accounts for the proper motion of the system. Since a typical pulsar describes an arc of significant extension on the celestial sphere when observed from Earth, its orbital inclination with respect to the plane of sky is constantly changing.

The term $(\dot{x}/x)_{GR}$ accounts for other gravity effects that could cause the precession of the total angular momentum such as the quadrupole momentum of the WD and the Lense-Thirring effect. However, they are negligible unless the WD has a very high spin ($P_{WD} < 200 \, s$), which is not the case for the systems discussed here. From the combined analysis of *PSR J1012 +5307* and *PSR J1738 +03333* an upper limit was set on $\hat{\alpha}_2$

$$|\hat{\alpha}_2| < 1.8 \times 10^{-4} \qquad (12.22)$$

which is three orders of magnitude larger than the limit obtained in the Solar System.

A better constraint is obtained observing isolated pulsars. We still use Eq. 12.20 replacing the orbital period by the spin period. The precession of the angular momentum would change the direction of radio emission determining a change in the shape of the pulse profile. Observation *PSR B1937 +21* and *PSR J1744 -1134* for over 15 years shows no changes in the pulse profile, leading to a combined constraint (Wex 2014)

$$|\hat{\alpha}_2| < 1.6 \times 10^{-9} \qquad (12.23)$$

Modelling the pulse profile in order to quantify the changes is a difficult task. As we are considering isolated pulsars, no information about the radial value of $\vec{w}$ can be derived.

Limits on PPN parameter $\hat{\alpha}_3$

Tests of $\hat{\alpha}_3$ can be derived from both binary and single pulsars, using slightly different techniques. A non-zero value would imply a violation of local Lorentz invariance as well as non-conservation of momentum. It would cause a rotating body to experience a self-acceleration in a direction orthogonal to both its spin angular velocity[6] $\vec{\Omega}_p$ and its absolute velocity $\vec{w}$.

$$a_{CM} = -\frac{\hat{\alpha}_3}{3} U_p \, \vec{w} \times \vec{\Omega}_p \qquad (12.24)$$

where $U_p = 2Gm_p/r_p c^2$ is (up to factors of order unity) the compactness parameter and $\vec{\Omega}_p$ the angular velocity of the pulsar. The corresponding term for the companion can be neglected since $\hat{\alpha}_3$ test is performed on systems with white dwarfs, that have a significantly smaller compactness. It is interesting to note that the double nature of $\hat{\alpha}_3$ is shown, in this equation, by the presence of both a self-acceleration and the velocity w.

The self-acceleration determines a *polarization* of the orbit that, as in the case of $\hat{\alpha}_1$, can be read as adding a forcing term that modifies the eccentricity vector, again much like the Nordtvedt effect:

$$|e_F| = \hat{\alpha}_3 \, \frac{U_p \, |\vec{w}|}{24\pi} \, P_b^2 \, \Omega_p \, \frac{c^2}{G(m_p + m_c)} \sin\beta \qquad (12.25)$$

The best constraints on $\hat{\alpha}_3$ are obtained selecting a sample of binary systems with large orbital periods P_b and small eccentricity. Following this approach, the achieved limit is

$$|\hat{\alpha}_3| < 5.5 \times 10^{-20} \qquad (12.26)$$

This shows that $\hat{\alpha}_3$ is the most tightly constrained PN parameter.

---

[6] In previous sections we used f, but the spin vector is always associated to $\Omega_p = 2\pi f$.

12.5.2 Limits on the PPN Parameter $\hat{\xi}$

The Post Newtonian Parameter $\xi \neq 0$ also implies a violation of the Local Position Invariance (LPI). As a consequence, the dynamics of a self-gravitating body could depend on its position with respect to the gravitational field of the galaxy and therefore on the overall distribution of matter.

A non-vanishing ξ would cause the precession of the total angular momentum around the direction of the external gravitational field

$$\Omega_p = \hat{\xi} \left(\frac{2\pi}{P_b} \right) \left(\frac{v_G}{c} \right)^2 \cos \chi \tag{12.27}$$

where χ is the angle between the direction pointing to the galactic centre and the orbital velocity $\vec{v}_G$ due to the galactic gravitational field. This equation is formally identical to Eq. 12.20: therefore the same kind of analysis can be performed (Wex 2014) yielding to

$$|\hat{\xi}| < 3.1 \times 10^{-4} \tag{12.28}$$

In this case as well, we can get more stringent limits by looking at isolated pulsars: then P_b is replaced by the pulsar spin period and we get

$$|\hat{\xi}| < 3.9 \times 10^{-9} \tag{12.29}$$

12.5.3 Limits on PPN Parameter $\hat{\zeta}_2$

A non-vanishing $\hat{\zeta}_2$ would induce a self-acceleration of the centre of mass in the direction of the periastron (Will 2018)

$$\vec{a}_{cm}(t) = \frac{\hat{\zeta}_2}{2} \, cT_\odot \, \frac{m_p \, (m_p - m_c)}{(m_p + m_c)^2} \left(\frac{2\pi}{P_b} \right)^2 \frac{e}{(1 - e^2)^{3/2}} \hat{e}_p(t) \tag{12.30}$$

Because of the rotation of $\hat{e}_p$, the unit vector pointing from the centre of mass to the periastron, this self-acceleration could be detected via the change in the pulsar frequency f, due to the changing Doppler shift between SSB and BSB. This rate of change is given by

$$\frac{\dot{f}}{f} \approx -\vec{a}_{cm} \cdot \frac{\hat{n}}{c}$$

where $\hat{n}$ is a vector aligned with the line of sight.

The measurement of $\dot{f}$ does not effectively constrain the value of $\hat{\zeta}_2$ since $\dot{f}$ is largely caused by the pulsar spin down due to the dipole emission of electromagnetic waves.

The second time derivative of f, instead, is a better quantity to test the existence of a self-acceleration term since it is much less sensitive to other factors.

Table 12.3 Upper bounds for the PPN parameters (as of 2020), measured in the *strong or quasi-strong field regime*, in isolated or binary pulsars. Since the values of all these parameters are consistent with zero, we report the experimental error, as a bound to each value. Data from: Imperi et al. (2018), Will (2018) and references therein

Parameter	σ	Limits from compact stars
$\hat{\xi}$	$3.9 \cdot 10^{-9}$	PSRs B1937+21 and J1744+1134
$\hat{\alpha}_1$	$3.7 \cdot 10^{-5}$	Pulsar-White dwarf binaries PSRs J1012+5307 and J1738+0333
$\hat{\alpha}_2$	$1.6 \cdot 10^{-9}$	PSRs B1937+21 and J1744-1134
$\hat{\alpha}_3$	$4 \cdot 10^{-20}$	Binary ms pulsars
$\hat{\zeta}_2$	$4 \cdot 10^{-5}$	Binary motion $\ddot{P}$ in PSR 1913+16
η	$1.8 \cdot 10^{-6}$	Triple system PSR J0337+1715
$\dot{G}/G$	$1.1 \dot{1} 0^{-12}$	21-year timing of J1713+0747

Considering the second time derivative, and neglecting higher-order terms like $O((\dot{f}/f)^2)$, $O(\dot{P}_b)$, $O(\ddot{P}_b)$, we can derive the relation

$$\frac{\ddot{f}}{f} = \frac{\hat{\zeta}_2}{2} \, c \, T_\odot \, \frac{m_p \, (m_p - m_c)}{(m_p + m_c)^2} \left(\frac{2\pi}{P_b}\right)^2 \frac{e}{(1 - e^2)^{3/2}} \sin i \cos \omega \frac{d\omega}{dt} \tag{12.31}$$

The contribution due to e.m. emission is roughly two orders of magnitude smaller than the term of interest and that due to the galactic acceleration is, for the observed pulsars, as much as three orders of magnitude smaller. A third possible contribution could arise due to the acceleration from a nearby third body. In the worst case, this acceleration could cancel that due to $\hat{\zeta}_2$, but it is statistically unlikely that this would happen for every observed pulsar.

We finally note that in Eq. 12.31 the term $\cos \omega$ is time dependent, due to precession. In the data analysis, two different approaches have been pursued: use a reference value (usually $\cos \omega$ at half of the observation time) or use a value for $\cos \omega$ uniformly distributed between all values. The two methods provide very similar limits for $\hat{\zeta}_2$:

$$|\hat{\zeta}_2|_{reference} < 2.7 \times 10^{-5} \qquad |\hat{\zeta}_2|_{distributed} < 2.6 \times 10^{-5} \tag{12.32}$$

Note that Solar System experiment and observation set a much weaker limit on $\xi < 10^{-3}$, while no bounds have been set on ζ_2 (Table 12.3).

12.6 Pulsars as Test Masses to Detect Gravitational Waves

Assuming that all the pulsar timing parameters have been accurately estimated and all the noise sources (described below, in Sect. 12.7) have been minimized, from the

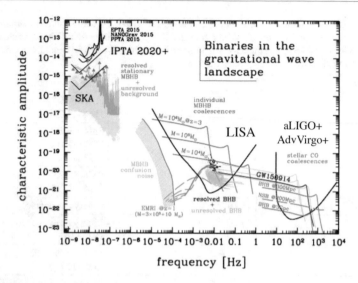

Fig. 12.10 Sensitivity curve of GW experiments to different sources of gravitational waves. Courtesy of Alberto Sesana

so-called timing residuals we can extract a signal due to the metric change of a GW crossing the radio signal path (Lynch 2015). In practice, the radio wave emitted by the pulsar and received on the Earth probes the distance between these two bodies in a similar way of the light bouncing between two mirrors set at enormous distance. This technique is substantially identical to that proposed in the 1970s to detect GW by Doppler tracking of artificial satellites with radio beams (Estabrook and Wahlquist 1975), and it is not too different from that used to measure the Shapiro time delay (see Sect. 5.4.1). The use of pulsar signals offers an enormously longer baseline, and an extremely stable radio signal, making the Pulsar-Earth system an attractive detector.

Just as in e.m. antennas, a GW detector works best for wavelengths of the order of its dimensions, $\lambda \gtrsim 2L$; else, the signal is reduced, because the two mirrors move in phase and the differential motion is smaller.

The Earth-based interferometers, as LIGO and Virgo, with their arms of 3/4 km, detect the GWs in the frequency range of $10 - 10^4$ Hz. The Laser Interferometer Space Antenna (LISA) will have arms of 2.5 millions of kilometres and is aimed at detecting GWs in the frequency range $10^{-4} - 10^{-1}$ Hz. By analyzing the timing signals emitted by an array of pulsar (Pulsar Timing Array - PTA), it is possible to probe the space metric on distances of the order of parsec and detect gravitational waves of very low frequencies, $10^{-9} - 10^{-7}$ Hz. LIGO and Virgo detect GWs from stellar-mass objects just before their merging, while PTAs is sensitive to GW emitted by supermassive black holes in the early stage of their inspirals. Therefore, PTAs provide a view of the gravitational wave sky complementary to the earth-based and space-based interferometers (see Fig. 12.10). This makes them suited to investigate the galaxy merger rate and the population of supermassive black hole binaries in the Universe. PTAs also provide an opportunity to test the theory of gravitation in the

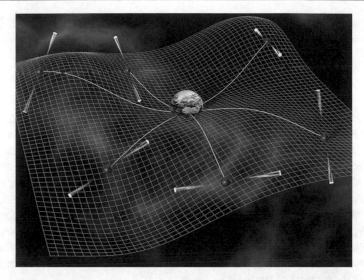

Fig. 12.11 Artist's rendition (not to scale) of pulsar timing: PTAs detect the changing time-of-flight of the electromagnetic beam emitted by a pulsar as a GW passes through the beam propagation path. Courtesy of David J. Champion

nanohertz regime (Dahal 2020). Just as in the case of ground-based interferometers, the electromagnetic signals that travel along the *arms* of the detector take up varying delays, owing to the influence of the GW. In the case of LIGO-Virgo, the e.m. signal is a laser light, while for pulsar timing the e.m. signal is the radio beam from a pulsar. Figure 12.11 shows only few pulsars, but PTA scientists actually monitor many dozens of isolated pulsars (Lommen 2017).

12.7   Sensitivity and Noise in Pulsar Timing

There are several issues to be considered when we plan to use pulsars as gravitational wave detectors:

- **The accuracy of the Pulsar Timing model**: given an initial set of parameter estimates, the accuracy of the model is improved by acquiring new data. However, if we analyze a single pulsar, we would be unable to distinguish between data that are actually improving our model and the ones that are affected by a GW crossing the beam path. The main solution of this problem, in all detection schemes, is to combine the signals of several pulsars. The drawback of this approach is that the response of each pulsar to a GW depends on two terms: one associated with the interaction of GW with the Earth, common to all the pulsars, and one due to the interaction of the GW with the pulsar beam, different for each pulsar. We will address this problem in Sect. 12.8.
- **Wide-bandwidth**: The second key issue in PTAs is the consequence of the enormous gains in the receiver bandwidth of the radio telescopes in the present config-

urations. This fact brings both advantages and disadvantages. The SNR in radio telescopes, when the receiver thermal noise dominates, follows a simple scaling rule (Lommen 2015)

$$SNR \propto \frac{A \sqrt{\tau \Delta f}}{T_{sys}} \qquad (12.33)$$

where A is the collecting area of the telescope, τ is the integration time, Δf is the bandwidth and T_{sys} is the temperature of the receiver. Therefore, quadrupling the bandwidth is the same as doubling the collecting area of the telescope, or quadrupling the observation time, the latter two of which come at a very high cost: this is a great advantage. However, the signal profile of a pulsar changes with frequency: as seen in Sect. 12.2, the scintillation effects and the dispersion are inversely proportional to the signal emission frequency. Indeed the integrated pulse profile is stable, but because of these frequency-dependent effects, we have variation in a very large bandwidth. To address this issue, astronomers divide the observation band into sub-bands and perform independent analysis for each of these sub-intervals of frequency.

- **Noise**: timing errors are usually divided into those pertinent to the time tagging of pulses (i.e. TOA measurement by template fitting) and those relative to the physical properties of pulsar or interstellar medium. Among the latter, there is an intrinsic noise of the pulsars, with characteristics both white (flat spectrum) and red (peaked around a given frequency). The white noise contribution is associated with the emission mechanism of the signals and depends on the unmodeled variation of the rotation frequency of pulsars: it causes a jitter effect, i.e. a deviation from true periodicity of the signal. Indeed, individual pulse can jitter in phase at the level of single pulse width, and its amplitude can change more than 100 % from pulse to pulse, thus requiring to average over several hundreds of pulses. Jitter currently only affects a relatively small number of the PTA pulsars, but in the upcoming era of larger and more sensitive telescopes, jitter will become a major limitation for PTA research.

  The red noise is instead associated with *star-quakes* or interactions between the crust of the pulsars and the innermost superfluid: they take place with a maximum frequency of $\sim 1/year$ and they cause sudden changes in the rotation period, *pulsar glitches*, and consequently large variations in TOAs.

12.8 Timing Residuals Due to Gravitational Waves

When a GW crosses the line of sight from the Earth to the pulsar, the TOAs change: to extract the GW information they are measured and compared with predictions for the arrival times based on a pulsar timing model. The differences between the predictions and the measurements are known as the pulsar timing residuals. We can connect the residual in a pulsar signal due to the GW, R_{gw} to the amplitude h of the

GW causing it through the relation

$$R_{gw}(t) = \frac{1}{2}[1 + \cos\mu][r_+(t)\cos(2\psi) + r_\times(t)\sin(2\psi)] \qquad (12.34)$$

where μ is the angle between the GW source and the pulsar, as seen from the Earth,[7] while ψ is the GW polarization angle. The subscripts $+$ and $\times$ refer to the two polarization of the GW. The amplitudes r_+ and $r_\times$ are each composed of two terms:

$$r_{+,\times}(t) = r^e_{+,\times}(t) - r^p_{+,\times}(t) \qquad (12.35)$$

one related to the gravitational strain $h^e_{+,\times}(t)$ at the Earth (the Earth term) and one for the gravitational strain at the source (the pulsar term):

$$r^e_{+,\times}(t) = \int_0^t h^e_{+,\times}(\tau)\,d\tau \qquad (12.36)$$

$$r^p_{+,\times}(t) = \int_0^t h^p_{+,\times}\left[\tau - \frac{d}{c}(1 - \cos\mu)\right]d\tau \qquad (12.37)$$

where d is the distance between Earth and the pulsar. Note that the pulsar term $h^p_{+,\times}(t)$ has the same functional shape as the Earth term but evaluated at a delayed time. The delay is the time between two events: the first is when the GW arrives at the Earth, while the second is when the Earth receives the information that the GW has arrived at the pulsar.

The amount of the delay depends on the angle μ between the pulsar, the Earth and the GW source and on the distance d to the pulsar emitting the e.m. signal. Thus, in the case of a GW source emitting a monochromatic signal, the Earth term is coherent (fixed phase relationship) for all the pulsars in an array, while the pulsar terms are all delayed by different amounts. Therefore, in general, the pulsar term is an unknown quantity assuming different values for each observed pulsar and in practice appears as noise in the measurement, unless the delay can be determined and accounted for all the elements of the pulsar array. Unfortunately, only few pulsar distances have been measured, using VLBI, to better than 20% (20–200 pc), while the needed accuracy is a fraction of GW wavelength, i.e. a fraction of a pc. Ideally, a single very bright GW source would allow us to solve for the distances to all the pulsars.

The amplitude of the Earth term depends on the wave direction as $[1 + \cos(\mu)]$: for $\mu = 0$ the pulsar and the GW source are aligned and the term has a maximum; for $\mu = \pi$ the two test masses (Earth and pulsar) are anti-aligned in the sky and we have zero response, while for $\mu = \pi/2$, the amplitude is half of the maximum. However,

---

[7] Care must be taken in defining the angle μ: (Maggiore 2008, Chap. 23), offers a detailed derivation of Eq. 12.34, but considers the direction from the GW source to the Earth, opposite to our choice, resulting in a change of sign in front of $\cos\mu$.

when we also consider the pulsar term, Eq. (12.35) shows that the difference between the two contributions equals zero, $R_{gw}(t) = 0$, when we have a perfect alignment of the GW source with the line of sight of the e.m. signal. This agrees with the transverse character of the GW: it cannot give a non-zero response from pulsars oriented along the direction of propagation. Thus, we need a misalignment large enough for the GW to pass some fraction of a wavelength away from the pulsar (say, about a light year) in order to have $R \neq 0$. We can conclude that the optimum detection is when the pulsar is *nearly* aligned with the GW source.

12.9   Sources of Gravitational Waves at Low Frequencies

The nature of gravitational waves depends on the sources producing them and this determines the structure of timing residual. Figure 12.10 shows the sensitivity curve of the PTA and its complementary with those of the LISA and LIGO detectors. In the same figure, we also show candidates of GW sources detectable by PTAs that we review shortly:

- **Binary systems of super-massive black holes** ($M > 10^8 \, M_\odot$): our current understanding of galaxy evolution suggests that every galaxy contains a **super-massive black hole** (SMBH). The *hierarchical or bottom-up* formation scenario predicts that larger galaxies are generated by the merge of smaller galaxies at high red-shift. When two galaxies merge, we expect that the two SMBHs at their centres form a binary system. Binary systems are characterized by a non-vanishing second time derivative of their mass quadrupole moment, thus they are continuous source of GWs. The expected number of SMBH binaries (SMBHB) is extremely large, up to 10^6 depending on the red-shift and the mass range of the involved BHs. The incoherent superposition of the GW signals emitted from such a population of SMBHBs causes a **stochastic background of GWs**, usually considered isotropic. This is the most likely GW signal that PTA hopes to detect.
- **Cosmic strings**: they are the most controversial sources in the PTA band. They are hypothetical one-dimensional topological defects that may have formed during a symmetry-breaking phase transition in the early universe. If they exist, the cosmic strings will oscillate and emit GW signals of short duration and large amplitude.
- **GW cosmic background**: it consists of the relic GWs from the early evolution of the universe. The Big Bang is expected to be a prime candidate for the production of the many random processes needed to generate this stochastic GWs, which therefore may carry information about the origin and history of the universe. If these GWs were actually originated in the Big Bang, they should have stretched as the universe expanded and they can tell us about the very beginning of the universe. They would have been produced between approximately 10^{-36} to 10^{-32} seconds after the Big Bang, whereas the CMB was produced approximately 300.000 years after the Big Bang. Though widely believed to exist, a relic background will probably not be detected, since it is several orders of magnitude below the unresolved background due SMBHB (Maggiore 2000).

12.10   Detection Procedure

As mentioned in Sect. 12.9, detection of the stochastic background of GWs (SGWB) is much more likely than detection of GWs from an individual SMBHB. When we consider a single pulsar, the SGWB signal cannot be detected: it is embedded in noise processes such as timing noise, or masked by ephemeris errors. However, a GW emission would differently affect a set of many pulsars, depending on their position in the sky. Thus, the PTA detection principle is based on correlating the timing residuals of an array of pulsars. A seminal study on this method in the presence of a perturbation by an isotropic SGWB was carried out by R. Hellings and G. Downs. Given a pair of pulsars, separated by an angle γ in the sky, they showed that the correlation C takes the functional form, known as Hellings and Downs curve (Hellings and Downs 1983):

$$C(\gamma) = \frac{3}{2}\left[\frac{1-\cos\gamma}{2}\log\left(\frac{1-\cos\gamma}{2}\right) - \frac{1}{6}\frac{1-\cos\gamma}{2} + \frac{1}{3}\right]. \qquad (12.38)$$

The correlation C, shown in Fig. 12.12, starts out at a maximum when the two pulsars are seen in the same direction (zero angular separation); it has a zero when $\gamma = 0.855$ rad $= 49°$; then it reaches a minimum at 1.431 rad $= 82°$ and finally grows up to half of its initial zero-separation value at π rad. The Hellings and Downs curve is computed using the Earth term only (Eq. (12.36)), because the pulsar terms are not correlated, as discussed in Sect. 12.8. The geometry shown in Fig. 12.13 can help us understand the physics of the Hellings and Downs curve. For $\gamma = 0$, i.e. for a given pair of pulsars at the same "sky location", aligned on the same line of sight, the correlation between their TOAs will be maximum: the timing residuals due to crossing of GW (Eq. (12.34)) will be the same. On the other hand, when the angular separation is $\gamma \sim \pi/2$, we have negative correlation: as explained in Sect. 7.2, GWs alternatively stretch and compress space-time along orthogonal directions. Therefore, when the TOA from pulsar 1 is affected by the space-time dilation due to the GW, the TOA from pulsar 2 is shrunk and viceversa. Searches for stochastic gravitational-wave backgrounds using pulsar timing arrays effectively compare the measured correlations with the expected values from the Hellings and Downs curve to determine whether or not a signal from an isotropic, non-polarized background is present in the data.

Fig. 12.12 The Hellings and Downs correlation curve. Reproduced from Jenet and Romano (2015), with the permission of the American Association of Physics Teachers

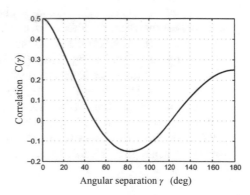

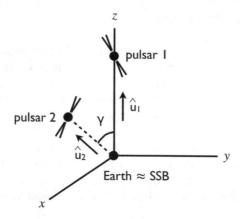

Fig. 12.13 Geometry for the Hellings and Downs correlation function. The Earth (rather, the Solar System Barycenter, but the difference is irrelevant: $1\ AU \ll 1kpc = 2 \cdot 10^8 AU$) is located at the origin, pulsar 1 on the z-axis and pulsar 2 in the xz-plane. The angle γ is defined by the two directions from the Earth toward the two pulsars. Reproduced from Jenet and Romano (2015), with the permission of the American Association of Physics Teachers

12.11 PTA Collaborations

Foster and Backer (1990) initiated the observing program called *Pulsar Timing Array—PTA program* to observe three pulsars using the US National Radio Astronomy Observatory telescope (Hobbs and Shi 2017). Nowadays, the PTA approach is pursed by different groups and the number of the pulsars in the array has expanded (Tiburzi 2018; Dahal 2020). At present, there are three main collaborations:

- The North American Nanohertz Observatory for Gravitational Waves (**NANOGrav**) is composed of researchers from institutions in the United States and Canada. Since 2007, they mainly use the Green Bank Telescope, to monitor 45 millisecond pulsars on a monthly basis. The Green Bank Telescope (GBT), in West Virginia, is the world's largest fully steerable radio telescope. The Arecibo Telescope was a 305 m spherical reflector radio telescope built into a natural sinkhole in Puerto Rico. However, the telescope was damaged by a hurricane in 2017 and by earthquakes in 2019 and 2020. Following others considerable damage of the supporting structure, the National Science Foundation decided to decommission it. Less than a month after the decision, on December 1, 2020, the suspended receiver collapsed and crashed onto the dish.
- The European Pulsar Timing Array (**EPTA**) was founded in 2007 but uses data collected by its members since the 1990s. The collaboration is composed of five telescopes, the Effelsberg Radio Telescope in Germany, the Lovell Telescope in England, the Sardinia Radio Telescope in Italy, the Westerbork Synthesis Radio Telescope in the Netherlands and the Nancay Radio Telescope in France. EPTA uses these telescopes to monitor 42 millisecond pulsars, among the many observed, at a higher rate than other collaborations;

- The Parkes Pulsar Timing Array (**PPTA**): this project began in 2004 making use of the Parkes Observatory in Australia. It is currently taking observations of 25 pulsars. As the southernmost telescope currently used for PTA observation, Parkes is able to see many Southern Hemisphere millisecond pulsars that are not visible at NANOGrav and EPTA telescopes.

The three PTAs collaborate in the framework of the International Pulsar Timing Array (**IPTA**), which aims to facilitate GW scientific research through the sharing of data and software codes.

Efforts to establish new PTA collaborations are ongoing in India, China and South Africa. The Indian PTA observes millisecond pulsars with the Ooty Radio Telescope and the Giant Metrewave Radio Telescope The Chinese PTA had a kick-off in 2007: it operates several 100-meter class telescopes (e.g. NSRT, Kunming, Tianma), together with the two most important facilities: the world-largest Five hundred meter Aperture Spherical Telescope (FAST) and the QiTai Radio Telescope. Finally, MeerTIME is an approved proposal dedicated to pulsar timing, which will use the MeerKAT telescope (South Africa).

12.12 Current Results and Prospects

No detection of gravitational waves has been made to date with the Pulsar Timing Array technique, but PTA upper limits already contributed to rule out some models of galaxy formation. Continuous efforts are made to improve the sensitivity and to understand the astrophysical implications of the (so far) null detection (Hobbs and Shi 2017). The primary limitation is simply the small number of pulsars are in the arrays. In 2019, the IPTA second data release (IPTA dr2) was published (Perera et al. 2019): it consists of regularly observed high-precision timing data for 65 millisecond pulsars (see Fig. 12.14), 16 more than the previous data release.

The primary goal of the IPTA is to detect and characterize low-frequency GWs using high-precision pulsar timing. The collaborations (Lommen 2015) confidently state that PTA *will* detect GWs: the question is when, and what needs to be done now to take full advantage of the array in the future.

Obvious steps toward increasing the detection probability of GW are:

Fig. 12.14 The Galactic distribution of 65 pulsars analyzed in the second IPTA data release, including 49 pulsars from the first IPTA data release (red dots) and 16 new pulsars (blue dots). Galactic latitude and longitude are expressed in degrees. Credits: Perera et al. (2019)

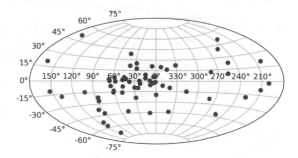

- To increase the SNR of the detected signals, using larger radio telescopes.
- To increase the observation frequency of the *ms* pulsar array, i.e. coordinating more radio telescopes together. In this respect, IPTA is a crucial tool for collaboration.
- To discover more and more millisecond pulsars, in order to have larger and larger arrays and increase the accuracy in the measurement of time residual caused by GW.
- to identify and correct for corrupting effects on the TOAs. At present, most of the ongoing research focuses on the study of several long-period processes affecting the residuals, such as inaccuracies in the Solar System ephemerides, intrinsic pulsar-timing noise and instrumental instabilities.

The new generation of radio telescopes that are now coming online will greatly increase our sensitivity to low-frequency GWs. The most sensitive instrument, the Square Kilometer Array (SKA), should begin operations in 2025. It will be the world largest telescope, with the potential of discovering all the pulsars in our galaxy with the beam shining on the Earth. In preparation for SKA, several pathfinders have been deployed, such as MeerKAT and the MWA (Western Australia), LWA (New Mexico, USA), and the LOw Frequency ARray (LOFAR, in Europe). They are essential tools to tackle some of the mentioned main challenges. For example, the low-frequency facilities such as LOFAR, LWA and MWA, are useful instruments to monitor the turbulent ionized interstellar medium and its effects on pulsar timing. In particular, DM variations are among the main sources of red noise in the TOA time series. Ionized interstellar medium studies at low frequencies will provide precious insights to improve the red noise model, and to disentangle its contribution from intrinsic timing noise generated from instabilities in the pulsar spin The science reward of this effort, detection by Pulsar Timing Arrays at low frequency, will be an invaluable insight into galaxies mergers and black hole dynamics, the exploration of fundamental physics problems like the cosmological constant, and high sensitivity tests of General Relativity. Over the next decades, PTAs will open a window over a new and unique regime of the GW universe, nicely complementing ground and space-based interferometers.

References

Baade, W., Zwicky, F.: On super-novae. Proc. Nat. Acad. Sci. USA **20**, 254–259 (1934). Baade, W., Zwicky, F.: Cosmic rays from super-novae. Proc. Nat. Acad. Sci. USA **20**, 259–263 (1934)

Becker, W., Kramer, M., Sesana, A.: Pulsar timing and its application for navigation and gravitational wave detection. Space Sci. Rev. **214**, 30 (2018)

Breton, R.P., et al.: Relativistic spin precession in the double pulsar. Science 321, 104 (2008)

Burgay, M., Perrodin, D., Possenti, A.: Timing neutron stars: pulsations, oscillations and explosions. In: Belloni, T.M., Méndez, M., Zhang, C. (eds.) General Relativity Measurements from Pulsars. Springer, Berlin, Heidelberg (2021)

Dahal, P.K.: Review of pulsar timing array for gravitational wave research. J. Astrophys. Astr. **41**, 8 (2020)

Detweiler, S.: Pulsar timing measurements and the search for gravitational waves. Ap. J. **234**, 1100–1104 (1979)

Estabrook, F.B., Wahlquist, H.D.: Response of Doppler spacecraft tracking to gravitational radiation. Gen Relat. Gravit. **6**, 439–447 (1975)

Foster, R.S., Backer, D.C.: Constructing a pulsar timing array. Ap. J. **361** 300 (1990)

Hellings, R.W., Downs, G.S.: Upper limits on the isotropic gravitational radiation background from pulsar timing. Astrophys. J. Lett. **265**, L39-42 (1983)

Hewish, A., Bell, S.J., Pilkington, J.D.H., Scott, P.F., Collins, R.A.: Observation of a rapidly pulsating radio source. Nature **217**, 709–713 (1968)

Hobbs, G., Shi, D.: Gravitational wave research using pulsar timing arrays. Natl. Sci. Rev. **4**, 707 (2017)

Hulse, R.A., Taylor, J.H.: A high-sensitivity pulsar survey. Ap. J. **191**, L59 (1974)

Imperi, L., Iess, L., Mariani, M.J.: An analysis of the geodesy and relativity experiments of Bepi-Colombo. Icarus **301**, 9–25 (2018)

Jenet, F.A., Romano, J.D.: Understanding the gravitational-wave Hellings and Downs curve for pulsar timing arrays in terms of sound and electromagnetic waves. Am. J. Phys. **83**, 635 (2015)

Lattimer, J.M.: Introduction to neutron stars. AIP Conf. Proc. **1645**, 61 (2015)

Lommen, A.: Pulsar timing arrays: the promise of gravitational wave detection. Rep. Prog. Phys. **78**, 124901 (2015)

Lommen, A.: Pulsar timing for gravitational wave detection. Nat. Astron. **1**, 809 (2017)

Lorimer, D.R.: Binary and millisecond pulsars. Living Rev. Relativ. **11**, 8 (2008)

Lorimer, D.R., Kramer, M.: Handbook of Pulsar Astronomy. Cambridge University Press, Cambridge (2005)

Lynch, R.S.: Pulsar timing arrays. J. Phys. Conf. Ser. **610**, 012017 (2015)

Lyne, A.G., et al.: A double-pulsar system: a rare laboratory for relativistic gravity and plasma physics. Science **303**, 1153 (2004)

Maggiore, M.: Gravitational wave experiments and early universe cosmology. Phys. Reports **331**, 283–367 (2000)

Maggiore, M.: Gravitational Waves, vol.2: Astrophysics and Cosmology, chap. 23. Oxford University Press, Oxford (2008)

Manchester, R.N.: Pulsars and gravity. Int. J. Mod. Phys. D **24**, 1530018 (2015)

Pacini, F.: Energy emission frome a neutron star. Nature **216**, 567–568 (1967)

Perera, B.B.P., et al.: The international pulsar timing array: second data release. MNRAS **490**, 4666–4687 (2019)

Perrodin, D., Sesana, A.: Radio pulsars: testing gravity and detecting gravitational waves. In: Rezzolla, L., et al. (eds.) The Physics and Astrophysics of Neutron Stars. Astrophysics and Space Science Library, vol. 457. Springer, Cham (2018)

Sazhin, M.V.: Opportunities for detecting ultralong gravitational waves. Astronomicheskii Zhurnal **55** 565–568 (1978). Translated in: Sov. Astron. **22**, 36–38 (1978)

Shao, L., Wex, N.: New tests of local Lorentz invariance of gravity with small-eccentricity binary pulsars. Class. Quantum Grav. **29**, 215018 (2012)

Shapiro, S.L., Teukolsky, S.A.: Black Holes, White Dwarfs, and Neutron Stars: The Physics of Compact Objects. Wiley-Interscience Publication. Wiley, New York (1983)

Stairs, I.H.: Testing general relativity with pulsar timing. Living Rev. Relat. **6**, 5 (2003)

Taylor, J.: Pulsar timing and relativistic gravity phil. Trans R. Soc. Lond. A **341**, 117–134 (1992)

Tiburzi, C.: Pulsars Probe the Low-Frequency Gravitational Sky: Pulsar Timing Arrays Basics and Recent Results. Publications of the Astronomical Society of Australia **35** e013 (2018). Astronomical Society of Australia 2018; published by Cambridge University Press

Voisin, G., et al.: Optimization of long-baseline optical interferometers for gravitational-wave detection. A&A **638**, A24 (2020)

Weisberg, J.M., Huang, Y.: Relativistic measurements from timing the binary pulsar PSR B1913+16. Ap. J. **829**, 55 (2016)

Weisberg, J.M., Nice, D.J., Taylor, J.H.: Timing measurements of the relativistic binary pulsar PSR B1913+16. Ap. J. **722**, 1030 (2010)

Wex, N.: Testing relativistic gravity with radio pulsars. In: Kopeikin, S.M. (ed.) Applications and Experiments, Frontiers in Relativistic Celestial Mechanics, vol. 2. Walter de Gruyter GmbH, Berlin/Boston (2014). arXiv:1402.5594

Will, C.M.: Theory and Experiment in Gravitational Physics. Cambridge University press (2018)

Further Reading

Lyne, A.G.: A Review of the double pulsar–PSR J0737–3039–Chin. J. Astron. Astrophys. Suppl. 2, 162 (2006)

Miao, X. et al.: Tests of conservation laws in post-Newtonian gravity with binary pulsars. Ap. J. 898, 69 (2020)

Sesana, A., Vecchio, A., Colacino, C.N.: The stochastic gravitational-wave background from massive black hole binary systems: implications for observations with Pulsar Timing Arrays. MNRAS 390, 192–209 (2008)

The Sagnac Effect

13

13.1 Introduction

Galileo first observed that linear uniform motion does not influence mechanical phenomena, concluding that it is not possible to distinguish the relative motion of (inertial) observers by studying of phenomena such as the free fall of bodies or the swing of the pendulum. Non-inertial references are inherently distinguishable. In them a free body describes non-linear trajectories, the pendulum oscillation plane rotates in space and the massive body falls moving from the vertical direction.

Although these examples refer to the motion of bodies, which follow the laws of mechanics, electromagnetism-induced phenomenology also allows to discriminate between inertial and accelerated reference frames. In 1913, the French physicist Georges Sagnac devised an interferometric device that was able to highlight the rotational state of the reference system in which the apparatus rests. He noticed that the rotation of the interferometer induces a delay, and consequently a phase difference, between two light waves that propagate in opposite directions along a closed optical path at rest in the rotating reference frame (Fig. 13.1). Its output is an interference figure from which it is possible to derive the angular velocity of the rotation. This is the Sagnac effect. Indeed, the first description of the Sagnac effect in the framework of special relativity was done by Laue (1911), two years before Sagnac performed his experiment (Sagnac 1913a), observing the correlation of angular velocity with the light phase-shift.

13.2 The Sagnac Effect

To derive the phase difference produced by the Sagnac effect, we consider a circular optical path of radius R, obtained for example by propagating the beam in a planar

© The Author(s), under exclusive license to Springer Nature Switzerland AG 2022   349
F. Ricci and M. Bassan, *Experimental Gravitation*, Lecture Notes in Physics 998,
https://doi.org/10.1007/978-3-030-95596-0_13

Fig. 13.1 The Sagnac
interferometer

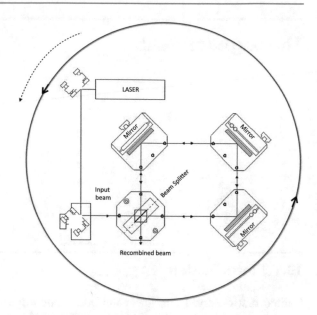

loop created in a platform that can be rotated around its symmetry axis, with angular
velocity Ω.

Referring to the Fig. 13.2, the light beam enters from point **A**, where it is divided
by a beam splitter in two different beams that propagate one in clockwise and the
other in counter clockwise direction. The two rays recombine at the interferometer
output, the point **A'**, which has moved from the position **A** due to the rotation of
the entire system. In the case of a simple circular loop, the path length is $L = 2\pi R$,
along which the light propagates in a time $\Delta t = L/c = 2\pi R/c$ if $\Omega = 0$, i.e. when
the platform does not rotate. As usual, c is the propagation velocity of the laser light.

When the platform rotates at Ω, we must compute different travel times for the
beam propagating in the clockwise direction (same as the platform) Δt_1 and that of
the beam travelling counterclockwise, Δt_2. At the recombination point A' (Fig. 13.2)
we have

$$c\Delta t_1 = 2\pi R + R\Omega\Delta t_1 \qquad c\Delta t_2 = 2\pi R - R\Omega\Delta t_2 \qquad (13.1)$$

$$\Delta t_1 = \frac{2\pi R}{c - R\Omega} \qquad \Delta t_2 = \frac{2\pi R}{c + R\Omega} \qquad (13.2)$$

Therefore, the time delay between the two rays is

$$\Delta T = \Delta t_1 - \Delta t_2 = \frac{4\pi R^2 \Omega}{c^2 - R^2\Omega^2} \simeq \frac{4\pi R^2}{c^2}\Omega \qquad (13.3)$$

neglecting $R^2\Omega^2 << c^2$. The corresponding optical path difference is:

$$\Delta L = c\Delta T \simeq \frac{4\pi R^2}{c}\Omega \qquad (13.4)$$

Fig. 13.2 Schematics of a
Sagnac interferometer: the
two rays, entering in **A**,
recombine in **A′** due to the
rotation of the platform

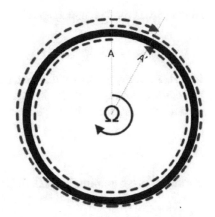

It can be proven that the above formula, derived for a circular path normal to the
rotation axis, is valid for all closed optical paths. Therefore, the Sagnac effect is
represented in terms of the general equation:

$$\Delta T = 4\frac{\vec{\Omega} \cdot \vec{S}}{c^2} \tag{13.5}$$

where $\vec{S}$ is the vector oriented perpendicular to the surface S enclosed by the optical
path.

This time difference is associated with a phase difference Ψ between the two
waves, given by the following expression:

$$\Psi = \omega \Delta T = \frac{\omega}{c}\Delta L = \frac{4\omega}{c^2}\left(\vec{\Omega} \cdot \vec{S}\right) \tag{13.6}$$

where ω is the angular frequency of the laser light.

The difference in optical path that is generated by the rotation of the interferometer
is generally a very small effect. Consider, as an example, the simple case a loop of
$R = 10$ cm, rotating with $\vec{\Omega}$ parallel to $\vec{S}$, at slow pace of $\Omega \simeq 2$ revolutions per
hour: we have for ΔL a value of about $1.5 \cdot 10^{-12}$ m, $\sim$ a millionth of the typical
wavelength of a laser light ($0.4-1\,\mu$m). To increase ΔL, and with it the phase
difference Ψ, a long optical fibre, of the order of kilometre, is used as a guide for
light. The fibre is wound to form a coil, in order to increment the total area which
determines the Sagnac effect by the number of the coil turns. We can express the
area in terms of the total length L of the fibre:

$$\Psi = 4\pi\frac{RL}{c\lambda}\Omega \tag{13.7}$$

where $\lambda = 2\pi c/\omega$ is the light wavelength. To determine the intensity of the resulting
beam at the interferometer output, we assume, for now, that the waves at the input
A are monochromatic and in phase:

$$s_1 = s_2 = S_0 \sin(\omega t)$$

At the end of the two paths, ignoring possible losses, the light signals are:

$$s_1 = S_0 \sin \omega(t + L_1/c); \qquad s_2 = S_0 \sin \omega(t + L_2/c) \qquad (13.8)$$

In the case of two perfectly matched signals, i.e. waves linearly polarised in the same plane, the resulting wave is simply: $s = s_1 + s_2$. After a simple manipulation, and defining the Sagnac-induced phase difference

$$\Psi \equiv \frac{\omega}{c} \Delta L$$

the output wave amplitude takes the form:

$$s = 2S_0 \sin \left(\omega t - \frac{\Psi}{2} \right) \cos \frac{\Psi}{2} \qquad (13.9)$$

We can now compute the intensity of the resulting wave:

$$I = I_0 cos^2 \frac{\Psi}{2} = \frac{I_0}{2}(1 + cos\Psi)$$

I_0 is the maximum of the light intensity that we have for $\Psi = 0$ (signals in the phase and non-rotating interferometer). The minimum values of intensity ($I = 0$) are obtained for $\Psi = \pm m\pi$.

In synthesis, the phase change due to the rotation of the interferometer has the effect of inducing an interference figure and the Sagnac phase can be determined directly from the interference figure. However, this method is extremely sensitive to changes in the intensity of the light focused in the fibre. The usual technique that addresses this problem consists in phase modulating the signal at a frequency ω_m, by means of a moving mirror that adds a time varying phase:

$$\Phi(t) = \Phi_m \cos(\omega_m t)$$

The two waves that propagate in opposite directions undergo the same modulation but in different time instants, t and $t - \tau$ where τ is a fixed, but yet to be determined delay.

This induces an additional phase difference between the two rays of $\Phi(t)$ and $\Phi(t - \tau)$. The signal intensity at the interferometer output is now time dependent:

$$I(t) = I_0 \frac{1}{2} \{ 1 + cos[\Psi + \Phi_m \cos \omega_m t - \Phi_m \cos \omega_m (t - \tau)] \}$$

After a few trigonometric manipulations, and defining the *modulation depth* (see Appendix A) $\phi_c \equiv 2\Phi_m sin(\frac{1}{2}\omega_m \tau)$, we can rewrite the intensity as:

$$2I(t) = \frac{I_0}{2} \left[1 + cos\ \Psi \cos \left(\phi_c \sin \omega t \right) - \sin \Psi \sin(\phi_c sin\omega t) \right].$$

where we also neglected an uninteresting phase $\omega\tau/2$ in the time dependent functions.

We can expand this equation in series of Bessels functions

$$I(t) = \frac{1}{2}I_0\big[1 + J_0(\phi_c)cos(\Psi)\big]+$$

$$+I_0\,cos\,\Psi\Big(\sum_{k=1}^{\infty}J_{2k}(\phi_c)cos(2k\omega t - k\omega\tau)\Big)+$$

$$+ I_0\,sin\,\Psi\left(\sum_{k=1}^{\infty}J_{2k-1}(\phi_c)sin((2k - 1)\omega t - \Big(k - \frac{1}{2}\omega\tau\Big)\right) \tag{13.10}$$

When we compare the terms of Bessel expansion with those of the Fourier one

$$I(t) = S_0 + \sum_{k=1}^{\infty}S_k cos(k\omega t + \alpha_k)$$

where $\phi_c = 2\phi_m sin\big(\frac{1}{2}\omega_m t\big)$, we end up with the following identities:

$$S_0 = \left[\frac{1}{2}I_0\big[1 + J_o(\phi_c)cos\,\Psi\big]\right]$$

$$S_1 = \Big[I_0 J_1(\phi_c)sin\,\Psi\Big]$$

$$S_2 = \Big[I_0 J_2(\phi_c)cos\,\Psi\Big]$$

$$S_3 = \Big[I_0 J_3(\phi_c)sin\,\Psi\Big]$$

$$S_4 = \Big[I_0 J_4(\phi_c)cos\,\Psi\Big]$$

The even harmonics are proportional to the cosine of the Sagnac phase, while the odd ones are proportional to the sine. In the absence of rotation ($\Psi = 0$) we have $S_{2k-1} = 0$ and at the interferometer output we get a signal at twice the modulation frequency. At the photodiode output we have a broadband signal that includes only the even harmonics (2ω, 4ω, 6ω, ...) whose amplitudes are determined by that of the modulation. When the gyroscope rotates all harmonic components are present (ω, 2ω, 3ω, 4ω, ...).

In the above expressions, the constant argument ϕ_c, that depends on τ, is still undetermined and must be estimated. To this purpose, we calculate the ratios between consecutive even (or odd) harmonics. For example:

$$\frac{S_1}{S_3} = \frac{J_1(\phi_c)}{J_3(\phi_c)}$$

Each of these ratios provides an independent estimate of ϕ_c that is not affected by fluctuations in the laser intensity. Once ϕ_c is known, a similar procedure is used to obtain the Sagnac phase. In this step, however, the ratios must be computed between two consecutive odd and even harmonics:

$$\frac{S_1}{S_2} = \frac{J_1(\phi_c)}{J_2(\phi_c)} tan(\Psi)$$

13.3   Sagnac Gyroscopes—Technologies and Applications

The first commercial devices, based on the Sagnac effect, were the Ring Laser Gyroscopes (RLG), first demonstrated in the early 1960 (for a review, see Anderson 1994): a sealed ring-cavity with very high-quality mirrors, filled with He-Ne, operates as an active resonator for two counter-propagating laser beams. These can then re-circulate a great number of times, depending on the cavity finesse. The two waves resonate at different frequencies, because the storage time is different, and the output has a beat note that is simply related to the rotation frequency. To avoid issues with too low beat frequencies, when $\Omega < 1^o/$ min, a *dithering* is applied to the cavity, modulating the signal to higher frequency. This degrades the performance and adds mechanical noise.

More recently, Fibre Optic Gyroscopes (FOG) have become popular: in a FOG, the ring is not a part of the laser. Rather, an external solid-state semiconductor laser injects counter-propagating beams into a fibre ring, where rotation causes a relative phase shift between those beams when interfered at the exit. The Sagnac effect can be enhanced by using a fibre coil with many turns. However, in such a passive interferometer, the signal is a phase (and not frequency) difference between the counter-propagating waves. This implies some additional signal processing and a lower sensitivity, because the phase difference does not grow with time. While based on the same physical principle, and having similar theoretical performance, the technology for the two devices has important differences, mostly to FOGs advantage: A FOG consists of few components, standard in the communication industry, there is no dithering (no moving parts, no mechanical noise), there is no gas as active medium (no leakage, better durability and reliability), it is scalable just by adding turns to the fibre coil. Commercial FOGs have now thousands of turns in a diameter of few mm. These devices have sensitivity of 10 nrad/s/$\sqrt{\text{Hz}}$ in the rotation range 0.01–100 Hz. The sensitivity can be enhanced at will by using longer and longer fibres, but with a trade-off: increased thermal phase noise and increased shot noise (due to fibre attenuation) limit the long term stability and portability of these devices. FOGs have replaced mechanical (spinning) gyroscopes in most applications of inertial navigation.

Today, the Sagnac interferometer, in either version, is a tool of current technology. Its foremost use is in inertial guidance systems, where high sensitivity to rotations is of foremost importance. Global navigation satellite systems (GNSSs), such as GPS, GLONASS, COMPASS or Galileo, need to account for the rotation of the Earth in the procedures for synchronising clocks by multilateral exchanges of radio signals (see Chap. 14).

13.4 The Sagnac Effect and Gravitation

At present, the best sensitivity, below $20\,\mathrm{prad/s/\sqrt{Hz}}$ is achieved by RLG (Di Virgilio 2020) and an effort is underway in several nations to develop a network of RLG for geophysical applications but also with the aim of measuring the Gravitomagnetic effects of GR on the Earth.

The Sagnac configuration may have a crucial role in future gravitational wave detectors. Shaddock proposed (2004) a phase-locking configuration for the LISA space interferometer. that should significantly simplify the mode of operation. The proposed scheme is based on one Sagnac signal readout inherently insensitive to laser frequency noise and optical bench motion for a non-rotating LISA array. This Sagnac output is also insensitive to clock noise, requires no time shifting of data, nor absolute arm length knowledge. As all measurements are made at one spacecraft, neither clock synchronisation nor exchange of phase information between spacecraft is required. Note that, given the size of the LISA constellation, the Sagnac effect amounts to $\Delta L = 3.5$ km.

Although this implementation of the LISA readout will probably not be implemented, LISA will have a *Sagnac channel* among its science output: indeed, the sum of the phases along the three sides of the constellation is insensitive to g.w. and thus provides a useful veto signal.

In principle the Sagnac configuration can be used also on the ground-based GW detectors. The Sagnac interferometer is generally used as a rotation sensor, but it is also sensitive to displacements of its mirrors (except for that located at half the round trip length).

By designing the interferometer such that the area enclosed by the two counter-propagating beams is zero, the interferometer can be made insensitive to rotations though keeping it sensitive for displacements. This interferometer could be insensitive to laser frequency variation, mirror displacement at dc, thermally induced birefringence, and reflectivity imbalance in the arms, allowing a simplified control system and reduced optical tolerance requirements. The peak response of the Sagnac interferometer should be competitive with that of a Michelson interferometer with the same storage time and power on the beam splitter (Fig. 13.3). Although a first prototype showed no advantage in comparison with a classical Michelson (Sun et al. 1996), it was later discovered that a zero-area Sagnac interferometer, by its very

Fig. 13.3 a Conventional
Sagnac interferometer. **b**
Zero-area Sagnac
interferometer for
displacement sensing with a
geometry that meets the
requirements for a ground
based g.w. interferometer.
Credits: Reprinted figure
with permission from Sun et
al. (1996). Copyright 1996
by the American Physical
Society

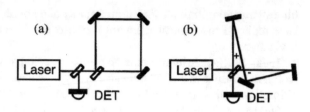

nature, can suppress the quantum back-action noise for displacement measurements
over a broad frequency band (Chen 2003).

13.4.1 The Sagnac Effect and Gravitomagnetism

There is a close connection between the Sagnac effect, seen in previous sections, and
Gravitomagnetism, that we discussed in Sect. 5.7. We shall show here that an effect,
quite similar to Sagnac, is generated if the light beam propagates in regions of space
where a gravitomagnetic field, generated by rotating source masses, exists.

We consider an interferometer where counterpropagating e.m. beams describe a
close path in the space-time, at a mean distance R from a central source M, endowed
with angular momentum J. The closed path ℓ needs not encircle the source. This
situation can apply to the next generation of GPS and Galileo satellites orbiting the
Earth, that will exchange radio signal to synchronize their clocks, or to the LISA
constellation orbiting around the Sun. We shall assume that the interferometer path
is in free fall at a distance R from the source: this implies a circular orbit around the
source.

We start from a general expression for the metric line element and recall that, for
a photon, such interval is null:

$$ds^2 = g_{00}(cdt)^2 + 2g_{0i}cdtdx^i + g_{ij}dx^idx^j = 0 \qquad (13.11)$$

We can solve this equation for t, i.e. the coordinate time interval for a space displacement vector $\{dx^i\}$ along a light ray. The result is:

$$cdt = -\frac{g_{0i}dx^i}{g_{00}} + \frac{\sqrt{(g_{0i}dx^i)^2 - g_{00}g_{ij}dx^idx^j}}{g_{00}} \qquad (13.12)$$

We have chosen the sign so that the propagation always be to the future, $dt > 0$,
irrespective of the sign of the dx^i's, i.e. of the direction of propagation along the
space path.

We can integrate Eq. 13.12 along the closed three-dimensional path ℓ to find the time of flight for a loop travel from event A (injection) to event B (extraction).

$$t_{AB} = -\oint \frac{g_{0i}}{cg_{00}} \frac{dx^i}{d\ell} d\ell + \oint d\ell \frac{\sqrt{(g_{0i}dx^i)^2 - g_{00}g_{ij}dx^i dx^j}}{cg_{00}} \tag{13.13}$$

Notice that the first integral depends on the direction of propagation, while the second does not. Therefore, when considering the time of flight for two beams counter-propagating along the same closed path, the first integral gives different contributions, and we obtain two different results, say t_+ for signal co-rotating with the reference frame, and t_- for the counter-rotating signal.

$$\delta t = |t_+ - t_-| = -\frac{2}{c} \oint_\ell \frac{g_{0i}(\vec{x})}{g_{00}(\vec{x})} d\ell^i \tag{13.14}$$

The difference is due to the off-diagonal components g_{0i} of the metric tensor, those related to the source rotation.

Solution of Eq. 13.14 is rather complex in the general case: we outline here the necessary steps to evaluate the time difference in the WFSM limit (Chap. 5), referring to Ruggiero and Tartaglia (2019) for details.

1. Choose the metric to work with. For instance, the WFSM limit of Kerr's metric of Eq. 5.30. We allow ourselves the additional, simplifying liberty of considering only motion in the equatorial plane ($\theta = \pi/2$):

$$ds^2 = (1 - \frac{2GM}{rc^2})c^2 dt^2 - \frac{dr^2}{(1 - \frac{2GM}{rc^2})} - r^2 d\varphi^2 - \frac{4G\,J}{c^3 r} cdt d\varphi$$

from this, we read the metric elements

$$g_{00} = 1 - \frac{2GM}{rc^2} \qquad g_{0i} = -\frac{2G\,J}{c^3 r} \delta_{i\phi} \tag{13.15}$$

2. Convert from coordinate time to the observer's proper time, where the beams are recombined: $\delta\tau = \sqrt{g_{00}(\vec{x}_{obs})}\delta t$.
3. Convert to the rotating frame where the interferometer is at rest: $\varphi^{'} = \varphi - \Omega t$, where $\Omega = \sqrt{GM/R^3}$
4. Apply a Lorentz boost at the Keplerian speed $\vec{V} = R\Omega\hat{\varphi}$: it affects both the time t and the azimuthal distance $R\varphi$
5. Expand $g_{00}(\vec{x})$, $g_{0j}(\vec{x})$, to be integrated along the closed path ℓ, as well as $g_{00}(\vec{x}_{obs})$, to first order in $\frac{2G\,J}{c^3 r^2} \ll 1$ and $\frac{GM}{rc^2} \ll 1$.

The calculation generates a number of "first order terms": among them we can extract the classical Sagnac effect previously discussed. We focus here, instead, on the gravitomagnetic term due to g_{0j} that, to first order, simply reduces to

$$\delta\tau = \frac{4}{c^4} \oint \frac{GJ}{r} d\varphi \qquad (13.16)$$

For the LISA constellation, rotating around the Sun with a one year period, $\Omega = 2\pi/1\,year$, this amounts to

$$\delta\tau_{GM} = 1.5ns \qquad \delta\ell_{GM} = 0.45\,\text{m}$$

where we have assumed the angular moment of the Sun $J_\odot = 2 \cdot 10^{41}$ kgm^{2}s^{-1}, that generates a gravitomagnetic field $\vec{B}_g = 4G\vec{J}/c^2 R^3 = 1.7 \cdot 10^{-19} \text{m}^{-1}$ at the distance $R = 1\,AU$. This must be compared with a much larger classical Sagnac effect, that is due to rotation of the constellation around its center with the same period of one year:

$$\delta\tau_{Sagnac} = 24\,\mu s \qquad \delta\ell_{Sagnac} = 7200\,\text{m}$$

Finally, it is worth noting that the line integral of Eq. 13.14 can be converted, by virtue of Stokes' theorem, into a surface integral of the gravitomagnetic field B_g given by the curl of the vector potential $-2g_{0i}/c^2$ (Eq. 5.64). In this way, the connection with the encircled area typical of Sagnac phenomena is brought to evidence.

References

Laue, Max von (1911), *On an Experiment on the Optics of Moving Bodies* Münchener Sitzungsberichte, 405–412

Sagnac, G. *L' éther lumineux démontré par l'effet du vent relatif d' éther dans un interféromètre en rotation uniforme (The demonstration of the luminiferous aether by an interferometer in uniform rotation) - Comptes Rendus Académie de Sciences, Paris*, **157**, 708-710 (1913)

Anderson R., Bilger H. R., and Stedman G. E. *Sagnac effect: A century of Earth-rotated interferometers - Am. Jou. Phys.*, **62**, 975 (1994)

Di Virgilio A. et al. *Underground Sagnac gyroscope with sub-prad/s rotation rate sensitivity: Toward general relativity tests on Earth - Phys. Rev. Research* **2**, 032069(R) (2020)

Shaddock D. A. , *Operating LISA as a Sagnac interferometer - Phys. Rev.* **D 69**, 022001 (2004)

Sun K.-X., Fejer M.M. , Gustafson E., Byer R.L. *Sagnac Interferometer for Gravitational-Wave Detection - Phys. Rev. Lett.* **76**, 3053 (1996)

Chen Y., *Sagnac Interferometer as a Speed-Meter-Type, Quantum-Nondemolition Gravitational-Wave Detector - Phys. Rev. D* **67**, 122004 (2003).

Ruggiero, M.L., Tartaglia, A.: Test of gravitomagnetism with satellites around the Earth - Eur. Phys. Jour. Plus **134**, 205 (2019)

GPS and Relativity

14

14.1 Introduction

In 1973, the U.S. Department of Defence approved a project to create a global satellite navigation system: it was the act of birth of the Navigation System for Timing And Ranging (NAVSTAR) network, called in a simpler way Global Positioning System (GPS). The system was designed to determine the position of a vehicle on the globe with sufficient accuracy to avoid collision hazards and to guide it up to its destination.

A receiver of the GPS navigation signals measures, on a local clock, the arrival times of signals from four or more different satellites. The receiver then uses these measurements along with information in the navigation messages to solve for four unknowns: user position (x, y, z) and the receiver clock offset from GPS time. In this computation, a GPS receiver must apply two relativistic corrections in order to provide time or position to the user: the first due to special relativity, the second to GR. In the following we present the GPS system and discuss these corrections, with a main focus on estimating the geometric range delay, i.e. the time that GPS signals take to propagate from the transmitter to the receiver.

A constellation of satellites orbiting the Earth is the backbone of the GPS system. The constellation consists of 24 active satellites, arranged in 6 almost circular orbits inclined 55° relative to the equatorial plane (Fig. 14.1). All satellites orbit in a period of half sidereal day, $T = 43082$ s. Kepler's laws tell us that they orbit at an average altitude of 20184 km and an average speed $v_s = 3873.8\,\mathrm{m/s}$. There are 4 satellites in each orbit, placed at regular distances. This choice of orbits responds to precise criteria: it allows an even distribution of satellites around the globe, with orbits that are close enough to the poles to ensure that the system works properly even in those remote regions of the Earth, even if with a slight loss of precision. Moreover, with the current arrangement, a receiver anywhere on Earth is exposed to receive signals from at least five satellites.

F. Ricci and M. Bassan, *Experimental Gravitation*, Lecture Notes in Physics 998,
https://doi.org/10.1007/978-3-030-95596-0_14

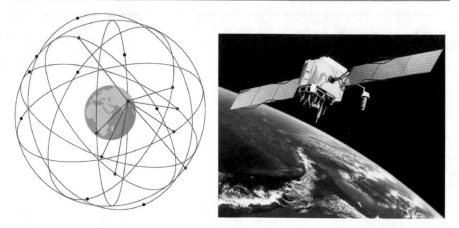

Fig. 14.1 Left: a drawing of the orbits of the 24 GPS satellites, arranged in pairs on the same orbit. The red segments indicate the signal propagation to a receiver on Earth. Right: artistic rendering of a GPS satellite. Credits: public domain images

All satellites experience disturbances such as gravitational fields and the solar wind, that make their control challenging. In addition to the 24 satellites that constitute the operating network, there are now 7 spares in orbit, plus several of previous generations or malfunctioning.

The first generation of satellites was launched between 1978 and 1985 and formed the system called GPS I. In that stage, the orbits of the satellites were tilted by 63° relative to the equatorial plane, the remaining characteristics were the same as the current ones, i.e. 24 satellites arranged, in groups of four, on six orbital planes. The satellites were 5.3 m wide, weighing 759 kg, each equipped with one atomic Caesium (Cs) clock and two Rubidium (Rb) clocks. Their expected operational life was 5 years, although many remained in service longer. They could be operated with 3 control axes and a hydrazine (N_2H_4) propulsion system. The energy was derived from two solar panels capable of providing ∼400 W of power and NiCd batteries when orbiting in the shadows. The navigation data were transmitted on two L-band frequencies, chosen because they are able to penetrate clouds, fog, rain, storms, and vegetation. They are called L1, at 1575.42 MHz and L2-P5, at 1227.60 MHz, both generated by upconverting, 154 and 120 times respectively, the local oscillator frequency $f_0 = 10.23$ MHz.

The second generation (1990–1997) satellites were equipped with one additional Cs atomic clock. In the last generations, the satellites have larger sizes and mass and a life expectancy of about 12 years. They are equipped with a hydrogen MASER as time reference.

Finally, we have the Block III satellites, the third generation of GPS satellites (GPS III). They began operation in 2009 and are expected to remain in service until 2030 and beyond.

14.2   Clock Fluctuations and the Allan Variance

The role of clocks is crucial in the use of GPS, and it is worthwhile to briefly sum-
marise their noise characteristics. Crystal oscillators and atomic clocks are affected
by phase noise, which has few different spectral components; the two main contributions are a white frequency noise and flicker frequency noise. The last one is a type of
electronic noise with a $1/f$ power spectral density and it is particularly troublesome,
being more and more significant as we go to lower frequencies. At very low frequencies, we can think of the noise as becoming a drift, although the mechanisms causing
drifts are usually different from flicker noise. There is a practical consequence to
this behaviour: if we try to evaluate the noise using traditional statistical tools, such
as standard deviation which implies the averaging operation, this estimator will not
converge. To overcome this difficulty, and to assess the frequency stability of a clock,
a suitable tool is the clock Allan variance, also known as *two sample variance*. We
note that the term *Allan variance* is a bit ambiguous, because there exist several
different definitions. Here we refer to the following:

*The Allan variance is one half of the time average of the squares of the differ-
ences between successive measurements of the frequency deviation sampled over the
sampling period.*

To understand this definition, we must first introduce the concept of M-variance

In mathematical terms, assume we sample the clock reading every T seconds;
being $x(k\,T)$ and $x(k\,T + \tau)$ the clock reading done at $k\,T$ and $k\,T + \tau$, we define
the incremental ratio computed at the delay time τ

$$y_k = \frac{x(k\,T + \tau) - x(k\,T)}{\tau} \tag{14.1}$$

Suppose we collect M measurements. The M-variance is defined as (Fig. 14.2):

$$\sigma_y^2(M, T, \tau) = \frac{1}{M-1} \left\{ \sum_{k=0}^{M-1} \bar{y}_k^2 - \frac{1}{M} \left[\sum_{k=0}^{M-1} \bar{y}_k \right]^2 \right\} \tag{14.2}$$

The Allan variance is a particular case of M-variance, for $M = 2$, and $T = \tau$:

$$\sigma_y^2(\tau) = \sigma_y^2(2, \tau, \tau) \tag{14.3}$$

and it depends just on the τ variable. The Allan deviation is $\sigma_y(\tau)$, the square-root
of the Allan variance. To clarify the meaning of this quantity and give a quantitative
example of a clock frequency stability, let us assume that $\sigma_y(\tau) = 1.3 \times 10^{-9}$ at an
observation time $\tau = 1$ s. If the clock signal is an oscillation at 10 MHz, it means
that we are dealing with a clock frequency instability equivalent to 13 MHz RMS.

In Fig. 14.3 the stability of quartz and Caesium atomic clocks verses the observation time τ are compared.
From the plots we can see that quartz clocks are stabler for short observation times,
up to tens of seconds, while atomic clocks have a much better stability at longer
times.

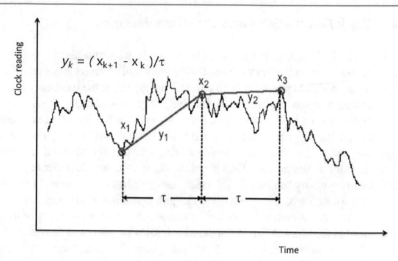

Fig. 14.2 Definition of the M-variance: x are the clock readings taken at constant time interval. The frequency fluctuation is evaluated by computing the M-variance of the y variable. This variance depends on the chosen observational delay τ, the sampling time T and the number of averaged measurements M, as shown in Eq. 14.2

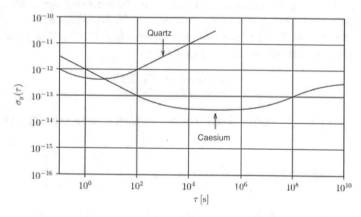

Fig. 14.3 Allan's deviation on the frequency of a quartz oscillator and a Caesium atomic clock vs the time interval τ between measurements. Credits: CC-BY Ashby (2003)-NC-ND/2.0

Once the satellites are in the correct orbit, it is necessary to carefully monitor their telemetry and operation to ensure the proper operation of the system.[1] For this purpose, many monitoring stations have been built and operated, first by U.S. military and later by other agencies, around the world (see Fig. 14.4).

---

[1] Factors such as the solar wind or attractions gravitational waves by other bodies in the universe can divert satellites from their intended trajectory. These are causes of error in the data of the navigation system.

Fig. 14.4 The network of Ground Stations for the control of the GPS system. Credits: GPS.gov

Finally, the signals of GPS satellites are detected on ground by the receivers, which consist of an antenna and a processing unit, capable to receive and interpret the signals sent by the satellites, plus a common quartz clock, which is constantly synchronised to the atomic clocks of the satellites.

At present, inexpensive receivers are available, capable of receiving and processing the signals of up to 12 satellites in parallel. There are two main types of receivers: for civil or military use. The former are only able to receive the L1 signal, while the latter receive and decode both L1 and L2 signals, thus providing higher accuracy. Indeed, one of the largest causes of errors is the slowing down of signals when crossing the atmosphere: military receivers can correct for this problem by comparing the propagation times of the two signals, the civilians receivers must instead rely on external systems. Even the L1 signals were degraded to a precision of 100 m, by the so called *Selective Availability*, implemented by the US military for security reasons. Many smart methods were then devised to circumvent this limitation, and finally, in 2000, the Selective Availability was turned off for good.

14.3 The Trilateration Method

The process that is used to determine the location is called *trilateration* and is based on the knowledge of the positon of some reference points to determine that of the point sought. We first give a conceptual approach. Assume any reference system with origin at the Earth center O. Consider a GPS receiver D (as Detector), located at an unknown point $\vec{r}_D$. The satellite S_1 sends a first signal, with encoded its emission time and its position $\vec{r}_1$, to the receiver. The distance $d_1 = |\vec{r}_D - \vec{r}_1|$ from the transmitter is calculated by measuring the travel time t_1 of the signal: $d_1 = ct_1$. The receiver can conclude that its position $\vec{r}_D$ is anywhere on the spherical surface of radius d_1, centered in $\vec{r}_1$. A second satellite S_2 sends another signal, whereby the receiver is

Fig. 14.5 A ring is obtained
from the intersection of the
two spheres. A third satellite
defines an additional sphere,
which reduces the
intersection to two points

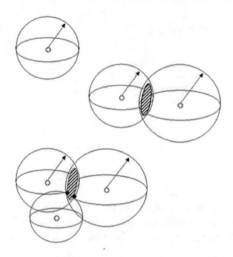

able to track a second sphere of radius $d_2 = |\vec{r}_D - \vec{r}_2| = ct_2$. The allowed positions
reduce now to the intersection of these two spheres, i.e. a circumference. With the
signal of the third satellite S_3 we finally reduce the intersections of the circle to just
two points, as shown in Fig. 14.5. This final degeneracy is easily removed because
one of the two points is normally at unacceptably high altitude.

In essence, all satellites simultaneously send a signal every millisecond that contains navigation parameters, from which their positions can be determined. The
receiver must detect at least four of these signals: for the jth satellite, the position
$\vec{r}_j$ correspond to the center of the sphere, while the radius, i.e. the distance from the
receiver, is given by

$$c\, t_j = |\vec{r_D} - \vec{r}_j| \qquad (14.4)$$

where t_j is the propagation time of the signal from the satellite to the receiver whose
position is located by the $\vec{r_D}$ vector.

The need to use at least four equations, and therefore four satellites, stems from
the fact that there are four unknowns in the problem: the time and the three space
coordinates that locate the event in space-time corresponding to the reception of the
signal simultaneously sent by the satellites.[2]

In order to obtain a position with a precision of the order of one meter, the travel
time of the signals must be measured with an accuracy of the order of 1 ns. For these
level of accuracy it is essential to consider and compensate relativistic effects that
characterise the space-time around the Earth, as well as the fact that the signals do
not reach at the same time the receiver, depending on the distance of their emitters
from the receiver. These effects are the consequence of fundamental features of

---

[2] The issue is actually more complex, because the timing of typical receiver (e.g. a smartphone)
is much less precise than the Caesium clocks on the satellites. A fourth satellite and additional
computation (see e.g. Ashby 2003) is thus required to remove the receiver clock bias.

signal propagation in the space-time, which is perceived, by an observer in motion, differently than by a stationary observer. Moreover, the gravitational field of all masses in the universe distorts the space-time.

Therefore, we have to compute three major relativistic corrections that affect the time of signal propagation:

- synchronisation: signals are not received at the same time by a moving (if nothing else, with the Earth) detector
- time dilation due to special relativity: satellites move at 3.9 km/s relative to Earth,
- time difference due to the general relativity: emitting satellites and receiver are at different heights and their clocks are in a different point of the gravitational field. Clocks on GPS satellites run faster than clocks at rest on the Earth's surface (gravitational redshift). Thus, GPS satellite clock frequencies need to be adjusted to compensate for this effect.

The first correction depends on the locations of the satellites and the receiver and, therefore, is not constant. Therefore, we need to correct the equations when we refer the result to the receiver. As for the remaining two effects, the satellites altitude as well as their velocity remain roughly constant along an orbit.

As a next step, we must choose the reference frame to compute the time intervals of the light propagation and, in doing that, we should keep in mind the need to include these relativistic corrections.

14.4   The Reference Frames

GPS time is defined using the principle of the constancy of c to synchronise an imaginary system of clocks everywhere in space in the neighbourhood of the Earth (the so-called Einstein synchronisation). GPS satellite clocks are in principle adjusted to agree with this imaginary reference system of clocks. The GPS network realises a coordinate time, a system of self-consistent time markers by which we label the events.

This definition of GPS time requires a locally inertial coordinate system. Therefore the time is defined relative to an Earth-Centred Inertial coordinate frame (ECI), but its pace is set to match the rate at which clocks would run on the geoid. An ECI is also used to simplify the paths of signals propagating from satellites, since, with sufficient accuracy for GPS,[3] light travels in Euclidean straight lines at the speed c in vacuum relative to such inertial frame.

The goal of the GPS system is to define the position of a point relative to the Earth, which is rotating. Thus, since the receiver is usually on the Earth, we also introduce the *Earth Centered—Earth Fixed* (ECEF) reference frame.

---

[3] The receiver must also account for ionospheric and tropospheric delay corrections, which we do not consider here.

This reference system has the origin in the Earth center and rotates with it. It is generally traced back to the reference system WGS-84(G873), with a constant angular velocity for the rotation around the Earth axis of $\Omega_\oplus = 7.292115 \cdot 10^{-5}$ rad/s (2π/sidereal day). This is clearly a non-inertial frame and, as a consequence, we cannot apply Eq. (14.4) used to determine the position and time of the receiver. We then need to compute these quantities using the ECI frame where the equations of special relativity are valid The frame is redefined by the receiver at each cycle of the calculation of the space-time coordinates, so it is *instantaneously inertial*, i.e. in free fall with the Earth in the gravitational field of the masses of the universe.

In essence, the computation steps are : (a) solve the signal propagation equation in the ECI reference, (b) transform the results into the ECEF to obtain the coordinates of the Earth-related position.

In the following, we shall run our calculations with the usual Spherical coordinates (r, θ, ϕ). However, GPS receivers output the position using the *Geographical coordinates*, and it is thus appropriate to recap the differences and the conversion between the two systems.

The Geographical coordinates are:

- Longitude λ: the angle calculated from the primary meridian defined as the semi-circonference that joins the two poles passing through the city of Greenwich (UK). Same as the azimuthal angle ϕ of spherical coordinates.
- Latitude φ or β [4] : is the polar angle, just as θ in spherical coordinates, but measured from the equator rather than from the North pole. In this notation, the North pole has latitude $+\pi/2$ and the South pole has $-\pi/2$. Actually, both latitude and longitude are usually expressed in degrees, rather than radians. We convert from one to the other with: $latitude = \pi/2 - \theta$.

- Altitude, or elevation h: is the height above the surface of the *Reference Ellipsoid*, a geometrical figure (an oblate ellipsoid of revolution) that best approximates the volume of the Earth.
  The altitude is thus defined as $h = r - R_\oplus(\theta)$, where r is the usual radial coordinate. $R_\oplus$ is the distance of the ellipsoid surface from the Earth center of mass (Table 14.1):

$$R_\oplus(\theta) = \sqrt{\frac{a^2 b^2}{a^2 \sin^2 \theta + b^2 \cos^2 \theta}} \qquad (14.5)$$

where $a = 6.378137 \cdot 10^6$ m is the equatorial radius (or semi-major axis)[5] and $b = 6.356752 \cdot 10^6$ m is the polar radius, or semi-minor axis (Fig. 14.6).

The ellipsoid is a geometrical approximation to the Earth, but we actually need a description of its gravitational potential: the *geoid* is defined as the locus of points,

---

[4] Unfortunately both symbols have different meaning in our text

[5] International Earth Rotation and Reference System Service (IERS) Conventions 2003.

Table 14.1 Symbols, range and units for various coordinates on a sphere. We also recall the cylindrical coordinates that will be used in the following

Coordinates	Radial	Polar angle	Azimuthal angle
Spherical	$0 < r < \infty$	$0 \leq \theta << \pi$	$0 \leq \phi < 2\pi$
Pseudo-spherical	$0 < r < \infty$	$-\pi/2 \leq \theta \leq \pi/2$	$0 \leq \phi < 2\pi$
Geographical	$R_\oplus \leq h < \infty$	$-90° \leq \varphi (or\,\beta) << 90°$	$0 \leq \lambda < 360°$
Cylindrical	$0 < r < \infty$	$-\infty < z < \infty$	$0 \leq \phi < 2\pi$

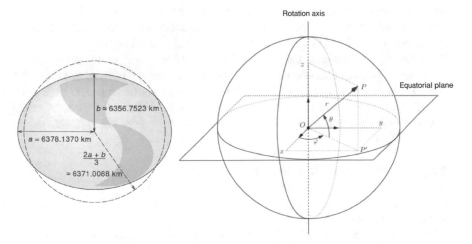

Fig. 14.6 Left: Graphical representation of the Reference Ellipsoid for the Earth. Credits: Creative Commons BY Cmglee-SA 4.0. Right: Graphical representation of pseudo-spherical and Cartesian coordinates of point P

on or near the Earth surface, with the same value of the potential $|\vec{g}|$. It has the shape that the free surface of water (if there were no tides, winds, streams etc.) would take. Because the geoid is defined in the rotating ECEF frame, we must also include the centrifugal potential. By truncating the multipole expansion of $|\vec{g}|$ to the quadrupole moment, a geoid is defined as the surface where

$$\tilde{\Phi}(r, \theta) = -\frac{GM_\oplus}{r}\left(1 - J_2\frac{a}{r}P_2(\cos\theta) + \ldots\right) - \frac{1}{2}\left(\omega_\oplus r \sin\theta\right)^2 \quad (14.6)$$

has a constant value $\tilde{\Phi}_0$. Such value can be computed on the equator:

$$\tilde{\Phi}_0 = \tilde{\Phi}(r = a, \theta = \pi/2) = -\frac{GM_\oplus}{a}\left(1 + \frac{J_2}{2}\right) - \frac{\omega_\oplus^2 a^2}{2}$$

In dimensionless unit, $\tilde{\Phi}_0/c^2 = -6.969 \cdot 10^{-10}$: the centrifugal term accounts for 1/500 of the monopole term, the quadrupole for 1/2000. Higher order terms are, obviously, even more negligible. We shall use the geoid later on in this chapter, dealing with GR timing corrections.

14.5   GPS and Special Relativity

Two main corrections concerning the GPS system directly derive from special relativity: the effect of time dilation and the problem of the simultaneous nature of events.

It is well known that clock in motion, with velocity u with respect to an ECI, runs slower than the time measured by a clock at rest in the ECI, i.e. the proper time,

$$\Delta t = \gamma(v_s)\Delta \tau \tag{14.7}$$

This is the time dilation effect due to the relative velocity, also called the second-order Doppler effect. To have an idea of the quantities involved, for the speed of GPS satellites $v_s = 3874$ m/s, we have $\beta = u/c = 1.3 \cdot 10^{-5}$ and $\gamma(v_s) - 1 = 8.23 \cdot 10^{-11}$. This is the fractional frequency offset needed to compensate for this effect, relative to the rate of clocks on the Earth. The main frequency of the clocks is 10.23 MHz (then upconverted to L band) and the frequency error accumulates over time. In other words, if we do not take this effect into account, within a day we accumulate an error of $\sim$7.2 μs, which results in an error on the receiver position of $\sim$2 km.

14.5.1 Simultaneity and Special Relativity

The revision of the concept of simultaneous events is a famous consequences of the invariance of the speed of light.

Two events, simultaneous in the **R** reference frame, are not such in the **R'** reference frame in motion with respect to **R**. In other words, *simultaneity* is not an absolute concept.

To estimate this relativistic correction, consider the simple case of two satellites A and B that moving along a straight trajectory with constant velocity. **R** is the rest frame of the receiver and **R'** that of the satellites (Fig. 14.7).

A and B now simultaneously, with respect to **R'**, send two radio signals, which will meet, at a later time, at the two satellites midpoint (for **R'**).

Suppose that in **R**, at the time of the emission, the detector was in the midpoint D between the two satellites: however, at the arrival time, the meeting point of the signals in **R** does not coincide with the position of the receiver. This effect is shown in Fig. 14.8, where we consider a simplified two-dimensional space-time diagram. If the reader wonders why the axes of **R'** are tilted with respect to those of **R**, consider the following: the world line of the origin O of **R** is the vertical axis $x = 0$. Conversely, the origin O' of **R'** is described by the equation $x' = 0$. Apply the Lorentz transformation to the spatial coordinate of O': $x' = \gamma(x - \beta ct) = 0$. This shows that its world line is described, in the (x, ct) plane, by the equation: $x/ct = \beta$, i.e. an axis at an angle $arctan(\beta)$ with respect to the ct axis. In the same way, the Lorentz transformation for $ct' = \gamma(ct - \beta x)$ shows that the x' axis $(ct' = 0)$ has inclination $arctan(\beta)$ with respect to the x axis.

In order to compute the detection time, we impose that the norm of $\vec{AD}$ of the Minkowski space be zero:

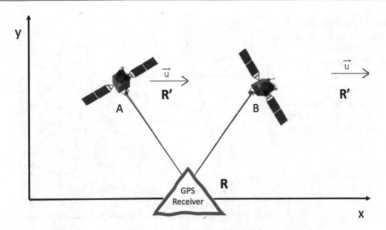

Fig. 14.7 Satellites A and B move along a straight trajectory at constant velocity u relative to the receiver. **R** and **R'** are the rest reference frames of the receiver and satellites

$$c^2(t'_D - t'_A)^2 = (x'_D - x'_A)^2 + (y'_D - y'_A)^2 \qquad (14.8)$$

from which we infer the detection time t'_D:

$$t'_D = t'_A + \frac{1}{c}\sqrt{\frac{1}{4}(x'_B - x'_A)^2 + h^2} \qquad (14.9)$$

where we have introduced the altitude $h = (y'_D - y'_A)^2$ and we have imposed that in **R'** the detection point is the midpoint of the satellites,

$$(x'_D - x'_A)^2 = \frac{1}{4}(x'_B - x'_A)^2 \qquad (14.10)$$

We now impose the Lorentz transformation, with $\beta = \frac{u}{c}$, $\gamma = \frac{1}{\sqrt{1-\beta^2}}$, and derive the position of the receiver in the Earth reference (**R**)

$$x_D = \gamma(x'_D + \beta c \, t'_D) =$$

$$= \gamma\left(x'_D + \beta c t'_A + \beta\sqrt{\frac{1}{4}(x'_B - x'_A)^2 + h^2}\right) \qquad (14.11)$$

This position differs significantly from that of the midpoint $x_M = \frac{(x_A + x_B)}{2}$. Indeed, we can express x_M using $t'_A = t'_B$ and the fact that the detection is at the midpoint $x'_D = x'_M = 1/2(x'_A + x'_B)$:

$$x_M = \frac{1}{2}\gamma(x'_A + \beta c t'_A + x'_B + \beta c t'_B) = \gamma(x'_D + \beta c t'_A) \qquad (14.12)$$

Fig. 14.8 Events represented in the simplified case of two dimension Minkowski space-time where we have drawn the two reference systems **R** (black) and **R'** (blue). A and B are the emission events generated by the satellites, *D* the receiver detection

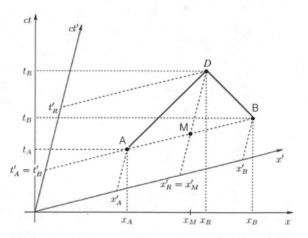

So, the difference between the space coordinates of the detection point x_D and the midpoint x_M is

$$x_D - x_M = \beta \sqrt{\frac{1}{4}(x'_B - x'_A)^2 + h^2}$$

In the case of $h \sim 20200$ km, $x'_A = 0$, $x'_B = 1000$ km, $u/c = 1.3 \cdot 10^{-5}$ $t'_A = t'_B = 0$ we get $x_D = 500.263$ km verses $x_M = 500.007$ km, for an error of the order of 250 m.

14.5.2 Synchronisation and Sagnac Effect

In the simplifying hypothesis that the speed of signal propagation remains constant while crossing the atmosphere, we must also take into account that the receiver is on the Earth, and moving with it. As mentioned before, the equations considered so far hold for an inertial frame: we need to compare the clock time of the satellites with that of the receiver, rotating around the Earth's axis with angular velocity $\omega_\oplus$. At the equator, the linear velocity due to the Earth rotation is $a \cdot \omega_\oplus = 0.464$ km/s, to be compared with the satellite velocity $u \sim 3.9$ km/s. In addition, the propagation time of the signal depends on the position of the satellite relative to the receiver. Localisation of the detector requires receiving signals from at least four satellites: because they arrive from different positions, these signals will not arrive in coincidence: the arrival time of the last signal can be delayed up to to 19 ms with respect to the first, as shown in Fig. 14.9. As discussed above, this is the basis of trilateration, but there is a catch: during this delay time, the Earth rotates approximately 1 μrad and the satellite moves by more than 30 m above a stationary receiver in the ECEF reference frame and more than 60 m in the ECI reference frame. This is a relevant effect, and we need to discuss its contribution.

We thus consider the photon travel time as measured in the non-inertial rotating frame. It is convenient to perform the conversion, from the inertial to the rotating

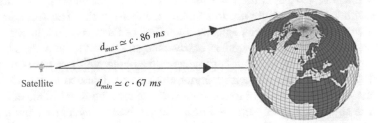

Fig. 14.9 A GPS satellite orbits at an altitude of 20200 km (67 ms). The transmitted signal takes between 67 and 86 ms to propagate to points the Earth surface in the satellite cone of visibility. Credits: public domain image by Rogibert

frame where the variables are marked as primed variables. In cylindrical coordinates, we have:

$$r' = r; \qquad z' = z; \qquad \phi' = \phi - \omega t \qquad t = t'$$

The time transformation $t = t'$ in Eqs. (3) is deceivingly simple. It means that the time variable t' for the rotating frame is really determined in the underlying inertial frame.

In the inertial frame ECI, we write the usual Minkowski metric as

$$ds^2 = (cdt)^2 - dr^2 - r^2 d\phi^2 - dz^2 \equiv (cdt)^2 - d\sigma^2$$

indicating by $d\sigma^2$ the square of the coordinate spatial distance. Upon change of coordinates, the metric becomes:

$$ds^2 = \left(1 - \frac{r'^2\omega^2}{c^2}\right)c^2 dt'^2 - d\sigma'^2 - 2r'^2\omega d\phi' dt' \tag{14.13}$$

Consider now two or more observers in the rotating frame who want to synchronize their clocks using the Einstein procedure (that is, using the principle that c is constant). Photons travel along a null worldline, $ds^2 = 0$. Neglecting second order terms in the small quantity $\omega r'/c$, we can rewrite the previous metric as:

$$c^2 dt'^2 - \frac{2\omega r' d\phi' c dt'}{c} - d\sigma'^2 = 0 \tag{14.14}$$

and solve for dt' to get

$$dt' = \frac{d\sigma'}{c} + \frac{\omega r'^2 d\phi'}{c^2} \tag{14.15}$$

The total time that a photon takes to traverse a given path ℓ is then:

$$\Delta t' = \int_\ell dt' = \int_\ell \frac{d\sigma'}{c} + \frac{2\omega}{c^2} \int_\ell \frac{r'^2 d\phi'}{2} \tag{14.16}$$

The first integral, $\Delta\sigma'/c$, is the time that naive observers, ignoring that they are rotating with the Earth, would use to synchronize their clocks. The second term is the error we would incur, if we neglected the effect of Earth rotation, in computing the propagation time. The integral is easily recognised as the area A_z swept by the position vector along the path ℓ, projected on a plane with $z = const$ (e.g., the equator). Observing from an inertial frame, this can be regarded as the additional travel time required by the light to catch up with the rotating frame. This difference can lead to significant errors: on the Earth, for an observer synchronizing a set of clocks around the equator ($\ell = 2\pi R_\oplus$), this amounts to:

$$\frac{2\omega_\oplus}{c^2} \int_\ell dA_z = \frac{2\omega_\oplus}{c^2}\pi a^2 = 207.4\,ns$$

This result echoes the Sagnac effect, discussed in Chap. 13: Eq. 13.3 has an extra facotr 2 due to the two counterpropagating beams that doubled the effect.

Some authors interpret the Sagnac effect as light not travelling at speed c, without violating the postulate of SR because the effect is analyzed in a non-inertial frame. It is however simpler to consider that the light does not travel, in the ECEF frame, in a straight path, but rather spiralizes due to rotation of the frame.

GPS can be used to compare times on two earth-fixed clocks when a single satellite is in view from both locations. This is the "common-view" method of comparison of primary standards, whose locations on Earth's surface are usually known very accurately in advance from ground-based surveys. Signals from a single GPS satellite in common view of receivers at the two locations provide enough information to determine the time difference between the two local clocks. The Sagnac effect is very important in making such comparisons, as it can amount to hundreds of nanoseconds, depending on the geometry. In 1984 GPS satellites 3, 4, 6, and 8 were used in simultaneous common view between three pairs of Earth timing centers, to perform an around-the-world Sagnac experiment. The centers were the National Bureau of Standards (NBS) in Boulder, CO, Physikalisch-Technische Bundesanstalt (PTB) in Braunschweig, West Germany, and Tokyo Astronomical Observatory (TAO). The size of the Sagnac correction varied from 240 to 350 ns. Enough data were collected to perform 90 independent circumnavigations. The actual mean value of the error, obtained after adding the three pairs of time differences, was 5 ns, which is less than 2% of the magnitude of the calculated total Sagnac effect (Allan 1985).

14.6   GPS and General Relativity

Satellites orbit at a mean altitude $h = 20184$ km and are subject to a weaker gravitational field than a receiver on Earth. In previous chapters, we discussed experiments that highlight how time slows down as the gravitational field grows. Therefore, time runs faster for a clock placed in a satellite than in a receiver on Earth. It is therefore necessary to slow down the clocks of satellites, in order to ensure that the frequencies of the signals sent correspond, at the reception, to those of the local receiver clock.

The difference in gravitational potential $\Delta\Phi$ between any satellite at altitude h and the receiver on the Earth is[6]

$$\Delta\Phi = -GM_\oplus \left(\frac{1}{R_\oplus + h} - \frac{1}{R_\oplus} \right) \tag{14.17}$$

According to General Relativity, the emission (on spacecrafts) time intervals Δt_e clock is related to the detection (on Earth) time intervals Δt_D as

$$\Delta t_e = \Delta t_D \left(1 + \frac{\Delta\Phi}{c^2} \right) \tag{14.18}$$

that is, a change in frequency

$$\frac{1}{\Delta t_e} \simeq \frac{1}{\Delta t_D} \left(1 - \frac{\Delta\Phi}{c^2} \right) \tag{14.19}$$

The relative change in frequency is then $-5.284 \cdot 10^{-10}$ for the $h = 20184$ km altitude.

This implies that, in a day, the atomic clocks of satellites, oscillating at 10.23 MHz, would slow down by a staggering 45.705 μs. In the absence of a correction, this we would cause a positional error of 13 km/day, cumulating with time.

We note that the Special Relativity correction related to time dilation (Eq. 14.7), and that associated with General Relativity, have opposite signs. Therefore, they partially compensate.

Summarizing, the total correction is:

$$\frac{1}{\Delta t_e} = 10.23 \cdot (1 - 5.284 \cdot 10^{-10} + 8.229 \cdot 10^{-11})$$

$$= 10.2299999954363 \text{ MHz} \tag{14.20}$$

In one day the clocks have to slow down by a staggering 38.54 μs to avoid a positional error of ~11.5 km.

In fact, to correct these two relativistic effects, it is enough to tune the atomic clocks, before launch, at a frequency of 10.2299999954363 MHz so that, once in orbit, this frequency is perceived on Earth as 10.23 MHz. This accuracy in the definition of frequency should not be surprising: it comes from the requirement of locating the position of objects to better than 1 m, and this implies having a timing error less than 3 nanoseconds.

---

[6] We neglect it here, for sake of simplicity, the contribution from the Earth quadrupole moment

$$\Phi(r, \theta) = -\frac{GM_\oplus}{r} \left(1 - J_2(\frac{R_\oplus}{r}) P_2(sin\theta) \right)$$

that is usually included in the computation.

We conclude by pointing out that, in principle, the time correction for gravitational effects is not constant, because on one hand, the Earth gravitational field is not uniform; on the other hand the satellites are also exposed to time-dependent gravitational fields of external bodies, such as the Sun and, to a larger extent, the Moon. However, the error associated to these contributions are of the order of a centimetre.

14.6.1 Relativistic Corrections: Another Approach

These results can also be derived using the simpler argument by Mehr Un Nisa. We will analyze the ticking of clocks using the Schwarzchild metric, seen in Chap. 5. Consider a coordinate time, at rest at infinity (with the distant stars), and the proper time ticking with the GPS satellite τ_s and with the receiver at rest on the Earth τ_D. For each of these we can write the metric as:

$$ds^2 = \left(1 - \frac{2GM_\oplus}{rc^2}\right)(cdt)^2 \left(1 - \frac{2GM_\oplus}{rc^2}\right)^{-1} dr^2 \qquad (14.21)$$
$$- r^2(d\theta^2 + \sin^2\theta d\phi^2) = (cd\tau)^2$$

Both the receiver D and the satellite S move at constant distance from the Earth ($R_\oplus$ and $r_s = R_\oplus + h$, respectively) on almost circular orbits, and therefore we can set $dr = d\theta = 0$. We can divide Eq. 14.21 by $(cdt)^2$ and write two equations, for both the satellite and the detector proper times:

$$\left(\frac{d\tau_s}{dt}\right)^2 = \left(1 - \frac{2GM_\oplus}{r_s c^2}\right) - r_s^2 \sin^2\theta_s \left(\frac{d\phi_s}{cdt}\right)^2 \qquad (14.22)$$

$$\left(\frac{d\tau_D}{dt}\right)^2 = \left(1 - \frac{2GM_\oplus}{R_\oplus c^2}\right) - R_\oplus^2 \sin^2\theta_D \left(\frac{d\phi_s}{cdt}\right)^2 \qquad (14.23)$$

In the last term, we easily identify $r \sin\theta \, d\phi/dt$ as the velocity along the circular orbit. Dividing the two above equations we get:

$$\left(\frac{d\tau_s}{d\tau_D}\right)^2 = \frac{1 - \frac{2GM_\oplus}{r_s c^2} - (\frac{v_s}{c})^2}{1 - \frac{2GM_\oplus}{R_\oplus c^2} - (\frac{v_D}{c})^2} \qquad (14.24)$$

This equation gives the ratio of the proper times of the clocks at two different positions from the centre of the Earth.

We can separately analyse the kinematic effect, or second order Doppler shift:

$$\frac{d\tau_s}{d\tau_D} = \sqrt{\frac{1 - (\frac{v_s}{c})^2}{1 - (\frac{v_D}{c})^2}} \simeq 1 - \frac{v_s^2}{2c^2} + \frac{v_D^2}{2c^2} = -8.23 \cdot 10^{-11}$$

and the gravitational blueshift, due to the lower potential in orbit:

$$\frac{d\tau_s}{d\tau_D} = \sqrt{\frac{1 - \frac{2GM_\oplus}{r_s c^2}}{1 - \frac{2GM_\oplus}{R_\oplus c^2}}} \simeq 1 - \frac{GM_\oplus}{r_s c^2} + \frac{GM_\oplus}{R_\oplus c^2} = +5.28 \cdot 10^{-10}$$

Thus, we end up with the same corrections derived in the previous sections.

To be more refined, we should substitute the last term (the Earth monopole potential) with the potential on the geoid $\tilde{\Phi}_0/c^2$ introduced in Sect. 14.4. Numerically, it makes little difference (0.2%) but it is important to remark that all clocks on the geoid tick at the same frequency.

As usual, life is more complicated than this: we did not take into account, in this simplified analysis, several issues like the ellipticity (small, but non zero) of the satellite orbits, the non-sphericity of the Earth, the velocity of receiver with respect to the ECEF (negligible for a hiker, less for a car, definitely not for an airplane), the Shapiro time delay and the ray bending due to Earth, time dependent perturbations due to the Moon and other planets, variations in the Earth rotation rate... All these perturbations, discussed in Ashby (2003) are responsible for corrections below one meter, and can be neglected in a primer like this chapter.

References

Ashby N. *Relativity in the Global Positioning System* - Living Rev. Relativity **6**, 1 (2003). http://www.livingreviews.org/lrr-2003-1

Allan, D.W., Weiss, M., and Ashby, N. *Around-the-World Relativistic Sagnac Experiment* - Science, **228**, issue 4695, pp. 69-70 (1985)

Correction to: Experimental Gravitation

Correction to:
F. Ricci and M. *Bassan, Experimental Gravitation*, Lecture Notes in Physics 998, https://doi.org/10.1007/978-3-030-95596-0

In original version of the book, the following belated corrections received from the author have been incorporated in respective chapters and Appendix at backmatter.

Chapter 2

Equation 2.27 has been removed and remaining equations are renumbered

Figure 2.8 has been replaced with revised figure

Chapter 7

In Equation 7.20 "(TT)" has been removed from equation

In page 161 the in line equation modified ($10^{11} - 10^{15}$ T) to ($10^7 - 10^{11}$ T)

Chapter 9

In Page 215, few lines has been replaced with updated as in below:

The updated versions of these chapters can be found at
https://doi.org/10.1007/978-3-030-95596-0_1,
https://doi.org/10.1007/978-3-030-95596-0_2,
https://doi.org/10.1007/978-3-030-95596-0_3,
https://doi.org/10.1007/978-3-030-95596-0_4,
https://doi.org/10.1007/978-3-030-95596-0_5,
https://doi.org/10.1007/978-3-030-95596-0_7,
https://doi.org/10.1007/978-3-030-95596-0_9,
https://doi.org/10.1007/978-3-030-95596-0_12

From

"In the DL case this difference may be due to different values of the curvature radii of the mirrors of the two cavities while in the FP interferometer, different values of finesse in the two FP cavities, or due to differences in either the radius of curvature or the reflectivity of the mirrors, so one can place even more demanding conditions on the reduction of frequency noise of laser light."

To

"This difference is due, in both cases, to asymmetries in the practical implementation of the two arms: unequal curvature radii for the DL, unequal finesse (that depends on both the curvature radius and the reflectivity of the mirrors) in the case of the FP cavities. This asymmetry places even more demanding conditions on the reduction of frequency noise of laser light." In page 218 line has been removed "Although an entire chapter of this text is devoted to it"

In addition, some minor corrections have been made throughout the book that does not change the basic facts. The correction chapters and the book has been updated with the changes.

The Correction chapters and the book has been updated with the changes.

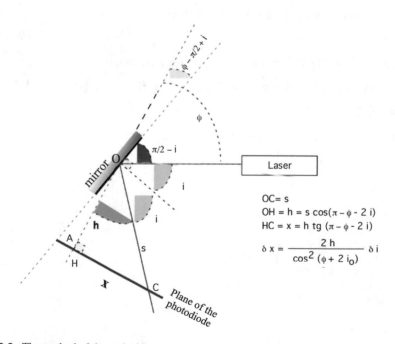

Fig. 2.8 The method of the optical lever

Modulation Techniques

A

A.1 Introduction

Modulation is a technique applied to electromagnetic signals before transmission, in order to adapt them to the characteristics of the communication channel, while keeping the information unchanged. For example, a human voice has a frequency content that does not exceed 5 kHz: if we tried to directly transmit the voice, we would transform it by a microphone into an electrical current with the same spectral components, irradiate it through an antenna and receive it on another, distant antenna. Transmission of electromagnetic waves of such low frequencies would have a number of unacceptable consequences:

- The size of the antennas would be enormous: the optimum length of an antenna is $\lambda/2$, and the wavelength corresponding to a frequency of 5 kHz is $\lambda = c/5$ kHz $\simeq 60$ km. Therefore the antennas, to have a good efficiency, would have to be 30 km long.
- The power required to power an antenna of this size would be enormous.
- The transmitter would be heavy and voluminous. The frequency band would be the same (0–5 kHz) for all users. In other words, the e.m. channel, in the absence of modulation, would be unique: all the users in the world would talk and listen at the same time on the same channel, making any communication absolutely impossible.
- The communications being de facto public, there would not be, and there could not be, any form of "IT privacy".

From these considerations it emerges the need to translate the signal in frequency, allocating in different channels the transmissions of different users, overcoming the above-listed disadvantages. As we will discuss in the following section, by transmitting in *Frequency Modulation* (FM), i.e. around 100 MHz, the length of the antennas is 75 cm, the required power is much lower and the size of the transmitter is minimal.

© The Editor(s) (if applicable) and The Author(s), under exclusive license to Springer Nature Switzerland AG 2022
F. Ricci and M. Bassan, *Experimental Gravitation*, Lecture Notes in Physics 998,
https://doi.org/10.1007/978-3-030-95596-0_A

Using different values of the frequency ratio factor (the ratio between the signal and the carrier frequencies), we can use different frequencies for each transmission, which can then occur at the same time.

In essence, the method of modulation allows the characteristics of the signal spectrum to be transmitted, adapted in such a way to pass well through the channel and, at the same time, allow the *multiplexing* process, i.e. the transmission of many signals on the same channel without interference.

A modulation process is characterized by two frequency components

- the information to be transmitted in the form of an electrical current or voltage: it is called *modulating wave*
- the other wave, at much higher frequency, called *carrier* allows the modulating signal to be translated in frequency

The modulation operation therefore needs a modulator, an electronic device capable of transferring the signal in frequency while keeping the information to be transmitted unchanged.

The operations associated with modulation can be carried out in a number of different ways, depending on whether the signal is analog or digital, and whether the channel is an electric cable, an optical fibre or the vacuum. A classification of the various types of modulations is shown in Fig. A.1.

In the following two paragraphs we will focus on the two main modes of analog modulation.

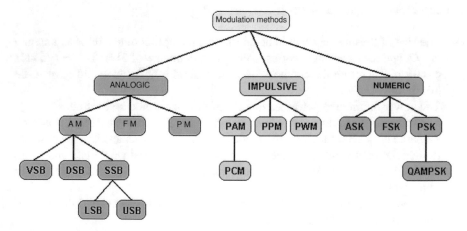

Fig. A.1 Classification scheme of the various modulation modes

A.2   Amplitude Modulation

The standard definition of amplitude modulation (AM) is:

The amplitude of the carrier signal varies in accordance with the instantaneous amplitude of a low frequency modulating signal. It is carried out using an electrical signal (typically a sine wave) with a frequency $f_c = \omega_c/2\pi$ in the typical range used in radio transmissions ranging from 140 kHz (long waves) up to 700 THz (optical fibre transmission). Each transmitting institution (e.g. a radio station) is assigned a carrier frequency and a bandwidth: this is called the channel.

The carrier propagates to a distant receiver the information contained in the low frequency signal, called the modulator. We shall assume, for ease of calculation and without losing generality, that also the modulator is a sine wave of frequency $f_m = \omega_m/2\pi$. A sinusoidal modulation is not very interesting, because it has only two bits of information: its amplitude and frequency; nevertheless, Fourier's theorem assure us that any periodic (or even aperiodic) signal can be decomposed in the sum of a number, often infinite, of sinusoidal waves.

In Fig. A.2 the three signals shown are, from top to bottom:

- the low frequency modulating signal
- the high frequency carrier and finally
- the modulated signal: it has the frequency of the carrier, but its amplitude varies according to the modulator. The periods and amplitudes of the three signals are also indicated.

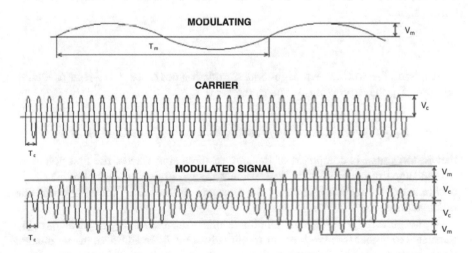

Fig. A.2 The three waveforms: the modulating signal, the carrier wave, the modulated signal

Fig. A.3 The three graphs represent the Fourier components of the three signals: the modulator, the carrier, the modulated signal

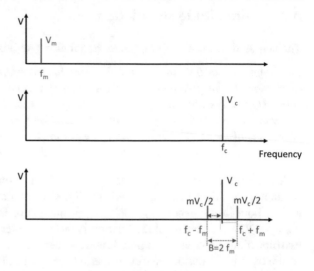

Formally we represent the AM modulation as follows

$$v_m(t) = V_m cos\omega_m t \quad v_c(t) = V_c cos\omega_c t \tag{A.1}$$

are the modulating and carrier signals, respectively. The resulting modulated signal is:

$$v_{AM}(t) = (V_m cos\omega_m t + V_c)cos\omega_c t = (mcos\omega_m t + 1)V_c cos\omega_c t \tag{A.2}$$

Here we introduced the modulation index m (also known as the modulation depth), defined as the ratio of the amplitude of the modulating signal to the amplitude of the carrier signal

$$m = \frac{V_m}{V_c}$$

Applying the well-known trigonometric relation $\cos \alpha \cos \beta = \frac{1}{2}[\cos(\alpha - \beta) + \cos(\alpha + \beta)]$, the expression A.2 takes the form

$$v_{AM}(t) = V_c \left[cos\omega_c t + \frac{m}{2} \cos(\omega_c - \omega_m)t + \frac{m}{2} \cos(\omega_c + \omega_m)t \right]$$

that is, the signal is composed of the sum of three sine waves: the first coincides with the unperturbed carrier, while the others are two sinusoids of amplitude $m\, V_c/2$ oscillating at frequencies given by the sum and difference of the carrier and the modulation frequencies (Fig. A.3).

We note that the process of modulation results in the modulating signal at f_m being translated to higher frequency by an amount given by f_c. In addition, the modulated signal extends over a bandwidth that is twice the f_m modulating frequency.

The modulation index m usually ranges between 0 and 1. Equation A.2 shows that $m = 0$, means no transmission of modulated signal, while still engaging the channel with the carrier. When $m > 1$ we observe at the output a strong distortion of the

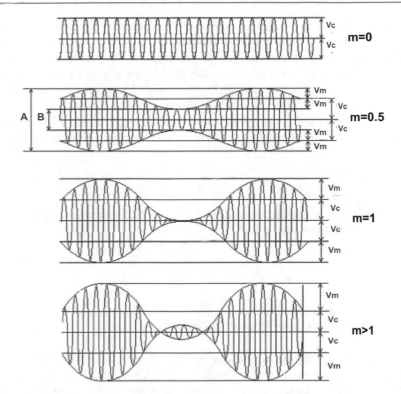

Fig. A.4 In the plots we show the signal in the time domain $v_{AM}(t)$ for the four cases $m = 0, 0.5, 1$, and for $m > 1$ (overmodulation)

modulated signal (*crossover*) as shown in Fig. A.4, since the amplitude modulated signal consists of three distinct components in the frequency domain, its total time averaged power is the sum of the powers of the three contributions:

$$P_{AM} = P_c + P_{c-m} + P_{c+m} \qquad (A.3)$$

where $P_i = V_i^2/2Z \quad (i = c, c + m, c - m)$ is the power of the carrier and of the right and left sidebands, and Z is the radiation impedance of the transmitting antenna. We get this simple form because the cross products have null time average. We observe that

$$P_{c+m} = P_{c-m} = P_c \frac{m^2}{4}$$

and deduce the overall power of the AM modulated signal, with a modulation index m:

$$P_{AM} = P_c \left(1 + \frac{m^2}{2}\right) = \frac{V_c^2}{2Z}\left(1 + \frac{m^2}{2}\right) \qquad (A.4)$$

Note that the strength of the AM modulated signal grows with the modulation index m. We also note that, of the three spectral lines composing the modulated signal, only the two sidebands contain the transmitted information (the same in both side-bands), while the third, more intense, is the carrier which has no information content. This observation suggests that amplitude modulation has a low informational perfor-mance: to transmit one of the two sidebands that alone contains all the information, we are forced to transmit two more lines, one of which, with much larger power.

We define the modulation yield as the ratio of the power of the transmitted infor-mation signal, contained in only one of the two side lines, to all the power that must be transmitted, which is due to all the three lines. By applying the relationship A.4, we have

$$\eta = \frac{P_{c-m}}{P_{AM}} = \frac{m^2}{2(m^2 + 2)} \tag{A.5}$$

Equation A.5 gives us the range of available performance as we vary m between its limit values of m, $0 < m < 1$:

$$0 < \eta < \frac{1}{6}$$

A.3   Frequency and Phase Modulations

A more modern method used in radio is the *Frequency Modulation* (FM).

In particular, transmission of stereophonic sound must use FM. The set of fre-quencies of the stereophonic band is conventionally defined by the FCC[1] : it corre-sponds to the interval $B = 30$ Hz-5 kHz, which widely overlaps with the sensitive band of the human ear $B \simeq 20$ Hz $- 20$ kHz. On the contrary, AM transmissions, which are progressively being phased out, are limited to a narrower bandwidth, with a maximum frequency of 5 kHz, similar to the band of telephone transmissions ($B = 300$ Hz $- 3.4$ kHz) that does not allow to faithfully transmit music: the sound of a violin, for example, has a spectrum reaching 9 kHz.

The FM (like the AM) features:

- a modulating signal, or modulator, $v_m(t)$
- a carrier $v_c(t) = V_c \cos \omega_c t$.

The modulator is usually a periodic signal that can be expanded in Fourier series. Just as we did in the previous case of AM transmission, we shall consider for simplicity's sake just the lowest harmonic of the modulating signal $v_m(t) = V_m \cos \omega_m t$, where we assume that $\omega_c >> \omega_m$.

In frequency modulation, the amplitude of the modulated signal is constant and equal to the amplitude of the carrier V_c. On the other hand, the frequency changes pro-portionally to the instantaneous value of the modulating signal; this can be achieved,

---

[1] FCC: Federal Communications Commission. It is an independent agency of the US Government.

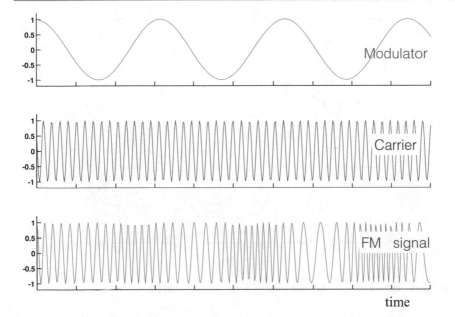

Fig. A.5 Working principle of frequency modulation

for example, using a Voltage Controlled Oscillator (VCO). Mathematically:

$$v_{FM}(t) = V_m \cos \omega_{FM}(t)t \tag{A.6}$$

where

$$\omega_{FM}(t) = \omega_c + K_F V_m cos\omega_m t = \omega_c + 2\pi\Delta f \ cos\omega_m t \tag{A.7}$$

where K_F is a normalization constant that renders $K_F V_m$ in units of [rad/s]. The *peak frequency deviation* Δf is the maximum distance from the carrier frequency:

$$\Delta f = \frac{K_F V_m}{2\pi} \tag{A.8}$$

in FM radio systems we have $\Delta f = 75\,\text{kHz}$ (Fig. A.5).

The modulated signal phase $\phi_{FM}(t)$ is a function of time, obtained by integrating over time the periodic function $\omega_{FM}(t)$:

$$\phi_{FM}(t) = \int \omega_{FM} dt =$$

$$= \int (\omega_c + K_F V_m cos\omega_m t) dt = \omega_c t + \frac{K_F V_m}{\omega_m} sin\omega_m t \tag{A.9}$$

This shows that the modulation index in this case is

$$m = \frac{K_F V_m}{\omega_m} = \frac{\Delta f}{f_m} \tag{A.10}$$

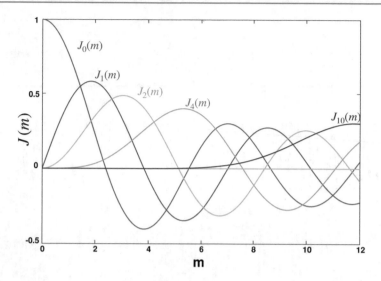

Fig. A.6 Bessel functions of the first kind, of order 0, 1, 2, 4 and 10

Summarizing, the FM modulated signal has the form:

$$v_{FM}(t) = V_c cos\phi_{FM}(t) = V_c cos(\omega_c t + m sin\omega_m t) \tag{A.11}$$

This periodic functions can be expanded in series of Bessel functions (Fig. A.6):

$$
\begin{aligned}
v_{FM}(t) = V_c \cos \phi_{FM}(t) = V_c cos(\omega_c t &+ m sin\omega_m t) = \\
V_c \Big\{ J_o(m) sin\omega_c t \quad &+ \\
J_1(m)[sin(\omega_c + \omega_m)t &- sin(\omega_c - \omega_m)t] + \\
J_2(m)[sin(\omega_c + 2\omega_m)t &+ sin(\omega_c - 2\omega_m)t] + \\
J_3(m)[sin(\omega_c + 3\omega_m)t &- sin(\omega_c - 3\omega_m)t] + \\
J_4(m)[sin(\omega_c + 4\omega_m)t &+ sin(\omega_c - 4\omega_m)t] + \cdots \Big\}
\end{aligned} \tag{A.12}
$$

where $J_i(m) \leqslant 1$ is the Bessel function of ith order.

A.3.1   The Spectrum of the Modulated Signal in FM

It appears from Eq. A.12 that the modulated signal spectrum is symmetric with respect to ω_c and consists of a series of lines with equal spacing ω_m (Fig. A.7). The amplitude

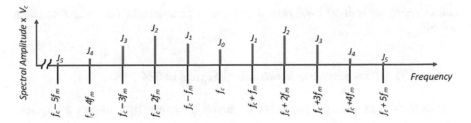

Fig. A.7 Typical spectrum of the FM modulated signal

of each line is given by the value $J_i(m)$ of the Bessel function of the ith order for the given value m of the modulation index.[2]

Let us consider the following practical example: we assume $f_c = 100\,\text{MHz}$, $f_m = 15\,\text{kHz}$, $\Delta f = 45\,\text{kHz}$, $V_c = 100\,\text{V}$, which corresponds to $m = 3$, as given by Eq. (A.10). From the Bessel function diagram of Fig. A.6 we find, in correspondence with the chosen value m, from the cross point with all the curves J_0, J_1, J_2, ..., multiplied by V_c, we obtain the corresponding voltage amplitude of the spectral lines.

In FM the bandwidth of a modulated signal is defined on the basis of the set of spectral lines having a appreciable amplitude. A typical threshold is to consider as detectable the lines as large as 1% of the carrier in absence of modulation: this has a normalized amplitude $v_{FM}^{max}/V_c = J_0(m = 0) = 1$.

In our example, it corresponds to exclude the contribution of all Bessel functions with order $i \geqslant 6$, leaving ± 5 sidebands separated by $f_m = 15\,\text{kHz}$. In general, the FM bandwidth results to be

$$B = 2N f_m \tag{A.13}$$

i.e. being $N = 5$ we have $B = 150\,\text{kHz}$.

In the case when the modulating signal has a continuous or broad spectrum of frequencies, the bandwidth required for transmission can be approximatively estimated by **Carson's bandwidth rule**

$$B_C = 2(\Delta f + f_{m_{max}}) \tag{A.14}$$

where Δf is peak frequency deviation, see Eq. A.8, and $f_{m_{max}}$ is the highest frequency in the Fourier spectrum of the modulating signal. This formula best approximates the value computed by Eq. A.13 for larger m values. Indeed, applying it to the previous example where we have $\Delta f = 45\,\text{kHz}$ e the $f_{m_{max}} = f_m = 15\,\text{kHz}$, the bandwidth calculated using the Carson rule is $B_C = 120\,\text{kHz}$, which is 20% lower than the number computed using Eq. A.13. A more fitting example is FM radio broadcast that, with $\Delta f = 75$ kHz allows transmission of frequencies up to 53 kHz. Equation A.14 then yields $B_C = 256\,\text{kHz}$. In principle, radio stations are assigned

---

[2] The Bessel functions (of the first kind) are a family of special functions, solution of the Laplace equation in cylindrical coordinates. The numerical values (just as the values of trigonometric functions) are tabulated but also directly available in modern computing tools.

carrier frequencies distant 400 kHz, in order to accomodate the full bandwidth, with margin.

A.3.2   The Power of the Modulated Signal in FM

Just as in the AM case, the power transmitted is computed by summing all squared amplitudes in the signal:

$$P_c = \frac{V_c^2}{2Z}\left[J_o{}^2(m) + 2\sum_i J_i{}^2(m)\right] \tag{A.15}$$

To this purpose, we can use an important sum rule for Bessel functions[3] :

$$J_o{}^2(m) + 2\sum_i J_i{}^2(m) = 1 \tag{A.16}$$

This relation, applied to Eq. A.12 shows that the energy of a transmitted signal is conserved, and the choice of m just changes the way it is distributed among the various frequency lines. It is therefore advantageous to minimize the amount of power contained in the carrier that transmits no information. This is achieved by choosing a value of m where J_0 vanishes: the first zeros of J_0 are found for $m = 2.4,\ 5.5,\ 8.7,\ 11.8\ldots\ldots$. In this condition of *carrier suppression*, the modulation yield (extending the definition Eq. A.5 to all the bands on one side of the carrier) reaches $\eta = 50\%$.

A.4   The Mixer

Mixers are non-linear devices, used in telecommunications systems, that accept two input signals of different frequency and produce, at the output, signals oscillating at various combination of the two frequencies:

- the sum of the frequencies of incoming signals
- the difference between the frequencies of incoming signals
- both original incoming frequencies (usually these latter components are undesired and filtered out).

The analysis of frequencies generated by a mixer can be used to translate signals between bands, or to modulate and demodulate them.

The incoming signals are, in the simplest case, sine waves of voltage of the type

$$v_i(t) = V_i sin(2\pi f_i t + \phi_i), \qquad i = 1, 2$$

---

[3] This relation is the "Bessel equivalent" to $sin^2(\alpha) + cos^2(\alpha) = 1$ for trigonometric functions.

and one way to obtain sum and difference of these frequencies f_i is to multiply the two signals together. In fact, by applying the trigonometric identity

$$\sin\alpha\,\sin\beta = \frac{1}{2}[\cos(\alpha - \beta) - \cos(\alpha + \beta)] \tag{A.17}$$

and assuming, for simplicity, that the two signals are in phase ($\phi_1 = \phi_2 = 0$), we have

$$v_1(t) \cdot v_2(t) = \frac{1}{2}[\cos 2\pi(f_1 - f_2)t - \cos 2\pi(f_1 + f_2)t] \tag{A.18}$$

producing both the sum frequency ($f_1 + f_2$) and the difference ($f_1 - f_2$).

In order to multiply two signals, we need a non-linear element. In the radio frequency band this is usually a simple diode. Diodes are characterized by a voltage-current $v - i$ relationship

$$i = i_o\left(e^{\frac{qv}{nkT}} - 1\right) \tag{A.19}$$

In the case of small voltage signals $v(t)$ applied to the diode input, we expand around zero the exponential in Eq. A.19

$$e^x - 1 \simeq x + x^2/2 + \cdots$$

If we now apply both signals at the device input $v(t) = v_1(t) + v_2(t)$, the output voltage (on a resistive load) will exhibit

$$v_{out}(t) = (v_1 + v_2) + \frac{1}{2}(v_1 + v_2)^2 + \ldots$$

$$= v_1 + v_2 + \frac{1}{2}(v_1{}^2 + v_2{}^2) + v_1 v_2 + \ldots \tag{A.20}$$

The two linear terms v_1 and v_2 contribute to the output with signals at the f_1 and f_2 frequencies, the quadratic terms v_1^2 and v_2^2 give rise to a (time averaged) constant term and terms at $2f_1$ and $2f_2$, while the product $v_1 v_2$ gives rise to signals at the sum and difference of f_1 and f_2. Then, the frequency band of interest must be filtered through an appropriate band pass circuit.

For example, in the case that $f_1, f_2 \gg |f_1 - f_2|$ (two signals with relatively close frequencies), the difference signal will be easily filtered compared to all other components because it oscillates at a much lower frequency than the others: in radio receivers, the mixer is used to extract this slow signal. In this way, a high frequency signal is demodulated down to audio frequencies.

A.5   Heterodyne and Homodyne Receivers

The *heterodyne*, also called the Armstrong regenerator, is a signal processing technique, largely used in radio equipment that shifts one frequency range into another (shifting up in transmission, down in reception). It was anticipated in 1896 by Nikola Tesla, developed by Reginald Fessenden in 1901 and perfected in 1913 by Edwin Armstrong during his studies on the operation of the triode. Modern implementations of this principle, taking advantage of the mixer, are called *super-heterodyne*.

The signal from the antenna is processed by a broadband radio frequency (RF) filter centred on a frequency f_s and then fed to the input of a mixer together with a signal generated by a local oscillator at a f_{LO} frequency such that their difference is a preset intermediate frequency:

$$f_i = |f_{LO} - f_s| \tag{A.21}$$

When the tuning command is activated on a super-heterodyne receiver, the value of both the frequency f_s of the first filter and that of the local oscillator f_{LO} are changed to keep Eq. A.21 true. In this way the mixer returns an output with the information content of the signal detected by the antenna, but with frequencies shifted around f_i. At this frequency, a narrow band filter operates by selecting the desired channel and sending the signal to the demodulators.

Another type of receiver is the *homodyne*, also known as *synchrodyne*, or direct-conversion receiver or zero-IF receiver. In this case, the receiver demodulates the incoming signals by mixing them with the local oscillator's signal **tuned to the carrier frequency**. The signal containing the useful information is obtained by filtering the mixer output with a suitable low pass filter. The homodyne was developed in 1932 by a group of British scientists looking for a method to replace super-heterodyne. This new type of receiver not only performs better, but also requires simpler circuitry, with lower power consumption. The potential limit of this method is due to the stability of the local oscillator signal. In fact, it may happen that the circuit becomes unstable in the long term if there is a slow drift of the reference frequency of the local oscillator.

In optical interferometry, we use the same expression: homodyne detection. In a perfect analogy with the case of radio frequency circuits, the method is based on the concept of deriving the reference signal (i.e. the local oscillator) from the same source generating the signal before the modulating process takes place. For example, in a measurement of laser light scattered by a sample, the laser beam is divided into two parts by a beam splitter. The first beam is the local oscillator, the other is sent to the diffusive sample. The scattered light is then recombined with the local oscillator on a photodetector. This configuration has the advantage of being insensitive to fluctuations in laser frequency.

In essence, the principle of optical homodyne detection is identical to that applied in the radio field: the frequency of the detector output is the same as the reference signal. Of course, the instruments and language are different.

Concerning the heterodyne system in optics, we have a frequency shift of the signal to be detected with respect to the reference. The reference radiation and the signal are superimposed in the mixer, which typically is a photodiode. This responds to the intensity of the incoming beam and therefore its response to the amplitude of the signal is non-linear.

The input electromagnetic field is

$$E_s(t) = E_s cos(\omega_s t + \phi)$$

and that of the local oscillator

$$E_{LO}(t) = E_{LO} cos \omega_{LO} t$$

The output of the detector is proportional to the light intensity that is the square the incident field:

$$I \propto \left(E_s cos(\omega_s t + \phi) + E_{LO} cos \omega_{LO} t \right)^2$$

By expanding this expression we have

$$\frac{E_s^2}{2}(1 + cos(2\omega_s t + 2\phi) + \frac{E_{LO}^2}{2}(1 + cos 2\omega_{LO} t) +$$

$$+ E_s E_{LO} \left[cos[(\omega_s + \omega_{LO})t] + cos[(\omega_s - \omega_{LO})t] \right] \quad (A.22)$$

and by collecting terms in the above equation, we identify the three different contributions:

- the static components $\frac{E_s^2 + E_{LO}^2}{2}$
- the high frequency components: $\frac{E_s^2}{2} cos(2\omega_s t + 2\phi) + \frac{E_{LO}^2}{2} cos(2\omega_{LO} t) + E_s E_{LO} cos[(\omega_{LO} + \omega_s)t + \phi]$
- the beat component: $E_s E_{LO} cos[(\omega_{LO} - \omega_s)t]$

In heterodyne detection, static and high frequency components are filtered, so that only the beat term survives. As it can be seen from Eq. A.22, the amplitude of the latter component is proportional to the amplitude of the input signal E_s.

A.6   The Lock-In Amplifier

A peculiar type of demodulator is the lock-in amplifier, also known as phase sensitive detector: it is an instrument capable of picking a signal of interest, with given frequency and phase, when it is buried in a large noise. It can achieve a Signal-to-Noise Ratio as large as 10^6. It is mainly used in an extended audio band (0–100 kHz), but there are also models working in the MHz range, while mixers are usually employed in radio frequency bands. The other, important difference with respect to a simple mixer is the phase sensitivity of the lock-in.

The lock-in building blocks are as indicated in Fig. A.8:

- an input amplifier, with an optional broad band-pass filter to reject most of the unwanted spectral components, and thus increasing the dynamic range
- a reference input, where a sine wave of the desired frequency is applied. This is fed to an adjustable phase shifter and then to
- an audio mixer, where the two signals are *multiplied* according to Eq. A.17.
- a low-pass filter
- an output amplifier

The input signal has, we assume, a component of interest at the frequency $f_i = \omega_i/2\pi$:

$$v_i(t) = A_i \sin(\omega_i t + \phi_i)$$

The amplitude A_i, much smaller than the rms value of the whole input, is constant, or slowly varying with time. The reference signal is a pure sine wave:

$$v_{ref}(t) = A_{ref} \sin(\omega_{ref} t + \phi_{ref})$$

where ϕ_{ref} can be changed at the operator's will. The instrument is usually adjusted so that the reference amplitude is included in the overall gain G, so that we can neglect it in the following. At the mixer, use of Eq. A.17, with unequal phases, yields

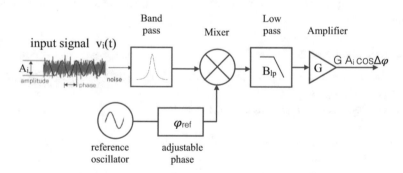

Fig. A.8 Block diagram of a lock-in amplifier. The input amplifier is not shown

$$v_{mixer}(t) = A_i \frac{1}{2}[\cos((\omega_i - \omega_{ref})t + \Delta\phi) - \cos((\omega_i + \omega_{ref})t + \phi_i + \phi_{ref})]$$

$$(A.23)$$

where $\Delta\phi = \phi_i - \phi_{ref}$. The low-pass filter, with a bandwidth B_{lp}, filters away the high frequency (sum) component, while the low frequency (difference) is almost unchanged if $|\omega_i - \omega_{ref}| < 2\pi B_{lp}$. Note that this difference frequency can also be negative: it just means that the phase evolves in the opposite (clockwise) direction. The bandwidth B_{lp} is chosen on the basis of the frequency content of $A_i(t)$, typically between 30 mHz and 10 Hz.

The original signal at ω_i is now translated down to a low frequency $\omega_i - \omega_{ref}$, that can be brought to zero by careful tuning of ω_{ref}. We are left with

$$v_{out} = \frac{G A_i}{2} \cos \Delta\phi \qquad (A.24)$$

Note that v_{out} is no longer an oscillating signal, but a constant voltage, as long as the input amplitude A_i is steady. Adjusting the phase ϕ_{ref} to a maximum of v_{out} will give us both the amplitude A_i and the phase, relative to the reference, of the input signal component $v_i(t)$.

There is a catch we overlooked so far: not always the reference signal is available as a clean sine wave. This can be a complex or noisy signal, derived from another apparatus. For this reason, the reference is conditioned within the lock-in into a square wave, that contains odd harmonics of the base frequency:

$$v_{ref} = A_{ref} \frac{4}{\pi} \left[\sin \omega_{ref} t + \frac{1}{3} \sin 3\omega_{ref} t + \frac{1}{5} \sin 5\omega_{ref} t + \cdots \right]$$

This has two consequences:

1. The signal v_{out} has a large series of terms like those in Eq. A.23, each with a different reference signal, odd multiple of ω_{ref}: however all these frequencies are easily removed by the output low pass filter.

2. If the input signal v_i has frequency components at one of the harmonics of v_{ref}, it will give a contribution at zero frequency, just like the base frequency. The band-pass filter at the input also serves for this purpose, nevertheless care should be applied, when assessing the amplitude A_i, that no tails of higher frequency peaks appear in the detection band.

In conclusion, calling $x(t)$ the final result, after the output low-pass filter, an integrator with time constant $\tau = 1/B_{lp}$, we have

$$x(t) = \frac{4}{\pi\tau} \int_{-\infty}^{t} dt' \, v_{in}(t') \cos(\omega_{ref} t') \, e^{-\frac{t-t'}{\tau}}$$

We can set up a second channel, identical to this, but with the phase of the reference signal shifted by $\pi/2$; we call this additional output $y(t)$:

$$y(t) = \frac{4}{\pi\tau} \int_{-\infty}^{t} dt' \, v_{in}(t') \sin(\omega_{ref} t') \, e^{-\frac{t-t'}{\tau}}$$

Therefore, we have at our disposal both the in-phase (cosine) and quadrature (sine) components of the input signal: $x(t)$ and $y(t)$ are to be considered as the two, time dependent, components of the complex vector, or *phasor*, representing amplitude and phase of the input signal $v_i(t)$. A more common representation of a phasor is in polar coordinates:

$$r(t) = \sqrt{x^2(t) + y^2(t)}; \qquad tg(\Delta\phi) = \frac{y(t)}{x(t)}$$

Lock-ins let the user choose between "cartesian" and "polar" output: in the polar form, the "r(t)" output always gives the magnitude of the input (at the desired frequency) regardless of the phase. This is useful if the phase jitters or drifts in time.

The invention of the lock-in is yet another contribution to experimental physics by the eminent scientist Robert Dicke[4] although he said he had found the idea in a previous paper.

A typical use is in solid-state experiments: the laser beam is modulated in a simple way by a *chopper*: a rotating disc with a slit, so that light is shone on the sample intermittently at the rotation frequency, and the response is expected with the same pace: the generator driving the rotating disc is fed as a reference to a lock-in amplifier, and the output only contains that frequency component.

Lock-ins are somewhat obsolete in this days, as Digital Signal Processing allows to perform the same task in software on a sampled signal. However, they are still used when frequencies are too high for real-time processing. Anyhow, software-based lock-ins work on exactly the operating principle described here.

A.7   Modulation and Controls

In laboratory experiments, modulation is often employed to move the signal of interest away from d.c.: if the experimenter manages to have the signal (e.g. a driving force) at a given frequency, this can then be detected by synchronous demodulation (that is what a lock-in amplifier does). A famous, historical example is the measurements of Equivalence Principle carried out by Roll, Krotkov and Dicke: while Eotvos was measuring the static pull of Earth the two test-masses of his torsion balance, comparing it with a not so well-defined rest position, Roll and collaborators measured the gravitational attraction of the Sun, that varies in the day. In this way, the signal was no longer static, but modulated with a 24 h period. Moreover, in order to move this signal in a more manageable frequency region, (electronic instruments, especially in the 1950s, were subject to relevant drifts), the signal was further modulated to 3 kHz

---

[4] We have encountered prof. Dicke in Chap. 3 (Roll-Krotkov-Dicke experiment of EP) and in Chap. 6 for the Brans-Dicke theory of gravity. He is also credited, among many other things, for predicting the existence of cosmic microwave background and for the correct interpretation of its serendipitous discovery by Penzias and Wilson (who were using a "Dicke radiometer"). He also founded Princeton Applied Research, the first company to produce commercial lock-ins.

by vibrating a wire in front of the photomultiplier. Also the Equivalence Principle experiment of Adelberger et al. (Schlamminger 2008) is modulated by rotating the whole apparatus containing the torsion pendulum.

Today, in most experiments involving optics, the laser light is phase modulated by an Electro Optical Modulator (EOM). EOMs are devices made of crystalline material, like Lithium Niobate, whose refractive index changes with the electric field applied to its sides: the molecules are distorted or change their position and orientation. The change in refraction index determines a change in the optical path length of the light crossing the crystal. Thus, the phase of the laser light exiting an EOM can be controlled the electric field in the crystal. This modulation, unlike anything mechanical, can be very fast: EOM can work at frequencies up to 2 GHz.

Recalling that the light intensity at the output of an interferometer depends on the phase difference of the recombined beams, we achieve amplitude modulation of a laser beam by inserting an EOM in one arm of an interferometer. Therefore, we have

$$\psi_1 = A \sin(\omega t); \quad \psi_2 = A \sin(\omega t + \Delta \ell_{12} + \ell_{eom} n_2(t))$$

where ψ_1 is the field in one arm and ψ_2 the modulated field in the second arm, $\Delta \ell_{12}$ is the static optical path difference, ℓ_{eom} is the length of the EOM, whose refractive index $n_2(t)$ is modulated in time. Therefore the intensity of the combined beams will be

$$I(t) = \frac{c \varepsilon_o A^2}{2}(1 + \cos \Delta \phi_0 + \phi(t))$$

where $\Delta \phi_0$ is a static phase difference and $\phi(t)$ the phase modulation.

Modulation is often a key element of control systems: it is an effective way to stabilize the working point of a non-linear instrument. Take for example the output of a Michelson interferometer (see Chap. 9): the optical path difference might slowly drift, moving the interferometer output away from the dark fringe condition. A fast (kHz to MHz) modulation of the path difference, using an EOM, by a small fraction of a fringe, will produce a symmetric light output, at twice the modulation frequency (the response around a minimum is quadratic) (Fig. A.9). If the working point moves away from the dark fringe, a light wave with a component at the modulation frequency appears: this can be used in a feedback loop to lock the working point onto the dark fringe. The topic of modulation in control system is dealt with in greater detail in Appendix B, dealing with controls.

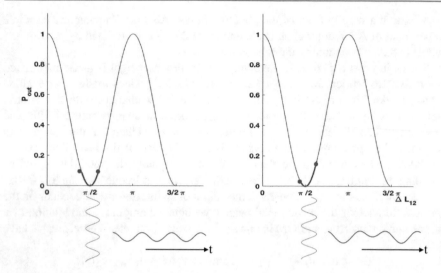

Fig. A.9 In a Michelson interferometer, modulating the optical path difference around the dark fringe ($\Delta\phi = \pi/2$) produces a light signal at twice the modulation frequency. If the operating point drifts (in our figure, to the right) a signal component at the modulation frequency appears that can be used as an error signal to stabilize operations on the dark fringe. In this figure, the amplitude of modulation $A_{mod} = 0.3$ rad is vastly exaggerated for clarity

Reference

Schlamminger, S., Choi, K.-Y., Wagner, T.A., Gundlach, J.H., Adelberger, E.G.: Test of the equivalence principle using a rotating torsion balance. Phys. Rev. Lett. **100**, 041101 (2008)

Feedback and Controls

B

B.1   Introduction

In this chapter we briefly outline the basic principles of feedback systems and the concept of stability, without any pretense of thoroughness. For that, we refer to textbooks of system theory and controls. For a comprehensive tutorial see Bechhoefer (2005).

To explain the meaning of a feedback process, we consider a very general system, called *plant*, which linearly transfers a signal from the input to the output port. This linearity relation is usually expressed in the frequency domain:

$$V^{(o)}(\omega) = A(\omega)V^{(i)}(\omega) \tag{B.1}$$

A is the *transfer function* that relates input and output of the system. $V^{(o)}$, $V^{(i)}$ can be physical quantities of any kind, but in the following we shall focus on electric circuits, so that A represents the *open-loop gain* of the network.

The plant is in a feedback configuration if a fraction of the output signal is measured, processed and added back to the signal input. If the feedback signal is applied to the input with opposite phase with respect to the incoming signal, we have a *negative* feedback.

The crucial advantage of the feedback scheme consist, as we shall see, in an increased stability of the input-output relation (the transfer function) that is, in general, subject to change due to fluctuations, e.g. thermal, of the plant components.

The feedback is a concept that goes far beyond its applications in Electronics. Let us consider an example: imagine you must deliver a glass of water without spilling its content: your eye (the "sensor") continuously monitors the non-horizontally of the water level (the "error signal"), and the brain (the "processor") sends a command to the supporting arm (the "actuator") to correct the tilting of the glass. In this way,

F. Ricci and M. Bassan, *Experimental Gravitation*, Lecture Notes in Physics 998,
https://doi.org/10.1007/978-3-030-95596-0_B

Fig. B.1 A simple scheme of feedback system. A is the open-loop transfer function of the plant; β is the transfer function of the feedback branch

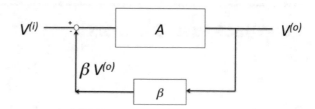

the attitude of the glass (the "plant") is continuously monitored and corrected, to maintain it horizontal and preserve its content.

Nature uses feedback widely and many functions of the human body are carried out by controlled systems (the regulation of body temperature, the rhythm of the heartbeat, the ear sensitivity, etc.).

In Fig. B.1 we show the scheme of a plant characterized by the transfer function A, inserted into a feedback loop. The feedback branch, characterized by the transfer function β, is then added, with inverted sign, to the input signal.

We infer the following equation:

$$V^{(o)} = A\big(V^{(i)} - \beta V^{(o)}\big) \tag{B.2}$$

As a result, the gain of the feedback amplifier A_f is given by the expression:

$$A_f = \frac{V^{(o)}}{V^{(i)}} = \frac{A}{1 + A\beta} = \frac{1}{\beta}\frac{1}{1 + \frac{1}{A\beta}} \tag{B.3}$$

where $1 + A\beta$ is the *feedback rate*.

In the common case $A\beta \gg 1$, we can approximate the closed-loop gain as $A_f = A/(1 + A\beta) \simeq 1/\beta$.

By applying feedback to an amplifier, significant changes in the overall characteristics of the system can be achieved. For example, we can

- stabilize the gain, i.e. make it less sensitive to changes in the values of circuit components (e.g. due to temperature changes)
- adjust the values of the input and output resistances of the amplifier stage, close to the ideal ones
- extend the frequency band of the system.

An example can be useful in showing the effect of feedback on the bandwidth of an amplifier.

Consider an amplifier that, at *open-loop* (without feedback), has a bandwidth of 100 Hz and a very high gain equal to $A = 10^5$. Moreover, for frequencies above 100 Hz, the gain of the circuit decreases by an order of magnitude per frequency decade, as shown in Fig. B.2). By applying an appropriate feedback, i.e. choosing an appropriate β value, we can reduce the gain by a factor 10^3, to get $A_f \simeq 1/\beta = 100$. This will expand the bandwidth by the same amount, i.e. up to 100 kHz. Consider the simple case of feedback with a frequency independent transfer function β. In

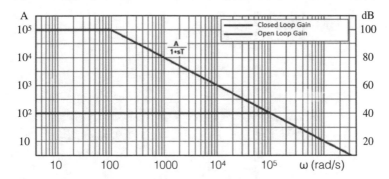

Fig. B.2 Gain curve of a system in open-loop (red) and with feedback (blue)

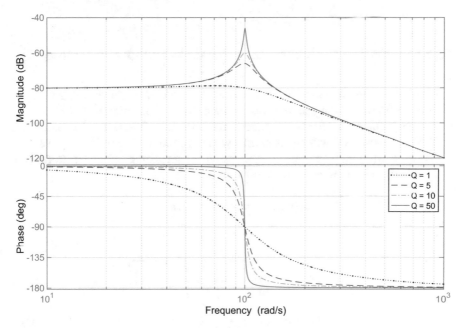

Fig. B.3 Bode plots for a harmonic oscillator, for various values of the quality factor Q: amplitude (top) and phase versus frequency. The plot is zoomed on the resonance frequency $\omega_0 = 100\,$rad/s

the closed-loop configuration the gain is reduced by a factor 1000 and the band is expanded by the same factor. This is because, lowering the gain, we also move the point where the gain starts to drop off, as shown in Fig. B.2: indeed the blue, "closed-loop" line crosses the red, "open-loop gain" curve at $\sim$100 kHz.

Before venturing further into the basics of controls, we need to recall a few concepts and math tools:

- In electronics, it is customary to express the gain in decibel ($A_{dB} = 20 \cdot log_{10}(A)$), to plot the A function over extended frequency ranges. With these units, in the

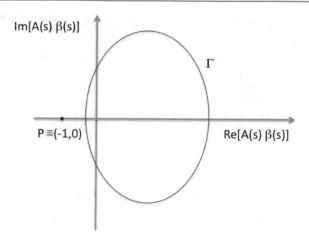

Fig. B.4 Diagram of the Nyquist criterion in the complex plan $\left(\mathrm{Re}[A(s)\beta(s)], \mathrm{Im}[A(s)\beta(s)] \right)$

above example we have open-loop gain $A_{dB} = 20\ log_{10}A = 100$ dB, $A_f \simeq 1/\beta = 40$ dB, and the gain decreases with a slope -20dB/decade.

- A transfer function is a complex function of the angular frequency ω. In particular, in what follows, we will be concerned with the denominator of Eq. B.3: we shall focus on $A(\omega)\beta(\omega)$: this can be expressed in cartesian coordinates $Re(A\beta) + i\ Im(A\beta)$ or in complex polar coordinates:

$$A(\omega)\beta(\omega) = G(\omega)exp[i\phi(\omega)]$$

with

$$G(\omega) = \sqrt{\left\{ \mathrm{Re}[A\beta] \right\}^2 + \left\{ \mathrm{Im}[A\beta] \right\}^2} \tag{B.4}$$

$$\phi(\omega) = arctan\frac{\mathrm{Im}[A\beta]}{\mathrm{Re}[A\beta]} \tag{B.5}$$

- The *Bode plots* is a graph of the frequency response of a transfer function (or, in general, of a linear system). It consists of two plots: magnitude versus frequency and phase versus frequency. Usually, the horizontal axis carries the frequency (either f [Hz], or ω [rad/s]) in logarithmic scale, the amplitude is expressed in dB and the phase in degrees. Figure B.3 shows the Bode plots for a simple harmonic oscillator.
- Other graphical representations of the *function* $A\beta$ exist: while a Bode graph draws two functions (magnitude and phase) versus frequency, one can condense the information in one single figure: the same variables can be plotted in polar coordinates, as shown in Fig. B.7, or in cartesian coordinates, with ϕ on the abscissa and G on the ordinate: this is called a *Nichols plot*. Note that in a polar Bode plot

we must abandon the *dB* units for the amplitude, because a magnitude cannot be negative.

- We can also plot $Im(A\beta)$ versus $Re(A\beta)$, in what is called a *Nyquist plot*, as shown in Fig. B.4. In this plane, varying the variable ω, the function $A(\omega)\beta(\omega)$ describes a path $\Gamma(\omega)$. The path is closed if we include negative frequencies $-\infty < \omega < \infty$.
  We shall see how the stability criteria have different statements depending on the graphical representation.
- In the Laplace transform, the conjugate variable is the complex frequency $s = \sigma + i\omega$. A function $f(t)$ is transformed according to

$$F(s) = \int_0^\infty f(t)e^{-st}dt$$

The Fourier transform is therefore a particular case of the Laplace transform, with $\sigma = 0$. The latter is more suited to describe both periodicity (ω) and damping (when $\sigma < 0$) as well as the response of the plant to transients. In the following we shall consider the function $A_f(s)$.

- A Transfer Function $A(\omega)$ is, most often, a rational function of ω, i.e. the ratio of two polynomials. A polynomials can be always be factored in terms of its roots and, as the roots are in general complex, it is convenient to replace $\omega \to s$. We thus write

$$A(\omega) = \frac{a_n\omega^n + a_{n-1}\omega^{n-1} + \cdots a_0}{b_m\omega^m + b_{m-1}\omega^{m-1} + \cdots b_0}$$

that becomes:

$$A(s) = \frac{(s - z_1)(s - z_2)\cdots(s - z_n)}{(s - p_1)(s - p_2)\cdots(s - p_m)} \tag{B.6}$$

In this expression, z_j $(j = 1\ldots n)$ are the zeros of A and p_k $(k = 1\ldots m)$ are its poles. A dynamical system is therefore completely defined by its zeros and poles. The case with multiple roots is a simple extension of this analysis, but for sake of simplicity we neglect it here. Converting this last equation to decibel we get

$$A_{dB} = \sum_{j=1}^{n} 20\, log_{10}|s - z_j| - \sum_{k=1}^{m} 20\, log_{10}|s - p_k|$$

- Zeros and poles are called the *break points* of the Bode plot: around these points the Bode plots exhibit the largest change in slope. Restricting again to $s = i\omega$, and using a rule of thumb, the amplitude Bode plot can be approximated with a series of straight segments connecting the frequencies of break points: the slope increases by 20 db/decade at each zero, i.e. for $\omega = |z_j|$, and decreases by -20 db/decade at each pole, i.e. for $\omega = |p_k|$.

B.2 Stability Criteria

A plant is stable if, due to the feedback, the output stays close to a given value (e.g. zero) independently of the value of the input. The issue of stability for a system with feedback can be traced back to the analysis of the complex poles of the function $A_f(s)$. A general result of complex analysis states:

to insure the system stability, $A_f(s)$ must not have poles with a positive real component.

Two stability criteria frequently used are those of **Bode** and **Nyquist**, because they are based on easily measured quantities.

We shall use an example to illustrate the criteria and their graphic representation. Consider the two following Transfer Functions:

$$H_1(s) = \frac{15}{(1+s) \cdot (1+s/2) \cdot (1+s/3)}$$

$$H_2(s) = \frac{2}{(1+s) \cdot (1+s/2) \cdot (1+s/3)}$$

they are identical, except for the low frequency gain.[5] In Figs. B.5, B.6 and B.7 the two functions are shown, H_1 in green and H_2 in blue, in the various representations discussed above.

The Nyquist criterion is based on the properties of the denominator of A_f, the analytic function $1 + A(s)\beta(s)$. This function has an obvious zero at $A(s)\beta(s) = -1$, and that is where the system will show instability.

We now plot the above functions $H_{1,2}(\omega) = A\beta$ in a Nyquist plot, see Fig. B.5, as described in the previous section, and state the criterion:

The system is stable if the path Γ does not contain the coordinate point $[-1, i0]$ within it.[6]

The Bode criterion.

We now consider the Bode plots for the function $A(\omega)\beta(\omega)$. As noted above, we have instability if this function takes the value -1, i.e. $G(\omega) = 1$, $\phi(\omega) = -\pi = -180°$. We introduce two particular angular frequencies:

1. The *phase crossover frequency* ω_π where the phase takes the value $-180°$, i.e. the feedback changes sign:
$\phi(\omega_\pi) = -180°$.

2. The *gain crossover frequency* ω_1 where the amplitude gain is unity:
$G(\omega_1) = |A(\omega_1)\beta(\omega_1)| = 1 = 0\,\text{dB}$

Bode stability criterion is expressed in terms of these two frequencies:

---

[5] Spoiler: H_1 is unstable, H_2 is stable, as we shall see.

[6] The criterion is actually a little more cumbersome, counting the number of turns in the clockwise and counter-clockwise direction. This statement, rigorous only for simple poles, contains the essence of the concept.

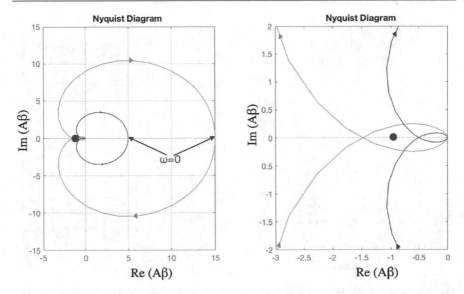

Fig. B.5 Nyquist stability criterion visualized in a Nyquist plot: the green Open-Loop Transfer Function is unstable, as it circles the instability point in clockwise sense as ω increases. In blue a similar, stable system. The symmetric branch at $Im(A\beta) < 0$ is described by negative frequencies. On the right, a zoom around the instability point $[-1, i0]$ (red dot)

If, at the phase crossover frequency, the corresponding gain is less than unity: $G(\omega_\pi) < 0$ *dB, then the feedback system is stable.*

In a stable system, we generally have $\omega_1 < \omega_\pi$.

These two frequencies are also used to define:

- The *gain margin*: find on the phase plot the frequency ω_π, where $\phi = -180°$. Then read on the amplitude plot the value of G at this frequency. The gain margin measures, in dB units, how far $G(\omega_\pi)$ is from the unity value:

$$\text{Gain margin} = 20\,log_{10}G(\omega_\pi) = 20log_{10}(|\beta(\omega_\pi)A(\omega_\pi)|)$$

- The *phase margin*: analogously, find on the amplitude plot the frequency ω_1 such that $G(\omega_1) = 1$. Then, read on the phase plot $\phi(\omega_1)$. If $\phi(\omega_1) > -180°$, the system is stable. The phase margin measures, in degrees, how far $\phi(\omega_1)$ is from $-180°$:

$$\text{Phase margin} = \phi(\omega_1) - (-180°) = 180° - |\phi(\omega_1)|$$

Both phase and gain margins are measures of how close the system comes to instability.

In these terms we can state: *a system is stable if both phase and gain margins are positive*.

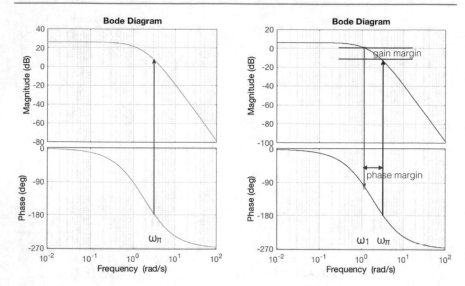

Fig. B.6 Bode stability criteria visualized in Bode plots: on the left the Open-Loop Transfer Function is unstable: the gain at ω_π exceeds unity. On the right a similar, stable system with the phase margin and gain margin. The only difference between the two functions is the low frequency gain

Fig. B.7 Bode stability criteria visualized in a Bode polar plot: the green Open-Loop Transfer Function is unstable, crossing the unit circle at an angle $\phi > -\pi$. The green curve starts (for $\omega = 0$) at a much larger amplitude ($G = 15$, $\phi = 0$, as seen in Fig. B.6), outside the plot margins. In blue a similar, stable system

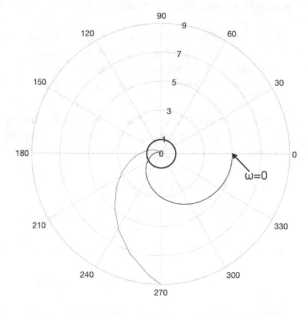

Finally, we can use a Bode polar plot, as in Fig. B.7. The pole of the Nyquist diagram, $[-1, i0]$ in Cartesian coordinates, corresponds to $[1, -\pi]$ in polar coordinates. The stability criterion can be stated as follows:

if the function $A\beta$ crosses the unit circle at a phase angle $-\pi < \phi < 0$, the system is stable.

Indeed, its gain will be less than one at $\phi = -\pi$.

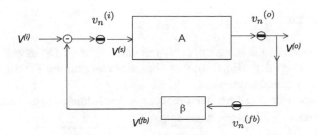

Fig. B.8 A scheme for evaluating noise in a feedback system. We are assuming that there are sources of voltage noise at the input $v_n^{(i)}$, at the output $v_n^{(o)}$ and on the reaction branch $v_n^{(fb)}$

B.3 Noise in a Feedback System

When we compose a feedback system, we add some circuit elements that, invariably, contribute to the overall system noise. To assess this effect, we refer to Fig. B.8, where the two blocks consisting of the amplifier A and the feedback circuit β are shown together with the noise sources that characterize each component. Thus, the circuit includes the noise voltage generators at the input $v_n^{(i)}$, at the output $v_n^{(u)}$ and on the reaction branch $v_n^{(fb)}$; we shall assume these random variables to be uncorrelated.

As before, $V^{(i)}$ is the input voltage, i.e. the signal to be detected, $V^{(o)}$ is the output signal, and $V^{(fb)}$ is the feedback signal. Moreover, $V^{(s)}$ is the voltage at the entry of the plant A, past the feedback node. We can now generalize Eq. B.2, to include the noise generators. We have

$$
\begin{aligned}
V^{(s)} &= V^{(i)} - V^{(fb)} + v_n^{(i)} \\
V^{(o)} &= A V^{(s)} + v_n^{(o)} \\
V^{(fb)} &= \beta (V^{(o)} + v_n^{(fb)})
\end{aligned}
\tag{B.7}
$$

From these equations we deduce the voltage at the output:

$$
V^{(o)} = \frac{A}{1 + \beta A} V^{(i)} + \frac{1}{1 + \beta A} [A v_n^{(i)} - A \beta v_n^{(fb)} + v_n^{(o)}]
\tag{B.8}
$$

The first term is the signal, conditioned by feedback as seen in Sect. B.1; the following terms show how the various noise voltages appear at the output, due to the effect of feedback. We now let $V^{(i)} = 0$, to focus on the total noise at the output:

$$
v_n^{(tot)2} = \left(\frac{A}{1 + \beta A} \right)^2 \left(v_n^{(i)2} + \frac{v_n^{(o)2}}{A^2} + \beta^2 v_n^{(fb)2} \right)
\tag{B.9}
$$

In the usual hypothesis $\beta A \gg 1$, the expression takes a simpler form:

$$
v_n^{(tot)2} = \frac{v_n^{(i)2}}{\beta^2} + \frac{v_n^{(o)2}}{\beta^2 A^2} + v_n^{(fb)2}
\tag{B.10}
$$

We conclude that

- The feedback has the same effect on the input noise of the amplifier $v_n^{(i)}$ and on the input signal V_i (Eq. B.8): therefore it cannot improve the Signal-to-Noise Ratio (SNR) of a measurement.
- By choosing a large value of the product βA, the feedback can reduce the contribution of noises $v_n^{(o)}$ acting at the output.
- The feedback branch contributes an additional noise $v_n^{(fb)}$, that appears with unity gain at the output.

Finally, we rewrite the total noise by referring it to the system input: this is the noise that is to be compared with the incoming signal, to assess the Signal-to-Noise Ratio of the measurement.[7] This can be done by solving Eq. B.7 with respect to $V^{(i)}$ or, more directly, by multiplying Eq. B.10 by the inverse of the input-output transfer function: in the same approximation $\beta A >> 1$, this is just $1/\beta^2$. We also introduce power spectral densities of the noise source: $S_{vn}^{(fb)}$ is the spectrum of $v_n^{(fb)}$ and similarly for $v_n^{(i)}$ and $v_n^{(o)}$. We have

$$S_{tot}^{(\text{at input})} = S_{vn}^{(i)} + \frac{S_{vn}^{(o)}}{A^2} + \beta^2 S_{vn}^{(fb)} \tag{B.11}$$

which highlights the contribution of the noise generator in the feedback branch to the total noise of the closed-loop system.

We conclude this section by mentioning a famous theorem, stating that feedback cannot, by any means, improve the Signal-to-Noise Ratio of a measurement. If applied correctly, feedback can improve other features (like the bandwidth, as we have seen, or the dynamics) and have little influence on the noise.

B.4   Controls and Digital Filters

The technological evolution of electronics allows signals to be processed at ever-increasing speeds. This is pushing for fully digital control system. The monitor signals are digitally converted by an ADC (*Analog to Digital Converter*) and processed by a computational unit, either a DSP (*Digital Signal Processor*) or a computer. This processes the error signals that is then converted back to analog via a DAC (*Digital to Analog Converter*), to drive the actuators and closing the control loop (Fig. B.9). Dealing with digital signals implies the discretization of the data, that is now sampled with a sampling time T. Computation of the filter consists of the *conditioning* of the error signal then of transforming the filtering analytic functions (the function $\beta(s)$ of previous section) into linear combinations applied to the discrete data series. These

---

[7] The SNR can be evaluated at any position of the detection chain. Computing the SNR at the input is often convenient because no conditioning (transfer function) needs to be applied to the signal.

linear combinations are referred to as *digital filters*. Given a discrete sequence of input data x_i, the output signal y_n, at time t_n, is generated by a linear combination of the past M input data and of the past Q output data:

$$
y_n = \frac{1}{a_0}[b_0 x_n + b_1 x_{n-1} + \cdots + b_M x_{n-M}
$$

$$
-a_1 y_{n-1} - a_2 y_{n-2} - \cdots - a_Q y_{n-Q}] \tag{B.12}
$$

or, in compact form:

$$
y_n = \frac{1}{a_0}\left[-\sum_{k=1}^{Q} a_k y_{n-k} + \sum_{h=0}^{M} b_h x_{n-h} \right] \tag{B.13}
$$

The following relation holds:

$$
\sum_{k=0}^{Q} a_k y_{n-k} = \sum_{h=0}^{M} b_h x_{n-h} \tag{B.14}
$$

If we apply a short impulse at the input, this type of filter can give a non-zero output even when the input has long returned to zero. For this reason it is called *Infinite Impulse-Response*—(IIR).

Conversely, for the so-called *Finite Impulse-Response*—(FIR) filters, Eq. B.12 only carries b_k terms. In other words, the output only depends on the input, and not on previous output values. For this reason, they are called filters with no feedback. FIR filters are by definition stable, while IIR can grow indefinitely

In most cases the feedback control loops are designed assuming to deal with continuous signals defined in the complex frequency domain s by their Laplace transform. Then, in the presence of discrete sequences of data x_k sampled at a time rate T, we need to introduce the Z transform defined as

$$
X(z) = \sum_{k=0}^{\infty} x_k z^{-k} \tag{B.15}
$$

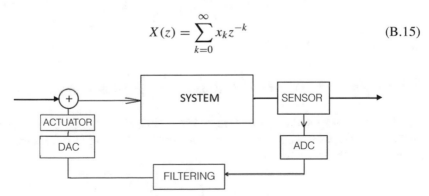

Fig. B.9 Digital feedback loop

Table B.1 Properties of the Z transform

Discrete domain	z space
$ax_n + by_n$	$aX(z) + bY(z)$
x_{n-k}	$z^{-k}X(z)$
$a^n x_n$	$X(\frac{z}{a})$
x_{-n}	$X(\frac{1}{z})$
$x_n{}^*$	$X^*(z^*)$
nx_n	$-z\frac{dX(z)}{dz}$
$x_n \otimes y_n$	$X(z)Y(z)$

where the complex variable z is linked to the Laplace variable s by the relationship

$$z = e^{sT} \tag{B.16}$$

The Z transform converts a discrete-time signal sequence into a complex frequency domain representation.

It can be shown that this transform has properties similar to those of the Laplace transform (Table B.1).

When we apply the Z transform to the input and output sequences of the linear filter of Eq. B.13, and properly use these properties, we obtain

$$\frac{Y(z)}{X(z)} = \frac{\sum_{k=o}^{Q} a_k z^{-k}}{\sum_{h=o}^{M} b_h z^{-h}}$$

If the filter function is known in the s domain, its transformation in the z space can be simply performed by creating a correspondence map based on a bi-linear transform derived from (B.16). We have

$$z = e^{sT} = \frac{e^{sT/2}}{e^{-sT/2}} \simeq \frac{1 + \frac{sT}{2}}{1 - \frac{sT}{2}} \tag{B.17}$$

and the inverse transformation is

$$s = \frac{1}{T}log(z) \simeq \frac{2}{T}\frac{1 - z^{-1}}{1 + z^{-1}} \tag{B.18}$$

In this way, if we know the transfer function $H_c(s)$ of a filter in the continuous domain of the s variable, we can infer the approximate function $H_d(z)$ of the equivalent digital filter through the relationship:

$$H_d(z) = H_c\left(\frac{2}{T}\frac{1 - z^{-1}}{1 + z^{-1}}\right)$$

Because of these approximations, the digital filter deviates from its analog equivalent. In order to assess the magnitude of this deviation, we take $s = i\,\omega T$ and then $z = exp(i\omega t)$. In this case, the function $T_d(z)$ is

$$H_d(exp(i\omega t)) \approx H_d\left(\frac{2}{T}\frac{e^{i\omega T/2} - e^{-i\omega T/2}}{e^{i\omega T/2} + e^{-i\omega T/2}}\right) =$$

$$= H_d\left(i\frac{2}{T}tan\left(\frac{\omega T}{2}\right)\right) = H_c(i\omega_c)$$

where $i\omega_c$ is the variable in the continuum space.

$$\omega_c \Rightarrow \frac{2}{T}tan\left(\frac{\omega T}{2}\right) \tag{B.19}$$

If we don't use the Z transform, a digital filter designed using continuous-time methods (Laplace transform) will produce an inaccurate transfer function: poles and zeros will be found at positions different from those desired.

B.5   The Control of the Michelson Interferometer

The control of optical systems is a field where optics, electronics and mechanics come together to achieve stability of the instrument: in a GW interferometer, lengths need to be controlled to much better than one wavelength, sometimes to nanometer level. The error signal is optical, processing is done by electronics and actuation is performed by micro- or nano-mechanical devices. The Fabry–Perot optical resonator has a dedicated appendix, with a detailed description of the sophisticated Pound–Drever–Hall technique to keep its length on resonance. Here we focus on the controls for the Michelson interferometer, crucial for the operation of the large GW optical detectors. We recall the basic relations for the simplest Michelson interferometer, as discussed in Chap. 9. A static difference in the length of the arms $\Delta L = l_1 - l_2$ causes a phase mismatch between the recombined beams on the beam splitter: $\phi = 2k\Delta L$. The power collected on the photodiode is

$$P_{d.c.} = \frac{P_{in}}{2}\frac{r_1^2 + r_2^2}{2}[1 + C cos(\phi + \phi_s)] \tag{B.20}$$

where the contrast $C = \frac{2r_1r_2}{r_1^2+r_2^2}$ in turn, depends on the reflectivity r_1, r_2 of the two end mirrors and $\phi_s(t)$ is an additive, time varying, phase factor due to a possible extra signal; for example, a gravitational wave. If high-reflective mirrors $r_1 = r_2 \approx 1$ are used, we have $(r_1^2 + r_2^2)/2 \simeq 1$. If we assume the deviation due to the signal to be small $\phi_s \ll 1$, we can expand the time dependent part of Eq. B.20 to get

$$\delta P_{d.c.} \simeq \left(C\frac{P_{in}}{2}\sin\phi\right)\cdot\phi_s(t) \tag{B.21}$$

For control purposes, we must choose the operation point by selecting a ϕ value. The choice is based on an optimization criterion. If we just want to maximize the signal, then it is obvious to choose $\sin \phi = 1 \rightarrow \phi = \pi/2$, that is, to lock the interferometer at half fringe, the intermediate position between the maximums of light and dark, where the slope of the instrument response is largest. However, in precision experiments, the correct quantity to maximize is the signal-to-noise ratio (SNR) and not simply the signal. We shall assume that the dominant noise is the shot noise of light power, with spectral density

$$S_{PP} = 2\frac{hc}{\lambda}\bar{P}_{d.c.} \tag{B.22}$$

and that the incoming GW signal has Fourier transform $H(f)$. It follows that the Fourier transform of the phase $\Phi_s(f) = 2k\,L\,H(f)$. Using Eq. B.22 for the noise and Eq. B.21, we have

$$\sqrt{SNR} = k\,L\sqrt{\frac{P_{in}\lambda}{hc}}\frac{C\sin\phi}{\sqrt{1+C\cos\phi}}H(f) \tag{B.23}$$

so that the optimum condition turns out to be: $\cos\phi = -C \simeq 1$, the dark fringe point.

Up to here, we have not introduced any modulation of the light field: the interferometer is at still on the chosen working point and ϕ_s produces the only time dependent output. This configuration is referred to as **D.C. detection:** it has the advantage of avoiding the phase noise of the oscillator generating the modulating signal.

However, laser sources are affected by the so-called $1/f$ noise, i.e. are much noisier at low frequencies (ϕ_s is expected to vary at $10 - 10^4$ Hz) than in the MHz region. In order to overcome this problem, the **frontal modulation** technique is used. The operational configuration is that of the *heterodyne* detector (Fig. B.10). The angular frequency ω of the incoming light is phase modulated at the angular frequency Ω. The resulting electric field is

$$E(t) = Ae^{i[\omega t + m\sin\Omega t]} = A\sum_{-\infty}^{\infty} J_n(m)e^{i(\omega+n\Omega)t} \tag{B.24}$$

where m is the modulation index. A more detailed account can be found in the Appendix A. We now focus on the first three frequency components: the carrier at ω and the two side bands at $\omega \pm \Omega$. The Michelson interferometer can then be configured in such a way that the length of the two arms has a static difference $\Delta L_S = l_1 - l_2$, called *Schnupp asymmetry*, chosen in such a way that the signal of the outgoing carrier verifies the condition of destructive interference (dark fringe) $\omega\Delta L_S/c = (2k+1)\pi$. In this way, the output signal of the carrier is suppressed, while the side bands are transmitted, as they do not verify the dark fringe condition. Indeed, in traversing the interferometer, the sidebands accumulate an additional phase

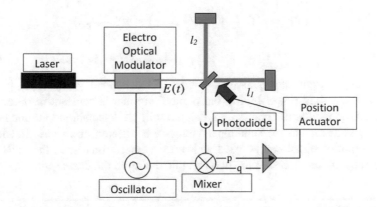

Fig. B.10 A simple Michelson control scheme based on the heterodyne method. Courtesy of the Virgo collaboration

difference $\pm 2\Delta L_S\Omega/c$. This allows us to choose the modulation frequency such as to maximize the sidebands amplitude:

$$\Omega = \frac{c}{2L}(n+1)\pi$$

If the reflectivities of the end mirrors r_1 and r_2 are different, the dark condition for the carrier is only partially verified, the light intensity is at a minimum, but non-zero: the output signal of the photodiode will also have a constant term proportional to $(r_1 - r_2)^2$. In fact, the wave function at the output of the interferometer is the sum of three fields:

$$\Psi_u(m, t) = \Psi_0 + \Psi_+\Psi_-$$

with

$$\Psi_0 = -J_o\left(e^{i\delta l\omega t/c} + e^{-i\delta l\omega t/c}\right)$$

$$\Psi_\pm = \pm J_1\left(e^{i\delta l(\omega\pm\Omega)t/c} + e^{-i\delta l(\omega\pm\Omega)t/c}\right)e^{\pm i(l_1+l_2)\Omega/c}$$

where we added to the static asymmetry of *Schnupp* a gravitational signal δl_{gw} so that $\delta l = \Delta L_S + \delta l_{gw}$.

It follows that the light power incident on the photodiode is proportional to

$$\begin{aligned}|\Psi|^2 =& |\Psi_0|^2 + |\Psi_-|^2 + |\Psi_+|^2 + \\ & + |\Psi_+\Psi_0{}^* - \Psi_0\Psi_-{}^*| \\ & + |\Psi_-\Psi_0{}^* - \Psi_0\Psi_+{}^*|\end{aligned} \tag{B.25}$$

neglecting terms in $|\Psi_+\Psi_-^*|$ and $|\Psi_+^*\Psi_-|$, that oscillate at 2Ω. We can rewrite this expression as

$$|\Psi|^2 = DC + A \, sin\left(\frac{\omega}{c}\Delta L_S\right) sin\left(2\frac{\omega}{c}\delta l_{gw}\right).$$

$$\left(A_1 \, sin\Omega t + A_2 \, cos\Omega t\right) \tag{B.26}$$

being A, A_1, A_2 constant quantities, while $DC = |\Psi_0|^2 + |\Psi_-|^2 + |\Psi_+|^2$ is the fraction of the light power hitting the photodiode that is constant in time. Equation B.26 tells us that the gravitational signal will be detectable at the modulation frequency Ω as long as the Schnupp asymmetry is different from zero. In addition, the same output signal can be used to keep the interferometer in the dark fringe condition by acting on one of the two end mirrors or on the *beam splitter*.

B.6   The Virgo Control of the Longitudinal Degrees of Freedom

We briefly recall here the complex optical scheme of Advanced Virgo, the European interferometric gravitational wave detector, described in greater detail in Chap. 9.

- The basic, simple Michelson consists of the two 3 km long arms, along the North and West directions, instrumented with a *beam splitter* (BS) and two end mirrors, North-End (NE) and West-End (WE).
- Each arm hosts an additional *input* mirror (WI, NI) that forms with the respective End mirror a Fabry–Perot (FP) cavity, where the light bounces back and forth. This increases the equivalent optical path and amplifies the gravitational signal. In order to maximize the signal-to-noise ratio, the length difference of two arms (North and West) is adjusted so that the rays destructively interfere at the interferometer output (dark fringe condition). In this set-up virtually all the light is reflected back towards the laser.
- By inserting an additional, partially reflecting mirror, the *recycling mirror* (RM), between the laser and the BS, a new optical cavity is formed, where light resonates. This is called *power recycling cavity* (RC) and consists of the RM mirror and all the interferometer that operates reflecting back towards the laser all the light, like a *virtual mirror*. By taking advantage of the resonance of the recycling cavity, the circulating light power can reach 1 kW on the beam splitter with only 20 W of laser power.
- Finally, we have the *Signal Recycling Mirror*, to enhance the interferometer response at a chosen frequency. For sake of simplicity, we shall not consider it here.

In this configuration (see Fig. B.11) the interferometer is a coupled cavity system: in addition to the dark fringe condition, all of these cavities must be *locked* in resonance in order to achieve a proper operating configuration for the detector.

Each mirror is suspended like a pendulum to isolate it from seismic noise: this behaves as a mechanical low-pass filter with a characteristic frequency of few Hz. However, the residual, unfiltered excitation at very low frequency will make the mirror oscillate around its equilibrium position, with an amplitude of the order of

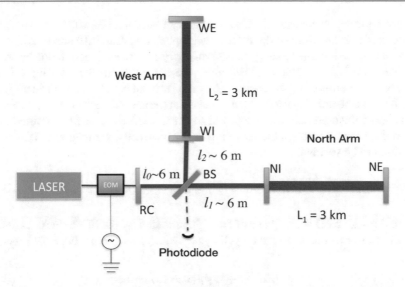

Fig. B.11 Simplified optical diagram of Virgo, without the signal recycling cavity. FP cavities are made up of pairs of mirrors: NI + NE and WI + WE. BS and RC are the beam splitter and the power recycling mirror. EOM indicates the electro-optical modulator used to frequency modulate the light. Courtesy of the Virgo collaboration

few tens of micrometres. A local control system, partly based on accelerometers and partly on displacement sensors anchored to the ground, reduces this amplitude to a fraction of a wavelength, still large enough to prevent stabilization of the resonance conditions for the optical cavities and Michelson dark fringe. In essence, purely local systems are not sufficient to keep the interferometer in the working position: the *locking* of the detector working point requires a system of global control of mirror distances, based on signals extracted from the interferometer itself.

Referring to Fig. B.11, there are four independent lengths that need to be controlled:

- the lengths of the two FP cavities (L_1 and L_2)
- the length of the recycling cavity $l_r = l_0 + \frac{l_1 + l_2}{2}$, where l_1, l_2 are the distances of the two input mirrors from the BS (see Fig. B.11)
- the arm length difference of the Michelson $\Delta l = l_1 - l_2$.

Thus, we need to extract four independent error signals for these four lengths, in order to implement actuation on the various mirrors. The control system must have a high gain at very low frequency (up to a few Hz) and should not introduce noise in the frequency bandwidth of the detector (10 Hz–5 kHz).

To get an idea of the challenge in the design of the control system, consider a numerical example relative to the initial Virgo configuration: the laser light has a wavelength of $\lambda = 2\pi c/\omega = 1.05 \,\mu$m and the Finesse of the various resonant optical cavities is $\mathcal{F} \sim 50$. In order to keep the 3 km long FP cavities in resonant condition,

L_1 and L_2 must be controlled with an accuracy better than $3 \cdot 10^{-7} \lambda \sim \cdot 10^{-13}$ m, or one-tenth of the width of the FP resonance peak (see Eqs. 9.19 and 9.20).

Here too, the locking strategy consists of applying a frontal modulation by adding an electro-optical modulator (EOM) along the input path of the laser light. In this case, the short arms of the Michelson (l_1 and l_2) are not equal, so that the dark fringe condition can be satisfied for the carrier, but not at the same time for the side bands. We then choose the modulation frequency Ω so that the sidebands are transmitted to the photodiode output through the recycling cavity. Formally, we require the following condition to be verified:

$$\pi + 2\frac{\omega \pm \Omega}{c}l_r + \Phi_{FP}(\omega \pm \Omega) = 2n\pi \qquad (n = 1, 2\ldots) \qquad \text{(B.27)}$$

where $\Phi_{FP}(\omega \pm \Omega)$ is the phase of the light reflected by the FP cavities. Assuming that the carrier resonates in the recycling cavity ($\omega l_r/c = 2m\pi$), Eq. B.27 becomes

$$\pi \pm 2\frac{\Omega}{c}l_r + \Phi_{FP}(\omega \pm \Omega) = 2n\pi \qquad \text{(B.28)}$$

A detailed analysis of these conditions shows that there is an optimum value for the Schnupp asymmetry, a value that maximizes the transmission of the sidebands at the photodiode output: (Vinet 2020)

$$cos\left(\frac{\Omega(l_1 - l_2)}{c}\right) = r_r r_{ITF}$$

where r_r is the reflection coefficient of the recycling mirror and r_{ITF} that of the second, *virtual* mirror of the RC, composed by the entire interferometer. The physical meaning of this condition is that the power extracted from the side bands should match that lost within the recycling cavity. In Virgo jargon, this is the *optimal matching condition at the output*. As mentioned above, in order to control the interferometer we need four independent signals related to the four lengths to stabilize. In practice, experimenters chose to control four degrees of freedom that are linear combinations of these lengths. In Fig. B.12 we sketch these motions:

- Differential mode of the Fabry–Perot arms: $\delta L_1 - \delta L_2$, named DARM—figure (a) of Fig. B.12,
- Common mode of the Fabry–Perot arms: $\delta L_1 + \delta L_2$, or CARM—figure (b) of Fig. B.12,
- Michelson's differential mode: $\delta l_1 - \delta l_2$, or MICH—figure (c) of Fig. B.12,
- Power Recycling Cavity Length common mode: $\delta l_1 + \delta l_2$, or PRCL—figure (d) of Fig. B.12.

The error signals used to control these quantities are the photocurrents produced in the photodiodes shown in Fig. B.13. The numbering of photodiodes, apparently bizarre, reflects the historical use adopted in the Virgo experiment.

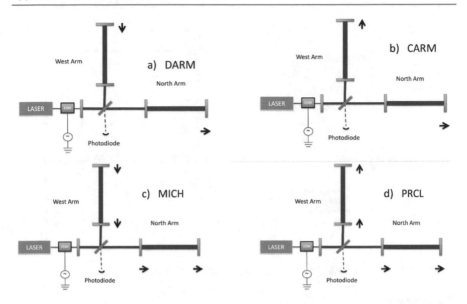

Fig. B.12 The four degrees of freedom of the Virgo interferometer: **a** DARM: Fabry–Perot differential mode. **b** CARM: Fabry–Perot common mode. **c** MICH: Michelson differential mode. **d** PRCL: Michelson common mode. Courtesy of the Virgo collaboration

The output signal at the photodiode 1, where the information about the GW signal is imprinted, is mainly determined by the differential mode of the 3 km FP cavity $(\delta L_1 - \delta L_2)$.

On the other hand, the signal from photodiode 2 mainly depends on the common mode of the FP cavities $(\delta L_1 + \delta L_2)$ and is much less influenced by the Michelson common mode $(\delta l_1 + \delta l_2)$.

The roles in the control strategy of the photodiodes 5 and 7 are similar: photodiode 5 controls the North arm FP cavity in reflection (and 7 in transmission); its output signal also strongly depends on the length of the West arm cavity. This effect is due to the coupling between the two FP cavities, induced by the recycling mirror. As a result, it is sensitive to the FP common mode and its dependence on the FP differential mode is much weaker. Photodiode 8 has an equivalent function in transmission of the West cavity.

In conclusion, most of the signals are more sensitive to changes in the length of the FP cavity than to that of the Michelson, whose arm lengths are smaller by a factor of $\sim 10^2$. To extract information about the differential motion of the Michelson and the length of the recycling cavity, we need to compute the difference between two or more signals. This reduces the SNR because the independent noises of the various photodiodes add up quadratically. In order to get around this problem in part, we can increase the information redundancy by using also higher harmonics of the modulation frequency, obtained by demodulating and filtering at the proper frequencies the output signal from photodiodes.

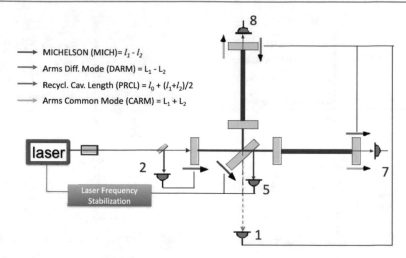

Fig. B.13 The Virgo longitudinal control. Courtesy of the Virgo collaboration

References

Bechhoefer, J.: Feedback for physicists: a tutorial essay on control. Rev. Mod. Phys. **77**, 783–835 (2005)

Vinet, J.-Y.: The VIRGO Physics Book, OPTICS and related TOPICS (revision 2020). https://artemis.oca.eu/rechercheartemis/projets/virgo/2081-the-darkf-optical-simulation-code

The Fabry–Pérot Cavity

C

C.1　Introduction

A Fabry–Pérot (FP) cavity is a linear optical resonator which consists of two highly reflecting mirrors, where the light bounces between the two reflecting surfaces, and is transmitted only for well-defined wavelengths. It is widely used in telecommunications, lasers technology and astrophysics. It is also employed as a high-resolution optical spectrometer, exploiting the fact that the transmission through such a resonator exhibits sharp resonances and is very small between those. As we shall see, these devices are used, in gravitational research, to extend the path of light in the interferometer arms, to keep their length equal to a few picometers, to stabilize the laser wavelength.

The FP consists of two mirrors positioned at a distance l from each other: as shown in the right pane of Fig. C.1, IM is the input mirror, with reflectivity, transmittivity and losses indicated by the symbols r_1, t_1 and p_1, respectively. EM is the second (end) mirror of the cavity, with corresponding quantities r_2, t_2 and p_2.

We assume to illuminate the cavity with a monochromatic plane light wave of frequency ν

$$\psi(x, t) = K\, e^{i(kx - 2\pi\nu t)} = A(t)\, e^{ikx}$$

where A is the complex (and time dependent) amplitude, $k = 2\pi/\lambda$ and $\lambda = c/\nu$ the wavelength. This wave, once propagated over a distance l, acquires an additional phase factor $i\,kl$. Moreover, at each reflection, the complex amplitude of the optical field changes by a factor $(i\,r)$ and at each transmission by a factor t. If we call ψ_{in} the field at the cavity input, the transmitted ψ_t and reflected ψ_r fields are deducted taking into account all these contributions. Figure C.1 shows how we combine the various contributions inside and outside the cavity. In high sensitivity cavities we

F. Ricci and M. Bassan, *Experimental Gravitation*, Lecture Notes in Physics 998,
https://doi.org/10.1007/978-3-030-95596-0_C

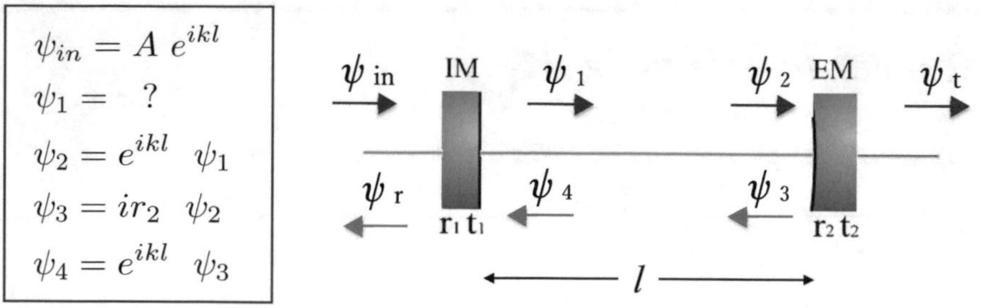

$$\psi_{in} = A \, e^{ikl}$$
$$\psi_1 = \; ?$$
$$\psi_2 = e^{ikl} \; \psi_1$$
$$\psi_3 = ir_2 \; \psi_2$$
$$\psi_4 = e^{ikl} \; \psi_3$$

Fig. C.1 Fields in the Fabry–Pérot cavity

have negligible losses: p_1, $p_2 << 1$, so that

$$t_i^2 + r_i^2 = 1 - p_i \qquad i = 1.2 \tag{C.1}$$

The resulting field is the superposition of an infinite series of waves, with geometrically decreasing amplitudes. Summing up all the contributions we have, e.g. the field ψ_1 inside the cavity:

$$\psi_1 = t_1 \psi_{in} + (ir_1)(ir_2)t_1 e^{2ikl} \psi_{in} + \cdots (ir_1)^n (ir_2)^n t_1 e^{2nikl} \psi_{in} + \cdots$$
$$= \psi_{in} \frac{t_1}{1 + r_1 r_2 e^{i2kl}} \tag{C.2}$$

Similarly, we can sum the transmitted field

$$\psi_t = \psi_{in} \frac{t_1 t_2 e^{-ik\,l}}{1 + r_1 r_2 e^{i2kl}}, \tag{C.3}$$

and the reflected field

$$\psi_r = i \, \psi_{in} \frac{r_1 + r_2(1 - p_1)e^{i2k\,l}}{1 + r_1 r_2 e^{i2kl}} \tag{C.4}$$

We can also derive the fields from their mutual relations, as shown in Fig. C.1. These relations show that when $e^{i2kl} = -1$, i.e. when $2l = (n + 1/2)\lambda$, we have a resonance phenomenon.

Suppose now we keep the mirrors at a fixed distance l and to change the light frequency $\nu = c/\lambda$. The cavity exhibits resonance peaks, one after the other, every time that the frequency equals

$$\nu_n = \left(n + \frac{1}{2}\right) \frac{c}{2l} \qquad n = 1, 2, 3 \dots . \tag{C.5}$$

The frequency difference between two adjacent resonances is the *Free Spectral Range* (FSR)

$$\Delta \nu_{FSR} = \frac{c}{2l} \tag{C.6}$$

The inverse of $\Delta \nu_{FSR}$ is the travel time of a round trip pass for the light inside the cavity, so that $\Phi = 2\pi\nu/\Delta\nu_{FSR}$ is the phase accumulated by the wave at each round trip. As an example, in the Virgo gravitational wave detector, where $l \simeq 3000$ m, we have $\Delta\nu_{FSR} \simeq 50$ kHz. Using a laser light with $\lambda = 1.06$ μm, i.e. $\nu \simeq 3 \cdot 10^{14}$ Hz, the resonance condition Eq. C.5 implies $n \sim 10^9$.

The ratio $S = \psi_1/\psi_{in}$, called *overvoltage factor*,[8] on resonance takes the value

$$S_{(max)} = \frac{t_1}{1 - r_1 r_2} \tag{C.7}$$

If the frequency is slightly detuned from the resonance condition, by an amount $\delta\nu << \Delta_{FSR}$ factor, we can rewrite the phase $\Phi = 2kl$ as

$$\Phi = (2n + 1)\pi + 2\pi \frac{\delta\nu}{\Delta\nu_{FSR}} \tag{C.8}$$

From Eq. C.2 we derive the magnitude of S:

$$|S|^2 = \frac{t_1^2}{(1 - r_1 r_2)^2 + 4r_1 r_2 sin\left(\pi\frac{\delta\nu}{\Delta\nu_{FSR}}\right)} = \frac{S_{(max)}^2}{1 + \left[\frac{2\sqrt{r_1 r_2}}{1 - r_1 r_2} sin\left(\pi\frac{\delta\nu}{\Delta\nu_{FSR}}\right)\right]^2} \tag{C.9}$$

Neglecting, for the moment, the losses p_i, we define the cavity *finesse* as

$$\mathcal{F} = \pi \frac{\sqrt{r_1 r_2}}{1 - r_1 r_2} \tag{C.10}$$

Note that $\mathcal{F}$ only depends on the combined reflectance $r_1 r_2$, just as $\Delta\nu_{FSR}$ only depends on the mirror distance l: so, $[\mathcal{F}, \Delta\nu_{FSR}]$ represent an equivalent description of the FP cavity.

Assuming $\delta\nu << \Delta_{FSR}$, Eq. C.9 rewrites in a simpler form:

$$|S|^2 = \frac{S_{(max)}^2}{1 + \left[\frac{2\mathcal{F}}{\pi} sin\left(\pi\frac{\delta\nu}{\Delta\nu_{FSR}}\right)\right]^2} \tag{C.11}$$

This transmittance function of a FP interferometer is also referred to as the *Airy* function, named after the British astronomer George Biddell Airy (1801–1892). In the vicinity of the resonance, it is well approximated by

$$|S| \simeq S_{(max)} \frac{1}{\sqrt{1 + \left[2\mathcal{F}\frac{\delta\nu}{\Delta\nu_{FSR}}\right]^2}} \tag{C.12}$$

---

[8] Similar definition is used in the case of resonant electrical circuits, hence the name.

From the relation (C.11), it is evident that the resonance of the Airy curve at half of its maximum value has a *Full Width Half Maximum*—(FWHM)

$$\delta\nu_{FWHM} = \frac{\Delta\nu_{FSR}}{\mathcal{F}} \tag{C.13}$$

This shows that a large finesse produces a narrow resonance: for this reason $\mathcal{F}$ is considered a sort of quality factor for a FP cavity.

C.2   The Fabry–Pérot as a Virtual Mirror

We now focus our attention on the field ψ_r reflected back by the cavity, as expressed by Eq. C.4. We consider the whole cavity as a *virtual mirror*, with a reflectivity $\mathcal{R}$ given by the ratio ψ_r/ψ_{in}. We compute this ratio substituting into Eq. C.4 the phase Φ expressed by Eq. C.8, and observe its behaviour versus the frequency shift $\delta\nu$ from resonance:

$$\mathcal{R} = i\,\frac{r_1 + r_2(1 - p_1)e^{i\Phi}}{1 + r_1 r_2 e^{i\Phi}} \tag{C.14}$$

$$|\mathcal{R}|^2 = \frac{1}{\left[1 + (r_1 r_2)^2 + 2 r_1 r_2 \cos\Phi\right]^2} \cdot$$

$$\cdot \left\{r_1\left[1 + r_2^2(1 - p_1)\right] + r_2\left[r_1^2 + (1 - p_1)\right]\cos\Phi\right\}^2 + \left\{r_2\left[(1 - p_1) - r_1^2\right]\sin\Phi\right\}^2 \tag{C.15}$$

$$\Phi_R \equiv arctan\{\mathcal{R}\} = \frac{\left[r_1^2 - (1 - p_1)\right]r_2\sin\Phi}{r_1\left[1 + r_2^2(1 - p_1)\right] + r_2\left[(1 - p_1) + r_1^2\right]\cos\Phi} \tag{C.16}$$

For $\delta\nu = 0$, i.e. $\Phi = (2n + 1)\pi$, the function has an absorption peak and its phase undergoes a swift change of 2π.

We can change the point of view: keep the light frequency ν (and the wavelength $\lambda = c/\nu$) constant, and allow the length of the cavity to change. The resonance condition Eq. C.5 expressed for cavity length becomes

$$l_n = \left(n + \frac{1}{2}\right)\frac{\lambda}{2} \tag{C.17}$$

We will then talk about of *Free Spectral Range length* $\Delta l_{FSR} = \lambda/2$ and *Full Width Half Maximum length*, $\Delta l_{FWHM} = \lambda/2\mathcal{F}$.

Consider again Eq. C.8, this time for l close to resonance condition:

$$\Phi = 2k\,l = \pi + 2\pi\frac{\delta\nu}{\Delta\nu_{FSR}} \quad \text{mod}(2\pi) \tag{C.18}$$

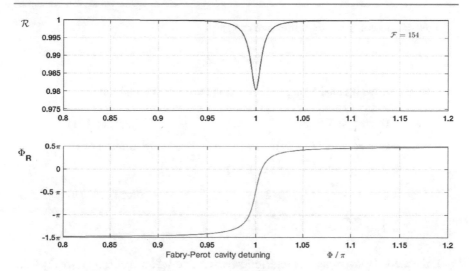

Fig. C.2 The reflectivity of a FP resonator versus the intra-cavity phase change $\Phi = 2kl$, normalized to π. Top: amplitude reflection, for a finesse $\mathcal{F} = 154$, achieved with $r_1 = 0.98$, $r_2 = 0.9998$. Bottom: phase of the light reflected by the FP resonator

To simply the notation we restrict Φ to the interval $[0, 2\pi[$ and introduce the reduced frequency $f \equiv \delta\nu/\Delta\nu_{FWHM} = \delta\nu/(\mathcal{F}\Delta\nu_{FSR})$, so that the phase takes the compact form

$$\Phi = \pi + 2\pi\frac{f}{\mathcal{F}}$$

For high-quality cavities, as in the case of gravitational experiments, the losses are small, $p_1, p_2 \ll 1$, and the finesse is high: $\mathcal{F} > 50$ for what discussed here (and up to 10^5 in special cases). As remarked above, $\mathcal{F}$ only depends on the product $r_1 r_2$. Inverting Eq. C.10 and neglecting terms $O(\pi/\mathcal{F})^2 << 1$, we get

$$r_1 r_2 \simeq 1 - \frac{\pi}{\mathcal{F}} \tag{C.19}$$

We substitute this relation into Eq. C.14 and define $(1 - p_1)r_2^2 = 1 - p$, where the parameter p includes all the cavity losses.

The cavity reflection coefficient $\mathcal{R}$ is well approximated with

$$r_2\mathcal{R} \simeq -\frac{1 - p\mathcal{F}/\pi + 2if}{1 - 2if} = -\frac{1 - \sigma + 2if}{1 - 2if} \tag{C.20}$$

In Eq. C.20 we have also introduced the *coupling parameter* σ of the cavity:

$$\sigma \equiv p\mathcal{F}/\pi \tag{C.21}$$

This parameter can vary within the range

$$0 < \sigma < 2$$

and we can prove it considering the following series of inequalities:

- $0 < r_1^2 < 1 - p_1$ (from Eq. C.1)
- $0 < r_1^2 r_2^2 < (1 - p_1)r_2^2 = 1 - p$ (definition of p)
- $0 < r_1 r_2 < \sqrt{(1 - p)} \simeq 1 - p/2$
- $0 < 1 - \dfrac{\pi}{\mathcal{F}} < 1 - p/2$ (use Eq. C.19)
- $0 < \dfrac{p\mathcal{F}}{\pi} < 2$

If the second mirror has very high reflectivity, $r_2 \simeq 1$, $\mathcal{R}$ at resonance (see Eq. C.20 computed at $f = 0$) is

$$\mathcal{R}(f = 0) = \sigma - 1 \qquad (C.22)$$

This relation shows the physical meaning of the coupling parameter σ:

- For $\sigma = 1$ we have maximum light storage in the cavity and no reflection. This is the condition of *optimum coupling*.
- For $0 < \sigma < 1$ the cavity is *over-coupled*. As σ grows in this range, the stored light intensity increases.
- For $1 < \sigma < 2$ the cavity is *under-coupled* and the stored light intensity decreases to the point, for $\sigma = 2$, where the condition of total reflection takes place

In other words, as σ increases in the range $1-2$ the stored light increases and the reflected signal becomes progressively less sensitive to the cavity status. This can be seen by studying $\mathcal{R}$ versus f. Its square modulus is

$$|\mathcal{R}|^2 = 1 - \frac{\sigma(2 - \sigma)}{1 + 4f^2} \qquad (C.23)$$

and the phase

$$\Phi_{\mathcal{R}} = \pi + arctan\left(\frac{2f}{1 - \sigma}\right) - arctan(2f) \qquad (C.24)$$

For completeness we write, in these notations, the overvoltage coefficient S:

$$S = \frac{\sigma(2 - \sigma)}{p(1 + 4f^2)} \qquad (C.25)$$

and its maximum occurs at $\sigma = 1$. Inverting Eq. C.21, to express the coupling parameter in terms of the finesse $\mathcal{F}$, we obtain

$$S_{max} = \frac{1}{p} = \frac{2\mathcal{F}}{\pi} \qquad (C.26)$$

C.3   Comparison with the Delay Line

It is interesting to evaluate how rapidly the phase changes around $f = 0$: this is indeed the error signal used to keep the cavity on resonance. Equation C.24, expanded near resonance to first order in f, has the form

$$\Phi_{\mathcal{R}} \simeq 2\frac{2-\sigma}{1-\sigma}f = 2\frac{2-\sigma}{1-\sigma}\mathcal{F}\frac{\delta\nu}{\Delta\nu_{FSR}} \qquad (C.27)$$

Combining this last equation with Eq. C.21, we obtain the desired relation:

$$\frac{d\Phi_{\mathcal{R}}}{d\delta\nu} = 2\sigma\frac{2-\sigma}{1-\sigma}\frac{\pi}{p\Delta\nu_{FSR}} \qquad (C.28)$$

$$\frac{d\Phi_{\mathcal{R}}}{d\delta l} = 2\sigma\frac{2-\sigma}{1-\sigma}\frac{2\pi}{p\lambda} \qquad (C.29)$$

In the second equation we used the equivalence $\delta\nu/\Delta\nu_{FSR} = \delta l/\Delta l_{FSR}$ and $l_{FSR} = \lambda/2$. In the case of low losses and low cavity coupling, we can approximate with

$$\frac{d\Phi_{\mathcal{R}}}{d\delta l} = \frac{8\mathcal{F}}{\lambda} \qquad (C.30)$$

As a last step, we use this last simplified equation to compare the rate of phase change for the light reflected by a FP cavity with that of an *optical delay line*. The latter is a cavity of two mirrors where the light wave is bounced n_{DL} times, avoiding overlap of the beams. The phase change of the beam exiting the delay line is $\Phi_{DL} = (4\pi n_{DL}l)/\lambda$ and its derivative

$$\frac{d\Phi_{DL}}{dl} = \frac{4\pi n_{DL}}{\lambda} \qquad (C.31)$$

Comparing Eqs. C.31 and C.30 we infer that, as regards the phase delay, the FP cavity is equivalent to an optical delay line where the light is bounced a number of times equal to $n_{FP}{}^{eff}$

$$n_{FP}{}^{eff} = \frac{2\mathcal{F}}{\pi} \qquad (C.32)$$

This conclusion can be used to define an equivalent optical path of light in the case in which the FP cavity is used as the arm of interferometric detector of gravitational waves.

C.4   The Optical Resonator

The Fabry–Pérot (FP) resonator is usually presented, for sake of clarity, in the configuration composed of two parallel plane mirrors. However, the FP analysis of the previous sections is general enough to hold validity in the general case of cavities made by curved mirrors. The name *optical resonator or optical cavity* refers to this general case.

In practice, an optical resonator is composed of two or more mirrors configured in such a way to confine the light, which is reflected many times in the cavity, producing standing waves for given *resonant* frequencies. The patterns of standing wave produced are called cavity normal modes. The longitudinal modes differ only in frequency, while the transverse modes have also different intensity patterns across the beam transversal section. To have a stable configuration, the geometry of the cavity must be chosen such to prevent the beam transversal dimension to continuously grow with multiple reflections.

The stability depends on the geometrical parameters of the cavity, that are the curvature of the mirrors R_1 and R_2 and the intra-cavity distance L. In the case of an unstable cavity, the beam transversal dimension increases up to the point that it becomes larger than the mirror diameter, and the light escapes from the cavity. In an unstable resonator, the light, regardless of the initial direction, will leave the cavity, while in a stable configuration rays are reflected back towards the system centre. We state the stability condition as follows:

the light rays remain in the vicinity of the optical axis during the propagation when the numbers of light bounces in the cavity tends to infinity.

To study the stability of a two-mirror cavity, we define for each mirror the stability parameter

$$g_i = 1 - \frac{L}{R_i} \text{ with } i = 1, 2$$

We then apply the formalism of ray transfer matrix, or *optical matrix*[9]: by repeatedly applying the mirror-propagation-mirror matrices to our rays, we can derive the stability requirement, which states that the hyperbolic function $g_1 \, g_2 = \pm 1$ define

---

[9] This ray-tracing formalism works in the case of paraxial rays: each optical element is described by a 2×2 transfer matrix, called ABCD matrix, applied to a 2×1 vector describing position and inclination of the input light ray. The result is the output vector describing the output ray:

$$\begin{pmatrix} x_{out} \\ \theta_{out} \end{pmatrix} = \begin{pmatrix} A & B \\ C & D \end{pmatrix} \begin{pmatrix} x_{in} \\ \theta_{in} \end{pmatrix}$$

with the matrix elements A, B, C, D peculiar to each optical element (lens, mirror, propagation, etc.) The computation of the light in the cavity resonator is done by applying the mirror matrix a number of time equivalent to the number of light bounces in the cavity.

Fig. C.3 Stability plot of the two-mirror optical cavity. The areas marked in blue are stability regions

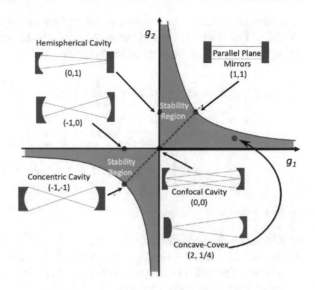

the boundary line in the g_1, g_2 plane where the cavity is stable:

$$0 \leqslant g_1 g_2 \leqslant 1$$

The region of stable resonator is limited by the coordinate axes and the two hyperbolas
Let us note a few special cases (Fig. C.3):

– When $L = R_1 + R_2$, the resonator is concentric.
– The straight lines $g_1 = 1$ and $g_2 = 1$ correspond to resonators with one plane mirror (infinite radius of curvature).
– The cavity made by two plane mirrors is represented by the point $g_1 = g_2 = 1$.
– For $R_1 = R_2 = L$, i.e. $g_1 = g_2 = 0$, the resonator is *confocal*.

As mentioned above, in a resonant cavity we have stationary electromagnetic waves, i.e. stable modes of oscillations classified as longitudinal or transversal to the cavity axis.

In the case of spherical mirrors we need to consider the transversal electromagnetic resonances, which are computed by solving the paraxial Helmholtz wave equation. Starting from the most general wave equation

$$\left(\nabla^2 - \frac{1}{c^2} \frac{\partial^2}{\partial t^2} \right) U(x, y, z, t) = 0 \tag{C.33}$$

and looking for stationary solutions for the electromagnetic field, separate the wave-function U in its time and spatial dependence: $U = A(x, y, z)T(t)$ and, because of the paraxial approximation, let $A(x, y, z) = u(x, y)e^{ikz}$, where z is the optical axis of the cavity. The solutions of the paraxial case of Eq. C.33 are given by superposition of Hermite–Gaussian modes when the amplitude profiles are described using

Cartesian coordinates, or a combination of Laguerre–Gaussian modes when using cylindrical coordinates around the cavity axis. The Hermite-Gauss representation of these transverse electromagnetic modes called TEM$_{nm}$, are (Fig. C.4)

$$A_{nm}(x, y, z) = E_0 \frac{w_0}{w(z)} H_n\left(\sqrt{2}\frac{x}{w(z)}\right) exp\left(-\frac{x^2}{w^2(z)}\right).$$
$$H_m\left(\sqrt{2}\frac{y}{w(z)}\right) exp\left(-\frac{y^2}{w^2(z)}\right).$$
$$exp\left\{-i\left[kz - (1+n+m)arctan\left(\frac{z}{z_R}\right) + \frac{k(r^2)}{2R(z)}\right]\right\} \qquad (C.34)$$

where:

- k is the modulus of the wave vector and $r^2 = x^2 + y^2$.
- H_n and H_m are the Hermite polynomials of order n and m respectively. The subscripts m and n correspond to the number of nodal lines, $A_{nm}(x, y) = 0 \ \forall \ z$, along the x- and y-axis, respectively.
- $w(z)$ is a linear measure of the transversal dimension of the beam: it is the radius at which the field amplitudes fall to $1/e$ of their axial values (i.e. where the intensity values fall to $1/e^2$ of their axial values), at the plane z along the beam. w_0 is the *beam waist*, i.e. the minimum transversal dimension of the beam.
- $R(z)$ is the curvature of the wavefront.
- z_R is the *Rayleigh range*, defined as the value of z where the cross-sectional area of the beam is doubled with respect to the waist.
- $\phi(x, y, z) = \left[kz - (1+n+m)arctan\left(\frac{z}{z_R}\right) + \frac{k(x^2+y^2)}{2R(z)}\right]$ is the phase of the wave.
- The term $arctan\left(\frac{z}{z_R}\right) \equiv \zeta$ represents the phase delay with respect to a plane wave.

For the first transversal mode $n = m = 0$, i.e. TEM$_{00}$, the intensity distribution along the z coordinate is

$$I(z) = I_0 \exp\left(-\frac{2r^2}{w^2(z)}\right)$$

This implies that the intensity profiles of the light, measured on all the planes perpendicular to the cavity axis, have a gaussian distribution, but with a width that varies along the axis, as shown by the presence of $w(z)$ in the variance. Due to diffraction, this Gaussian beam first converges to its minimum transverse size, at the *beam waist* $w(z) = w_0$, and then, from this section, it expands with a divergence angle $\theta = \lambda/(\pi w_0)$ as it is shown in Fig. C.5.

The diameter of the beam profile depends on z as

$$w(z) = w_0\sqrt{1 + \left(\frac{\lambda z}{\pi w_0^2}\right)^2}$$

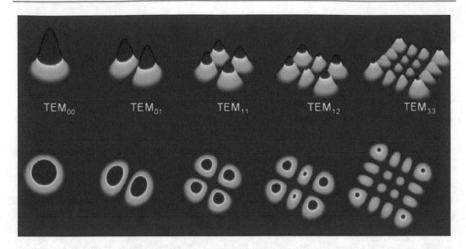

Fig. C.4 Hermite-Gauss optical modes TEM_{mn}. The subscripts m and n count the number of nodal lines in the x and y direction, respectively. *Credits* Creative Commons

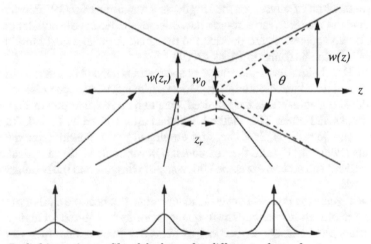

Radial intensity profile of the beam for different values of z

Fig. C.5 Profile of a Gaussian beam: it is defined by the waist w_0 and the Rayleigh range z_R

and, applying the definition of z_R, we conclude that

$$z_R = \frac{\pi w_0^2}{\lambda} \quad \text{and, as a consequence:} \quad w(z) = w_0 \sqrt{1 + \left(\frac{z}{z_R^2}\right)^2} \quad \text{(C.35)}$$

Finally we derive the resonance frequencies of the modes associated to the cavity. We recall that the phase of the transverse modes is

$$\phi(0, 0, z) = kz - (1 + n + m)arctan\left(\frac{z}{z_R}\right) = kz - (1 + n + m)\zeta(z) \quad \text{(C.36)}$$

Then, we impose the phase matching condition after a round trip of the light in the cavity of length L:

$$2kL - 2(1 + n + m)\Delta\zeta = 2q\pi \quad \text{with} \quad q = \pm1, \pm2, \cdots \quad (C.37)$$

where $\Delta\zeta = \zeta(z + 2L) - \zeta(z)$.

We conclude that the resonance frequencies are

$$\nu_q = \frac{c}{2L}\left[q + (1 + n + m)\left(\frac{\Delta\zeta}{\pi}\right)\right] \quad (C.38)$$

C.5 The Pound–Drever–Hall Method

In this section we describe the Pound–Drever–Hall (PDH) technique for stabilizing a FP cavity. As we saw, the resonance of the cavity is only determined by its length l: we might be interested in stabilizing the frequency, e.g. when the cavity is filled with the active medium of a laser, or the length, as in the arms of a GW interferometer. We assume the reader to have some knowledge of the basics of modulation and of feedback: two appendices are provided to that goal. A pedagogical introduction to this technique can be found in Black (2001).

The PDH technique has a wide range of applications beyond interferometric detectors for gravitational waves: atomic physics experiments where spectroscopic probes of individual quantum states are required, quantum optical computers, apparata for atomic clocks and other. This method is named after Robert V. Pound, Ronald W. P. Drever and John L. Hall: it was first developed by R. Pound[10] for microwave resonators (Pound 1946) and then extended (Drever et al. 1983) to optical cavities by Drever, Hall and coworkers at the University of Glasgow and the National Bureau of Standards.

In recent years the Pound–Drever–Hall technique has become a pillar of the laser frequency stabilization method. The frequency stability is required for high-precision experiments, but all lasers are subject to frequency fluctuations: they can be due to temperature changes, mechanical imperfections and laser gain dynamics, which change the lengths of the laser cavity, fluctuations in the supply voltage, frequency widths of atomic levels and many other factors. The PDH feedback method offers a solution to this problem by actively tuning the laser to match the resonance condition of a stable reference cavity.

To avoid degradation of performance, we shall make sure that the noise introduced by the feedback signal does not exceed the *shot* noise limit of the incident light on the detection photodiode. Under optimum control conditions, the line width depends on the absolute stability of the cavity and is ultimately imposed by the thermal noise

---

[10] This is the same R. Pound we encountered in Sect. 4.3 for the Pound–Rebka experiment. He is also credited for the discovery of Nuclear Magnetic Resonance.

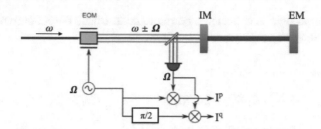

Fig. C.6 Schematics of the Pound–Drever–Hall stabilization technique. EOM is an electro-optic modulator that phase-modulates at frequency Ω the laser beam; the sidebands at frequency $\omega \pm \Omega$ are reflected off the FP cavity, detected and demodulated. The resulting d.c. signal I^p is proportional to the displacement from resonance. Adapted from (Vajente 2014)

of the mirrors. As an example, the frequency linewidth of an optical laser can be stabilized, applying this method, within 40 mHz, i.e. better than 2 parts in 10^{16}.

In gravitational wave interferometry, the PDH method is also used to keep the FP cavities of the interferometer arms on resonance (Fig. C.6). To describe the PDH technique, we now change notation, switching from ν to $\omega = 2\pi\nu$: for ease of writing the phases, and also for uniformity with the existing literature (Black 2001). As first, we inject into the cavity light of optical frequency ω, phase modulated[11] at frequency Ω in the radio band. We use the Jacobi-Anger expansion to express the light field in series of Bessel functions:

$$\psi_{in} = A_0 e^{i(\omega t + m\cos\Omega t)} = A_0 e^{i\omega t} \sum_{n=-\infty}^{\infty} i^n J_n(m) e^{in\Omega t} \qquad (C.39)$$

If the modulation index is small, $m \ll 1$, we only retain the carrier and the first order sidebands and approximate, to first order in m: $J_0(m) \simeq 1$ and $J_{\pm 1}(m) \simeq \pm m/2$, so that the relevant spectral components are limited to

$$\psi_{in} \simeq A_0 \left[e^{-i\omega t} + i\frac{m}{2} e^{-i(\omega+\Omega)t} + i\frac{m}{2} e^{-i(\omega-\Omega)t} \right] \qquad (C.40)$$

We can consider the light field as the superposition of three fields oscillating at (slightly) different frequencies. The reflected field is

$$\psi_r \simeq A_0 e^{-\omega t} \left[\mathcal{R}_0 + \frac{im}{2} \mathcal{R}_+ e^{+i\Omega t} + \frac{im}{2} \mathcal{R}_- e^{-i\Omega t} \right] \qquad (C.41)$$

where $\mathcal{R}_0$ is the cavity reflectivity at the carrier frequency ω and $\mathcal{R}_\pm$ the reflectivities at $\omega \pm \Omega$.

---

[11] Frequency and phase modulation are similar and yield the same results. But phase modulation is easier to implement, e.g. by varying the optical path with an electro-optical modulator like a Pockel cell.

The reflected light is collected by a photodiode whose output current is proportional to the intensity of the modulated light:

$$I \propto \psi_r \psi_r^* = I_0 \cdot \Big[\mathcal{R}_0 \mathcal{R}_0^* + \tag{C.42}$$

$$+ i\frac{m}{2}(\mathcal{R}_0 \mathcal{R}_-^* - \mathcal{R}_0^* \mathcal{R}_+)e^{+i\Omega t} - i\frac{m}{2}(\mathcal{R}_0 \mathcal{R}_+^* - \mathcal{R}_0^* \mathcal{R}_-)e^{-i\Omega t} +$$

$$+ \text{terms oscillating with } e^{\pm i2\Omega t} \Big]$$

We have set the amplitude of this signal to I_0 and will include in it, hence on, all constant terms introduced by the following stages, like conversion efficiency, amplifier gains, reference signal amplitude, etc.

The photo current is then processed by a mixer that extracts the amplitude contribution at frequency Ω.

Formally, the demodulation process consists of multiplying the incoming signal by the *reference* signal, a cosine at frequency Ω:

$$D(t) \propto \Big[e^{i(\Omega t + \theta)} + e^{-i(\Omega t + \theta)} \Big]$$

and then low-pass filtering the result, in order to reject higher harmonics. As shown, the process can introduce a phase delay θ, thus we must consider both the component *in phase* with the demodulating signal ($\theta = 0$) and that shifted by $\theta = \pi/2$, called *quadrature*. This is what a lock-in amplifier does, see Sect. A.6. The demodulated current is

$$I_{demod} = I(t) \times D(t) \tag{C.43}$$

$$= i\frac{m}{2}I_0 \Big[e^{i\theta}(\mathcal{R}_0^* \mathcal{R}_- - \mathcal{R}_0 \mathcal{R}_+^*) + e^{-i\theta}(\mathcal{R}_0^* \mathcal{R}_+ - \mathcal{R}_0 \mathcal{R}_-^*) \Big]$$

$$= \frac{m}{2}I_0 \, \Re e[ie^{i\theta}([\mathcal{R}_0^* \mathcal{R}_- - \mathcal{R}_0 \mathcal{R}_+^*)]$$

The argument in square brackets of the last equation is a complex number and the demodulation phase θ just rotates it in the complex plane: we pick the component, along the real axis by choosing $\theta = 0$ (the in-phase component) and that along the imaginary axis (quadrature) with $\theta = \pi/2$, recalling that $e^{i\pi/2} = i$:

$$I_{demod}^p = -\frac{m}{2}I_0 \, \Im m(\mathcal{R}_0^* \mathcal{R}_- - \mathcal{R}_0 \mathcal{R}_+^*) \tag{C.44}$$

$$I_{demod}^q = \frac{m}{2}I_0 \, \Re e(\mathcal{R}_0^* \mathcal{R}_- - \mathcal{R}_0 \mathcal{R}_+^*)$$

Fast modulation- Assume the Fabry–Perot cavity to be tuned on resonance, $\omega = 2\pi \frac{c}{2l}(n + 1/2)$ and allow the frequency, or the cavity length, to change by a small amount, $\delta\omega/(2\pi \Delta\nu_{FWHM}) \ll 1$. In the case of *fast modulation* we choose the modulation frequency such that the sidebands at $\omega \pm \Omega$ fall far away from resonance, ideally half-way between our resonance peak and the adjacent ones. The sidebands are then completely reflected, as shown by Eq. C.18 or by Fig. C.2. So, by choosing

$$\Omega \sim 2\pi \frac{c}{4l} \quad \Rightarrow \quad \mathcal{R}_\pm \simeq 1, \tag{C.45}$$

and dropping the subscript *demod*, Eq. C.43 simplifies to:

$$I^p = m I_0 \, \Im m(\mathcal{R}_0)$$
$$I^q = 0 \tag{C.46}$$

The reflectivity near resonance, derived in the previous section (Eq. C.20 with $r_2 \sim 1$), allows us to finally evaluate the PDH response:

$$\mathcal{R}_0 \simeq -\frac{1 - \sigma + 2if}{1 - 2if} \quad \Rightarrow \quad I_{demod}^p = -I_0 \frac{4m(2-\sigma)}{1+4f^2} f \tag{C.47}$$

We assumed small deviations from resonance, $f \ll 1$. Neglecting terms $O(f^2)$, the demodulated sideband output is thus proportional to f: it provides the desired error monitor, i.e. a signal proportional, in sign and amplitude, to the cavity displacement out of its resonant frequency, or length. This error signal is then fed to the feedback system that drives the actuator (piezo, electrostatic or magnetic) to exert the control force on one of the two mirrors.

The above relations are suited to handle the case with high reflectivity, small modulation depth, small displacement, of interest in most applications. A similar calculation yields the general case:

$$I^p = -2I_0 \, J_o(m) J_1(m) (t_1 t_2)^2 r_1 r_2 \frac{sin2kx}{1 + (r_1 r_2)^4 - 2(r_1 r_2)^2 cos4kx}$$
$$I^q = 0 \tag{C.48}$$

where x is the displacement from resonance. The slope of the linear region, near $x = 0$ can be evaluated:

$$I^p \simeq I_0 \frac{4m\mathcal{F}}{\pi} \frac{x}{\lambda}$$

having used Eq. C.10 that relates finesse and mirror reflectivities. A large value of $\mathcal{F}$ improves sensitivity, yielding a steep response, but on the other hand narrows the linearity range where the PDH method can be applied.

Figure C.7 shows how this function makes abrupt jumps when the condition $2l = n\lambda/2$ is met. In the lower pane, a zoom around the central resonance shows the linear response in a ± 2 nm interval.

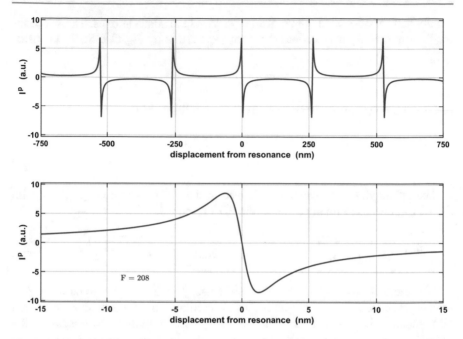

Fig. C.7 Simulated I^p response of a PDH scheme for $\lambda = 1.05\ \mu m$, $r_1 r_2 = 0.99$. The sharp segments occur every $\lambda/4 \simeq 250$ nm. The zoom on the resonance $x = 0$ shows the linearity region used for feedback

References

Black, E.D.: An introduction to Pound - Drever - Hall laser frequency stabilization. Amer. J. Phys. **69**, 79 (2001)

Drever R.W.P., et al.: Laser phase and frequency stabilization using an optical resonator. Appl. Phys. B **31**, 97–105 (1983)

Pound, R.V.: Electronic frequency stabilization of microwave oscillators. Rev. Sci. Instr. **17**, 490 (1946)

Vajente, G.: Readout, Sensing and Control. In "Advanced Interferometers and the Search for Gravitational Waves", Bassan M. ed. Springer Int. Publish (2014)

Printed in the United States
by Baker & Taylor Publisher Services